NAMING FOOD AFTER PLACES

Perspectives on Rural Policy and Planning

Series Editors:
Andrew Gilg, University of Exeter and University of Gloucestershire, UK
Henry Buller, University of Exeter, UK
Owen Furuseth, University of North Carolina, USA
Mark Lapping, University of South Maine, USA

Other titles in the series

A Living Countryside?
The Politics of Sustainable Development in Rural Ireland
Edited by John McDonagh, Tony Varley and Sally Shortall
ISBN 978 0 7546 4669 3

Rural Sustainable Development in the Knowledge Society
Edited by Karl Bruckmeier and Hilary Tovey
ISBN 978 0 7546 7425 2

Comparing Rural Development
Continuity and Change in the Countryside of Western Europe
Edited by Arnar Árnason, Mark Shucksmith and Jo Vergunst
ISBN 978 0 7546 7518 1

Sustainable Rural Systems
Sustainable Agriculture and Rural Communities
Edited by Guy Robinson
ISBN 978 0 7546 4715 7

Governing Rural Development
Discourses and Practices of Self-help in Australian Rural Policy
Lynda Cheshire
ISBN 978 0 7546 4024 0

Naming Food After Places
Food Relocalisation and Knowledge Dynamics
in Rural Development

Edited by

MARIA FONTE
University of Naples Federico II, Italy

APOSTOLOS G. PAPADOPOULOS
Harokopio University of Athens, Greece

ASHGATE

© Maria Fonte and Apostolos G. Papadopoulos 2010

All rights reserved. No part of this publication may be reproduced, stored in a retrieval system or transmitted in any form or by any means, electronic, mechanical, photocopying, recording or otherwise without the prior permission of the publisher.

Maria Fonte and Apostolos G. Papadopoulos have asserted their right under the Copyright, Designs and Patents Act, 1988, to be identified as the editors of this work.

Published by
Ashgate Publishing Limited
Wey Court East
Union Road
Farnham
Surrey, GU9 7PT
England

Ashgate Publishing Company
Suite 420
101 Cherry Street
Burlington
VT 05401-4405
USA

www.ashgate.com

British Library Cataloguing in Publication Data
Naming food after places : food relocalization and knowledge dynamics in rural development. -- (Perspectives on rural policy and planning)
1. Local foods--Economic aspects--Europe. 2. Rural development--Europe. 3. Traditional ecological knowledge--Europe. 4. Food supply--Europe. 5. Traditional ecological knowledge--Europe--Case studies. 6. Rural development--Europe--Case studies. 7. Food supply--Europe--Case studies.
I. Series II. Fonte, Maria. III. Papadopoulos, Apostolos G.
338.1'94-dc22

Library of Congress Cataloging-in-Publication Data
Fonte, Maria.
Naming food after places : food relocalization and knowledge dynamics in rural development / by Maria Fonte and Apostolos G. Papadopoulos.
    p. cm.
Includes bibliographical references and index.
ISBN 978-0-7546-7718-5 (hardback) -- ISBN 978-0-7546-9436-6 (ebook)
1. Agricultural diversification--Europe. 2. Local foods--Europe. 3. Rural development--Europe. 4. Traditional farming--Europe. 5. Agriculture--Technology transfer--Europe. I. Papadopoulos, Apostolos G. II. Title.
 S452.F66 2010
 338.109173'4--dc22

2010021625

ISBN 9780754677185 (hbk)
ISBN 9780754694366 (ebk)

Printed and bound in Great Britain by
TJ International Ltd, Padstow, Cornwall

Contents

List of Figures *vii*
List of Tables and Boxes *ix*
List of Abbreviations *xi*
Notes on Contributors *xiii*
Acknowledgements *xix*

Introduction: Food Relocalisation and Knowledge Dynamics for Sustainability in Rural Areas 1
Maria Fonte

PART I   REINVENTING LOCAL FOOD AND LOCAL KNOWLEDGE

1 'Local Food' as a Contested Concept: Networks, Knowledges, Nature and Power in Food-Based Strategies for Rural Development 39
Hilary Tovey

2 Creating a Tradition That We Never Had: Local Food and Local Knowledge in the Northeast of Germany 61
Rosemarie Siebert and Lutz Laschewski

3 The Reconstruction of Local Food Knowledge in the Isle of Skye, Scotland 77
Lorna Dargan and Edmund Harris

4 Local Food Production in Sweden: The Eldrimner National Resource Centre for Small-Scale Food Production and Refining 99
Karl Bruckmeier

PART II   VALORISING LOCAL FOOD AND LOCAL KNOWLEDGE

5 From the Local to the Global: Knowledge Dynamics and Economic Restructuring of Local Food 127
Isabel Rodrigo and José Ferragolo da Veiga

6	The Construction of Origin Certification: Knowledge and Local Food *Maria Fonte*	149
7	One Tradition, Many Recipes: Social Networks and Local Food Production – The Oscypek Cheese Case *Tomasz Adamski and Krzysztof Gorlach*	173
8	Traditional Food as a Strategy in Regional Development: The Need for Knowledge Diversity *Gunn-Turid Kvam*	197
9	Traditional and Artisanal Versus Expert and Managerial Knowledge: Dissecting Two Local Food Networks in Valencia, Spain *Almudena Buciega Arévalo, Javier Esparcia Pérez and Vicente Ferrer San Antonio*	215
10	Reclaiming Local Food Production and the Local-Expert Knowledge Nexus in Two Wine-producing Areas in Greece *Apostolos G. Papadopoulos*	237
	Conclusion: Europe's Integration in the Diversities of Local Food and Local Knowledge *Maria Fonte*	265

Index   *275*

List of Figures

I.1	Map of the study areas of the local food initiatives in the CORASON project	4
2.1	Case study region	66
3.1	Map showing location of Skye and Lochalsh	84
6.1	The construction of Aspromonte National Park certification as a hybrid forum	166
7.1	From local to adaptive knowledge	187
7.2	Map of the 'traditional' area of origin of oscypek cheese as social construction	190
9.1	Study area in the region of Valencia	217
10.1	The local production network of Mavro Messenikola VQPRD	247
10.2	The local production network of Nemea VQPRD	253
C.1	Agro-food contexts and strategies of re-localisation	267

List of Tables and Boxes

Tables

I.1	From the 'cold' negotiation of the market to the face-to-face relations of local food production	20
5.1	Production stages in the production of Barrancos Cured Ham	139
6.1	Examples from the ICEA table identifying typical products in Calabria	161
C.1	Main characteristics of the two strategies of food re-localisation	268
C.2	Local food and knowledge dynamics	270

Boxes

I.1	Knowledge forms and knowledge producers	15
I.2	Case studies in the reconnection perspective	19
I.3	Case studies in the origin of food perspective	19
4.1	The Eldrimner case study in the CORASON Project	101
4.2	Quality aspects of local food	113
4.3	Eldrimner's practice in facilitating local food production and processing	115

List of Abbreviations

AAFN	Alternative Agriculture Food Network
ACPA	*Associação de Criadores do Porco Alentejano* (Alentejo-Breed Pig-raisers' Association)
AGM	Annual General Meeting
AIAB	*Associazione Italiana per l'Agricoltura Biologica* (Italian Association for Organic Agriculture)
ANCPA	*Associação Nacional dos Criadores do Porco Alentejano* (National Association of the Alentejo-breed Pig-Raisers)
ANKA	Local Development Agency of Karditsa
ANP	Aspromonte National Park
AOC	*Appellation d'origine controllee* (Controlled designation of origin)
ASB	Association of Sheep Breeders
BSE	Bovine Spongiform Encephalopathy
CAP	Common Agricultural Policy
CEE	Central and Eastern Europe
CSA	Community Supported Agriculture
DLG	*Deutsche Landwirtschafts-Gesellschaft* (German Agricultural Society)
DE	Specific Designation
DECO	National Consumer Protection Association
DG	Generic Designation
DO	Designation of Origin
DORC	Regulatory Board (for a Label of Origin)
EDOAO	National Interprofessional Association of Vines and Wines
EEC	European Economic Community
ELOT	Hellenic Organization for Standardization
ENOAN	Nemea Union of Winemakers and Viticulture
ENOAP	Peloponnesian Union of Winemakers and Viticulture
EU	European Union
GDP	Gross Domestic Product
GDR	German Democratic Republic
GI	Geographical Indication
GM	Genetically Modified
GTS	Guaranteed Traditional Speciality
HIE	Highlands and Islands Enterprise
HACCP	Hazard Analysis and Critical Control Points
IFA	Irish Farmers' Association

IGEA	*Ispezioni e controlli per la Garanzia Ecologica dei processi agroalimentari* (Control Agency for the Ecological Guarantee of Agrofood Process)
INTERREG	Interregional Cooperation Community Initiative
IOFGA	Irish Organic Farmers and Growers Association
ISO	International Organization for Standardization
KEOSOE	Central Union of Vine and Wine Producing Cooperative Organisations of Greece
KRAV	Swedish certification system for organically grown products
LEADER	*Liaison Entre Actions de Développement de l'Économie Rurale*' ('Links between the rural economy and development actions').
LAG	Local Action Group.
LETS	Local Exchange and Trade Systems
LFA	Less Favoured Area
LFASS	Less Favoured Area Support Scheme
LFG	Local Food Group
LQC	Local Quality Convention
MATKULT	Food Culture (Swedish)
MS	Member State
NGO	Non-Governmental Organisation
NOK	Norwegian Crown (currency)
NUT	Nomenclature of Territorial Units for Statistics
OECD	Organisation for Economic Cooperation and Development
OPAP	*Onomasia Proelefseos Anoteras Piotitos* (Protected Designation of Origin of Superior Quality)
OPE	*Onomasia Proelefseos Eleghomeni* (Protected Designation of Origin)
PDO	Protected Designation of Origin
PGI	Protected Geographical Indication
PPP	Purchase Power Parity
R&D	Research and Development
RDP	Rural Development Programme
SAC	Scottish Agricultural College
SEO	Greek Wine Association
SLHDA	Skye and Lochalsh Horticultural Development Association
SPIN	*Senter for produktutvikling i næringsmiddelindustrien* (Centre for Product Development in the Food Processing Industry)
TSG	Traditional Speciality Guaranteed
UNIAPRA	*União das Associações de Criadores do Porco Alentejano* (Union of Associations of the Alentejo-breed pig-raisers)
VQPRD	Quality Wines Produced in Specific Regions
WTO	World Trade Organization
WIPO	World Intellectual Property Organization

Notes on Contributors

Tomasz Adamski is a Junior Researcher and PhD student at the Institute of Sociology, Jagiellonian University, Krakow. He has an MA in sociology from Jagiellonian University (2004) and an MA in European Studies from the University of Exeter, UK (2005); he is leading research assistant in 6FP project CORASON. He is a member and national correspondent of the European Society for Rural Sociology and Member of the Polish Association of Sociology. Research interests include use of local knowledge in rural development, transformation of EU rural policy and environmental sociology.

Karl Bruckmeier is Professor in Human Ecology at the School of Global Studies, Gothenburg University, Sweden. His main areas of research are societal change, environmental policy, rural development and natural resources management. His publications include *Innovating Rural Evaluation* (co-author and editor, Berlin 2001), *The Agri-Environmental Policy of the European Union* (co-author and editor, Basel 2002), *Ethik und Umweltpolitik: Humanökologische Positionen und Perspetiven* (co-author and editor, München 2008), *Rural Sustainable Development in the Knowledge Society* (co-author and editor with Hilary Tovey, Ashgate 2009).

Almudena Buciega Arévalo, PhD from the University of Valencia and Sociologist at the University of Alicante, Spain. She also attained a MSc in rural and regional resources planning at the University of Aberdeen, Scotland. Between 1998 and 2006 she worked at the University of Valencia (Department of Geography) as a main researcher in several EU projects. Her areas of interest include social dimensions of rural development, processes of social change and social capital. Currently she works as a practitioner for local development in a local municipality.

Lorna Dargan spent several years in the Global Urban Research Unit at Newcastle University UK, exploring issues around urban regeneration. Her specialist interests include social and urban policy and the way in which policymakers, practitioners and local residents define and rationalise key concepts in relation to regeneration policy. Recent publications include 'Participation and regeneration in the UK: The case of the New Deal for Communities' in *Regional Studies* 2009, 43(2); 'LEADER and innovation' (with Mark Shucksmith) in *Sociologica Ruralis* 2008, 48(3), and 'Conceptualising regeneration in the New Deal for Communities' in *Planning Theory and Practice* 2007, 8(3). She now works as a Careers Adviser for Newcastle University Careers Service.

Javier Esparcia Pérez is Professor at the University of Valencia (Spain) and currently head of the Department of Geography. Dr Esparcia has until recently been in charge of the Social Sciences Unit at DG Research of the Ministry of Science and Innovation. He also collaborates with the evaluation panels of the EU Framework Programmes. He is the Spanish representative in the Standing Committee for Social Sciences in the European Science Foundation. He has major experience as a team leader of EU and national-level research projects. His main scientific interests include change processes and socio-economic development policies in rural areas. Currently he is involved in the analysis of social capital and territorial development, with particular attention to social networks, power elites and leadership in the dynamics of rural areas.

José Ferragolo da Veiga, PhD in agronomic engineering from the Technical University of Lisbon and trained as an economist, is currently Senior Technician at the regional Directorate for Agriculture and Fishery in the Alentejo, Ministry of Agriculture, Rural Development and Fisheries, Portugal. He has been involved in the technical definition, implementation and assessment of agricultural and rural development policies at regional level. He was part of the research team of the Department of Agricultural and Rural Sociology at the *Instituto Superior de Agronomia*/Technical University of Lisbon in several research programmes. He is the author of *Desenvolvimento e Território* (Development and Territory) (2005).

Vicente Ferrer San Antonio is a Geographer at the University of Valencia (Spain). Between 2002 and 2006 he worked at the Department of Geography, involved in several projects in the field of rural development.

Maria Fonte, Associate Professor of Agriculture Economics at the University of Naples Federico II, Italy; e-mail: mfonte@unina.it. Her teaching and research topics include rural development, local food, innovation in agriculture, agro-biotechnology and property rights. She was the coordinator of the Italian team and the Work Package 6 on 'Local Food' in the FP6 EU research project CORASON (A Cognitive Approach to Rural Sustainable Development), with 12 participating European countries. She cooperates also with Latin American scientists and institutions in topics related to rural development and cultural identity.

Krzysztof Gorlach, Head of the Department of Rural Sociology and Agrarian Relations and Senior Lecturer in the Institute of Sociology of Jagiellonian University. Former fellow of the Kościuszko Foundation, the Fulbright Foundation, and the Andrew W. Mellon Foundation. Research experience at the University of Wisconsin in Madison (twice), at Harvard University (Center for European Studies), Oxford University (St. John's College) and the Institute for Human

Sciences in Vienna (*Institut für die Wissenschaften von Menschen*). He was awarded in 1992 with the Stanisław Ossowski Polish Association of Sociology and by the Minister of Education and Sport in 2002. Member of the editorial team of *Studia Socjologiczne* (Sociological Studies) and *Problemy Polityki Społecznej* (Problems of Social Policy). Advisory editor of *Sociologia Ruralis* and *Rural Sociology*. Member of advisory committee for the quarterly *Wies i Rolnictwo* (Countryside and Farming). Chairman of the Krakow branch of the Polish Association of Sociology and a member of the board of the same organisation. Member of the European Society for Rural Sociology. Coordinator of a number of international and national research projects.

Edmund Harris is a PhD student in the Graduate School of Geography at Clark University in Massachusetts. He holds undergraduate and Masters by Research degrees in human geography from the University of St Andrews and the University of Edinburgh, and his research interests focus on the spatialities of food systems and on alternative food politics. His recent research has explored the politics of localism in emergent alternative food networks in Fife, Scotland and addressed the constructions of place and scale mobilized by activists. He has published articles engaging with debates around alternative food politics in *Area* and *Geography Compass*, and has contributed book reviews to the *Scottish Geographical Journal* and *Agriculture and Human Values*. http://www.edmundharris.com.

Gunn-Turid Kvam is a Researcher at Centre for Rural Research (CRR) in Trondheim, Norway, where she has been working for the last 10 years. During this period she has also been working as a Coordinator of the research group Business Development, Innovation and Consumption. Her main research fields are local food systems and tourism. She has been author of a book, several chapters of books and articles on local food in Norway. Her PhD is from The Norwegian University of Science and Technology (1995), with a thesis on *Technology transfer from research and development organizations to small and medium sized firms*.

Lutz Laschewski is an independent Consultant and Researcher. His main research activities have been in the area of agricultural restructuring and rural change in East Germany, agro-environmental policies in Europe and policy evaluation. He has published numerous academic papers in German and English and has co-edited two books. Most recently he has co-edited a special issue on rural property rights in the *International Journal of Agricultural Resources, Governance and Ecology*, and published work on rural change and social capital formation in East Germany, the use of ICT based learning in rural areas, and agricultural education systems in Germany.

Apostolos G. Papadopoulos, Associate Professor at the Department of Geography, Harokopio University of Athens (Greece). He is trained sociologist (BA in Panteion University of Athens and MSc in London School of Economics and Political Science), while he received his PhD in geography from the University of Sussex. He has a long research and teaching experience in rural sociology and geography, social science methodology, migration and rural development. He has worked as senior researcher and project leader in numerous research programmes financed by the European Commission and by the Greek State. He has published a large number of peer-reviewed papers in international and in Greek academic journals, collected volumes and proceedings. He has edited, co-edited and/or co-authored four books in Greek and he has co-edited (with C. Kasimis) a book in English, entitled: *Local Responses to Global Integration* (Ashgate, 1999). He has received a research and writing grant from the MacArthur Foundation (USA) for a study entitled: 'The Multifunctional Role of Migrants in Rural Greece and Rural Southern Europe' (2004–2006).

Isabel Rodrigo, Associate Professor at the Agricultural Economics and Rural Sociology Department of the *Instituto Superior de Agronomia*/Technical University of Lisbon, Portugal. Trained as an agronomist, she has longstanding research and teaching experience in rural sociology. Her main areas of research include farming systems, social change and identities, environmental sociology and policy, rural development. She has worked as senior research and project leader in numerous research programmes financed by the European Commission and has published in international and Portuguese academic journals. She is author of *Percepção do Risco: a Seca no Baixo Alentejo* (*Risk's Social Perception: the Drought in the Lower Alentejo*).

Rosemarie Siebert, PhD from the Humboldt University, Senior Researcher and head of the 'Preferences and Conflict' section at the Leibniz Centre for Agricultural Landscape Research (ZALF). Her research interests include rural development, multifunctionality of land use, governance issues, farmers' attitude towards environmental issues, stakeholder involvement as well as participatory and action-oriented research methods. She works as independent evaluator for the EC and has been a member of the Executive Committee of the European Society for Rural Sociology and an Associate Editor of Rural Sociology. She has published several articles and is co-author and editor of the book *Agricultural Transformation and Land Use in Central and Eastern Europe* (with St. Goetz and T. Jaksch, Ashgate 2001).

Hilary Tovey, Department of Sociology, Trinity College Dublin, 1 College Green, Dublin 2, Ireland. email: htovey@tcd.ie. Hilary Tovey is a Senior Lecturer in sociology, a Fellow of Trinity College Dublin and a past president of the European Society for Rural Sociology. Her research interests address the general theme of rural societies in the age of the 'knowledge economy' (food production and consumption, environment, and civil society issues). She was the coordinator (2004–2007) of the EU FP6 research project CORASON (A cognitive approach to rural sustainable development), in which 12 participating European countries completed more than 80 case studies of local projects for rural sustainable development, with a focus on the dynamic relations and circulations of different knowledge forms within these. Recent publications include: *Rural Sustainable Development in the Knowledge Society* (Karl Bruckmeier and Hilary Tovey, eds.), Ashgate, 2009; *A Sociology of Ireland* – third edition (Perry Share, Hilary Tovey, Mary Corcoran), Gill and Macmillan, 2007; *Environmentalism in Ireland: Movement and Activists*, IPA Press, 2007.

Acknowledgements

We wish to thank Loukia-Maria Fratsea who collaborated in the formatting and editing of the book, Wayne Hall who took care of the language editing and Erasmia Kastanidis who elaborated the map of the study areas. We thank also Valerie Rose, Jude Chillman, Carolyn Court, Nikki Selmes, Katy Low and David Shervington of Ashgate Publishing Group who followed us through the production of this book with great patience and attention.

We want finally to acknowledge the European Union Sixth Framework Programme for the financial support to the CORASON (A cognitive approach to rural sustainable development: the dynamic of expert and lay knowledge) project.

We dedicate this book to all the people who shared their time and their knowledge with us during the field work and to our colleagues with whom we had the pleasure to work and to learn in the CORASON project

Introduction

Food Relocalisation and Knowledge Dynamics for Sustainability in Rural Areas[1]

Maria Fonte

Introduction

Since the 1980s the evolution of the agrofood economy and agrofood policy has experienced a profound change in Europe as indeed it has in other regions. Globalisation and liberalisation have led on the one hand to a reform of the agricultural policies of post-industrialised countries and on the other to a restructuring of production and markets in response to the application of new technologies and the emergence of quality as a new criterion for competitiveness. There has been a reversal of the previous tendency in the agrofood economy towards consolidation of a rigid vertically-integrated complex dominated by the processing industry and structured according to economies of scale and product standardisation. Global production has been re-organised into a flexible demand-driven value chain, ruled by standards of quality and co-ordinated by the retailing industries (Gereffi et al. 2004, Marsden et al. 2000).

On the side of this global system, though, a multitude of initiatives for the social and spatial re-embedding of the food economy have emerged and acquired new importance, pre-figuring features of an alternative model. A number of sub-types are included in this model, embracing pre-modern, non-modern and post-modern local food products, some of which were never detached from their socioeconomic and cultural contexts but were regarded by political economists and sociologists as peculiarities or 'irregularities' characteristic of backward or less favoured areas within the developed countries. Although the two models – conventional and alternative – are often considered autonomous, they operate in contiguous economic spaces, intersecting and overlapping with each other.

Whereas in the agro-industrial food complex, production processes are de-territorialised, placeless and centred around the commodification of food (*food from nowhere*), the alternativeness of the local food economies is contingent on their embeddedness in the social, cultural and territorial context (*food from somewhere*) as well as in affirmation of the importance of non-monetary values in food production and consumption. Socio-economic rights, rural citizenship, respect

---

1 I thank Les Levidow, Hilary Tovey and Apostolos G. Papadopoulos for comments on the draft of this introduction. Responsibility for content is, of course, entirely mine.

for the environment, fair trade and cultural identity all give the appearance of foreshadowing a new model of civic agriculture and food economy (Lyson 2004).

Local food is a focus of attention for many disciplines (rural sociology, anthropology, economic geography), but it has also triggered many controversies. After years of debate the scale of locality remains a critical factor, but it is still not clear what the optimum size might be for a locality. There is disagreement over whether local food is really alternative to the conventional food system, or whether by contrast it is merely a defensive, un-reflexive reaction against globalisation (Hinrichs 2003, DuPuis and Goodman 2005, DuPuis et al. 2006, Guthman 2007b). Conceived as consumer-driven, local economies are projected as in effect 'the progenitor of a neo-liberal anti-politics that devolves regulatory responsibility to consumers via their dietary choices' (Guthman 2007a: 264). Sharp political and academic battle-lines have been drawn around local food, with different practices accordingly understood – and legitimated or condemned – as good or bad, reformist or radical, alternative or conventional.

In the wealth of relevant literature that has emerged, the relation between the agro-industrial complex and the local food economy is often left implicit. According to some interpretations (e.g. Hendrickson and Heffernan 2002), the two are to a large extent interdependent. The pressures being exerted in the direction of homogenisation and standardisation also generate counter-pressures towards social and economic differentiation, which however involve only the 'interstices' (Renard 1999) of globalisation: the spaces left empty by the standardisation process of the agro-industry. The global and the local co-exist, the local being 'alternative' insofar as it is organised on different principles, without being a threat to the global.

The proliferation of initiatives and calls for relocalisation of food production over the last two decades or so have led many to imagine that local food might totally replace the dominant system of food provision. Rather than being seriously integrated into the local food debate, the subject is however left for political economists to discuss.

This volume represents an attempt to pursue further empirical investigation of food relocalisation, in conjunction with theoretical reflection on the findings. It emerged out of the CORASON project, CORASON being an acronym for 'A cognitive approach to rural sustainable development: the dynamic of expert and lay knowledge'. Funded through the EU VI Framework Research Programme and carried out in 12 European countries between 2004 and 2007, this research project was aimed at identifying the forms of knowledge and analysing the dynamics of their interaction in the economic development initiatives being carried out in the European rural areas, among which food relocalisation initiatives were being included.

A recent volume by Bruckmeier and Tovey (2009a) sheds light on the thinking and the organisation behind the CORASON project. Researchers from 12 European countries were involved, all of them belonging geographically to the European 'rim', the selection criteria deriving from – and representing an application of

– the 'Green Ring' hypothesis (Granberg, Kovach and Tovey 2001): Hungary, Poland, Czech Republic, Greece, Italy, Spain, Portugal, Ireland, Scotland, Sweden, Norway, Germany. What all these countries have in common is that agriculture and rural culture have played, and continue to play, an important part in their social, economic and political development. Bruckmeier and Tovey discuss the role and dynamics of local knowledge in initiatives pertaining to a non-agricultural economy, to innovatory development, nature protection and biodiversity. Here, from the same perspective, we present and analyse initiatives of food relocalisation. We have included 10 of the 12 CORASON partner countries because of their particular focus on the issue of interest (see Figure I.1). One specific contribution made by the present volume is that it presents a critique of modern science from the perspective of local food and the countryside.

Interest in knowledge dynamics in rural areas grew out of two social trends (Bruckmeier and Tovey 2009b: 3): the movement toward a knowledge society and the increasing emphasis on sustainable development. Both trends are significant in rural areas, but in their own particular way. Rural areas are often perceived as being rich in natural resources but lacking in the human capital and knowledge that are a necessary prerequisite for remaining competitive in a modern economy. It is not clear, on the other hand, what implications sustainable development might have for them. In several of its variants (Buttel 2000) ecological modernisation 'centres on the idea of rebuilding core industrial production processes using "clean technologies"' (Bruckmeier and Tovey 2008: 319), without this entailing any necessary concern for the social and economic conditions of rural sustainability. Inspiring EU and intergovernmental policies at the official level, this vision of sustainability, as an expert-dominated discourse employing rules of science for establishing percentages of allowable emissions, has the potential to block rather than promote rural development, excluding local actors and their knowledge from participation in its construction.

The CORASON project favours 'polycentric management' of local resources (Bruckmeier and Tovey 2008: 323), involving a new model of rural governance with the capacity to secure the participation of local people both as individual users and producers and as formal and informal groups and institutions. The model also creates opportunities for joint learning, collective formulation of principles and sharing of decision-making power.

While reflecting a variety of approaches, the local food case studies considered in the different chapters of this volume were all selected on the basis of common assumptions, which were discussed extensively among the researcher teams prior to crystallisation in a conceptual and methodological synthesis (Fonte and Grando 2006). Social and ecological embeddedness and the producer-consumer nexus were at the core of the analysis. As for local food, a broad differentiation of meanings soon emerged in the discussion, polarised around two main perspectives. In the first, 'local' was understood as denoting *socio-spatial proximity,* reconnecting producers and consumers in the same place (*the re-connection perspective*). In

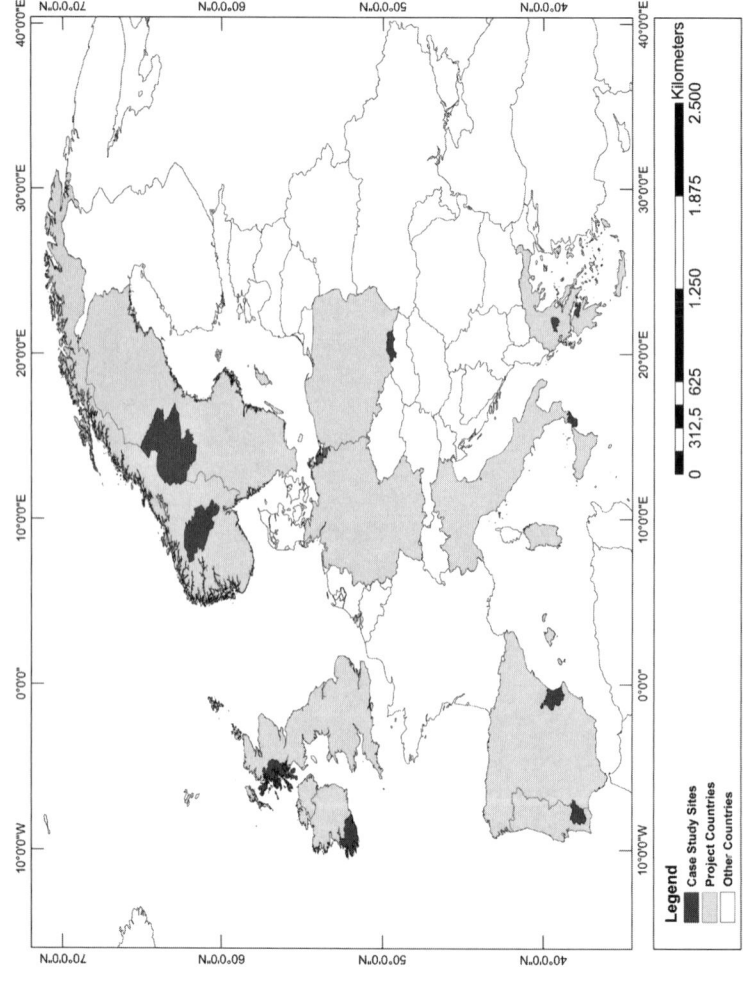

Figure I.1   Map of the study areas of the local food initiatives in the CORASON project

Source: Elaborated by Erasmia Kastanidis, Department of Geography, Harokopio University of Athens.

the second, the concept of 'localness' was also linked to the specific conditions of production in a *territory* (*the origin-of-food perspective*) (Fonte 2008).

Rather than privileging one discourse over the other, we decided to explore both of them, distinguishing – in the relation between producers and consumers – 'local production for local consumers' from 'local production for distant consumers'. For each perspective we agreed, through analysis of case studies in the CORASON research areas, to explore the characteristics of the network (including the actors and actants involved, the objectives and the strategies pursued) and to identify the forms of knowledge mobilised by the rural actors as well as the way they changed and interacted over time. Within this perspective the local is not only identified with a geographic location and a particular community but also constituted through 'its methods of producing situated knowledge' (Jasanoff and Martello 2004: 14).

The main research assumption was that the knowledge debate would enrich the analysis of local food. It is hard to exaggerate the role of science and technology in the constitution of the agro-industrial model of food production. The 'green revolution' is generally considered to have been the product of a breeding revolution brought about by scientists in the land-grant universities and diffused in the field by an army of extensionists and development agencies, persuaded that an increase in productivity would eradicate hunger and bring progress all over the world (Schultz 1964, Mosher 1966, Brown 1970, Evenson and Gollin 2003). Given that the political agenda of local food is to establish a new food economy offering an alternative to this model, the local food project will necessitate new ways of knowing and a new science (Kloppenburg 1991). Combined discussion of 'local food' and 'local knowledge' is seen as a first step in the construction of a new science of agriculture with a potential for elaboration in the various national and regional contexts. The attention paid to the dynamics of knowledge in developing local food is crucial both in addressing the objective of food re-localisation (as part of an attempt to construct alternatives to the dominant agro-food production model) and in gaining insights into the processes that may serve to legitimate different ways of knowing, in the process leading to new, democratic ways of generating knowledge.

In the following sections of this introduction we first consider some of the key insights that have contributed over the last decades to revitalising the debate on the role of local food. We pay particular attention to the spatial and socio-economic dimensions, subsequently touching on the question of knowledge dynamics in local food projects and always bearing in mind the most important findings of the case studies presented in this volume.

The chapters of the book are organised under two headings: 'Re-inventing Local Food and Local Knowledge', and 'Valorising Traditional Food and Local Knowledge', in accordance with what appears to us to be the differing economic and cognitive dynamics of the initiatives under analysis.

Local food and the political agenda

The local food movement has grown rapidly in the last decades, both in North America and Europe and so has academic debate about it. Local food is promoted as an alternative to the globalised industrial system of food production, whose products dominate the supermarket shelves. Shortening the food chain and the distance between producers and consumers is expected to have beneficial effects for the environment, the local economy and the rural community.

For its activists and proponents, local food represents a radical alternative for supporting food produced, retailed and consumed locally. There is an appeal to 'three aspects of sustainability: invigorating local economies; sustaining diverse environments; nourishing healthy communities'.² The political strategy constructed around local food proceeds to canvass the wider support of citizen-consumers, i.e. those who use their consumption choices as an expression of social agency and citizenship (Lockie 2009). For them the organisation of the global food value chain is based on unfair exchange relations favouring big intermediaries, above all the retailing industry, against the interests of agricultural producers – who do not earn a living income – and the final consumers – who pay too much for food (Patel 2007). The global food chain is moreover characterised by great paradoxes: overproduction and food shortages, systems of production that deplete the same natural resources that are necessary for future production and, most dramatically, the co-presence in the world of a billion people suffering from hunger and a billion suffering from obesity and related illnesses (diabetes and cardio-vascular disease). A recent international assessment (IAASTD 2009) recognises that the global food system is environmentally, socially and economically unsustainable.

The new movement for localising food production gained impetus in the 1990s as a result of growing dissatisfaction with the organic movement and its increasing 'conventionalisation' (Guthman 2003, Buck et al. 1997). There was a widespread perception that the organic movement had dropped its alternative/environmental ideological baggage and had been seduced by multinational retailing firms and the prospect of a mass market (Blythman 2005). Certification began to be seen as encouraging non-local food consumption, raising costs for producers and prices for local consumers. Accordingly, a 'post-organic' (Moore 2006) local food movement shifted the focus to direct sales to the consumer, specifically addressing the sustainability of the distribution system in the food chain.

Local food cannot challenge globalised industrial food production everywhere in the same way, for the simple reason that there is not, and could not be, either a generally accepted definition for local food production or a uniform practice of relocalisation. In its different guises, as community gardens, farmers' markets or community supported agriculture, as food circles or box schemes, as food fairs or certification programmes, the local food project emerges out of different contexts, is inspired by different values and may inspire different social practices

---

2   See the material at http://www.localfood.org.uk/.

and different social relationships. Because of the heterogeneity of the initiatives it inspires and the objectives it pursues, local food is, as Tovey convincingly argues in her chapter in this volume, also a 'contested concept'.

Localness is associated with space and short distance, but also with place, regions and territories; it is associated with small-scale farms, multifunctional agriculture, quality food, rural livelihoods, sustainable community agriculture. The most comprehensive list of the different objectives pursued by local food projects is provided by Pratt (2007: 288–289):

- alternative food movements promoting local produce for environmental reasons;
- localised food systems as part of a political project to construct local economies outside/against/opposing the capitalist system (locality is socially constructed as a space of resistance, autonomy, empowerment, sustainability);
- food system localisation as a strategy for increasing farmers' income in rural development policy (community-supported agriculture, rural development strategies);
- the connection between locality and quality (food quality as a territorial connotation);
- food sovereignty as the right of each society to establish its own food economy, an objective pursued by alternative global movements such as the small farmers' movement led by Via Campesina.

Each objective has to do with a different connotation of 'local': environmentally friendly, anti-capitalist, favouring small farms and marginal areas or food quality, food sovereignty. The different meanings and the different objectives may overlap, complement or contradict each other, so that one major issue for further investigation is whether all these different initiatives may be articulated with each other in a coherent trans-local project or whether they must remain partial, localised forms of resistance to the global food system. The question to be asked, in other words, is whether they are niche phenomena filling the spaces overlooked during globalisation of production and markets or whether they portend a paradigm shift leading to an alternative food economy (Morgan et al. 2006: 81–85). McMichael (2008: 95) suggests that they should be seen as an 'expression of transitional relations within/between food regimes in which both objective and subjective forces are at play'.

Environmental sustainability and the spatial dimension of the 'local'

In its immediate meaning 'local' has to do with the physical distance food travels from the place of production to the place of consumption, a distance expressed in miles, as in the '100-mile-diet' or kilometres, as in the '0 km restaurant'.

A sharp contrast is drawn between the short chain for local food and the long distance food is required to travel in the conventional, centralised, industrialised food system (Pretty et al. 2005). In the United States Pirog et al. (2001, 2003) analysed the transport arrangements for 28 fruits and vegetables to Iowa markets via local and conventional food distribution systems and calculated that produce in the conventional system travelled an average of 1,546 miles (about 2,500 kilometres) while by contrast locally sourced food travelled an average of just 44.6 miles (72 kilometres).

Attention to food miles links concern over food to environmental concern with climate change and emissions of carbon dioxide and other greenhouse gases (GHG) from transport. The environmental impact of the food economy, though, does not depend only on the distance 'from farm to fork', but also on *how* food is transported, grown, transformed and prepared. Only a life-cycle analysis of food can yield an accurate assessment of the total volume of gas emissions. Studies in the UK and the United States (Garnett 2007, Weber and Matthews 2008) revealed that it is agricultural production that accounts for the largest proportion of the food system's greenhouse gas emissions: between 50 per cent and 83 per cent of total emissions occur before food goes out the farm gate. Different food groups also differ widely in GHG-intensity; on average red meat is around 150 per cent more GHG-intensive than chicken or fish. It is thus evident that

> dietary shift can be a more effective means of lowering an average household's food-related climate footprint than 'buying local'. Shifting less than one day per week's worth of calories from red meat and dairy products to chicken, fish, eggs, or a vegetable-based diet achieves more GHG reduction than buying all locally sourced food. (Weber and Matthews 2008: 3508)

The difficulty of establishing well-defined boundaries for the notion of 'locality', taking into account the conditions for the entire life-cycle of production, appears to undermine the usefulness of 'localness' as a category for analysis of the sustainability of food systems. But there are two major considerations to corroborate the suggestion that the spatial dimension of local food remains important. First, most studies on GHG emissions from agriculture start from an assumption that there is no difference between 'long-distance' and 'short-distance' agriculture when it comes to production technique. If one were to take into account existing differences in farming practices and farming structure, this could lead to different results. Local food is better not only because it has travelled shorter distances, but also because it is grown differently, on farms of a different type, usually small and utilising more sustainable practices (DeWeerdt 2008). Harris (2008) also suggests that emphasis should be placed on scale as an important aspect of practice. He would like our attention to be directed to the ways in which 'scalar narratives, classifications and cognitive schemas constrain or enable certain ways of seeing, thinking and acting'. The 'local' becomes the space for enactment of a political agenda involving construction of a new, more equitable and more sustainable food system. 'The local' acquires a variety of complex meanings within this perspective, encompassing not only spatial but also political and social dimensions.

The socio-economic dimension of localness

Relations of production and relations of exchange in the food economy Local food is not only about short distances. From a sociological viewpoint local food is place-embedded, the opposite of the placeless food of industrial agriculture. This concept of embeddedness imparts social meaning to notions of place, social meaning to be elaborated by the rural communities inhabiting the 'places' in question. The shortness of the local food chains makes it possible to trace the food almost personally to the individual farmer who produced it, enabling relations of *trust* to be established in the local society. Food production is re-contextualised within the formal and informal social relationships that constitute the basis for community life. Geographical proximity is, then, important because it implies or makes possible social proximity, i.e. face-to-face interaction between producers and consumers. Such interaction has a significant impact on rural community life. Local food becomes part of a political project for keeping rural communities alive and constructing local economies with a degree of independence from the powerful forces of globalisation. Partially protected or separated from global competition, local economies encourage values other than the suffocating market law of profit: respect for natural resources, attention to cultural and biological diversity, economic sustainability for small farmers, social justice and food sovereignty.

Place-embeddedness of food may thus be conceived of as local society's *resistance* strategy against globalisation and neo-liberalism (Polanyi 1957). Place-embeddedness and differentiation of food comes forward as a cultural, individual and collective societal response to the commodification-of-everything (Strassen 2003) drive of the neo-liberal economy.

Not everyone agrees with this assessment. Place-embeddedness can be seen as having been co-opted by the same globalisation process that it is supposedly fighting. Valorisation of local foods – as 'commodities that embed ecological, social and/or place-based values' (Guthman 2007b: 456) – green labelling and 'fair trade' are nothing more than aspects of a 'third wave of marketisation' (Burawoy 2005a and 2005b) as it extends to the fictitious commodities of nature, land and natural resources. Commodification is able to embrace niche production and place-specific products, sweeping them up in a movement of appropriation by global capitalism that allows of no escape, with every action and every aspect of production susceptible of integration into the market mechanism. The 'commodification-of-everything' argument is evidently predicated on a linear conception of modernisation involving a progressive shift from non-market to market economy (through successive waves of commodification), absorbing everything and destroying cultures and society.

In the Marxian tradition commodities are associated with the capitalist mode of production and with the production of goods for sake of their exchange value.

They represent a *fetish* insofar as they conceal the fact that (surplus) value has its real source in labour and production relations are relations of labour exploitation. In the recent debate on commodification, references to relations of production are downplayed and circulation, i.e. the market, rather than production, becomes the source both of value (or rent) and of a new form of alienation deriving from excessive individualism and loss of sociality (not the alienated worker but the alienated consumer). In post-industrial society there has undeniably been an expansion of the commodity sphere, but cultural anthropology warns us against 'an excessively positivist conception of the commodity, as being a certain *kind* of thing' (Appadurai 1986: 13). Appadurai instead proposes to see things as having a social biography and a 'social life', in the course of which they may change status and switch from a condition of commodities to one of non-commodities. In this perspective the commodity and the gift are not something separate and the one does not exclude the other.

Gifts, as we know, are conceived of as a type of exchange involving both things and persons and embedding the flow of things in the flow of social relations; 'commodities represent [by contrast] the drive ... of goods for one another, a drive mediated by money and not by sociality' (Appadurai 1986: 11–12). But the term 'commodity' should not be taken as denoting a fixed category of thing. It is rather a socially specific situation in which things are exchanged in a certain regime of values. 'The commodity situation in the social life of any "thing" may be defined as the situation in which its exchangeability (past, present and future) for some other thing is its socially relevant feature' (Appadurai 1986: 13). Following the social life of things in their total trajectory from production to exchange to consumption 'we can see *things* moving into and out of the commodity state' (Appadurai 1986: 13), subject to various processes of commodification and de-commodification. Kopytoff (1986) identifies the former process (commodification) with homogenisation and the latter (de-commodification) with singularisation.[3]

In this reading the commodity is not born with the capitalist mode of production: the term denotes something that is cross-cultural and common to numerous modes of production. The 'tendency of all economies to expand the jurisdiction of commodification' is moreover counterposed to the tendency of 'all culture to restrict it' (Appadurai 1986b: 17).

The situational, contingent construction of things as commodities comes over as a contradictory process, which it is therefore important to investigate. There are in fact in any society culturally defined hierarchical spaces surrounding commodities and serving to establish which items are exchangeable. But, apart from this, individuals too have their own criteria for evaluation and their own need to discriminate between things, and these criteria do not necessarily coincide with those applying in general in the public sphere or in the society. Kopytoff

---

3 Appadurai (1986: 17) is less convinced of this opposition, noticing that the most interesting cases show a permanent commodifying of singularities. Flexible accumulation can be also seen as a process of appropriation of 'singularities' by capital.

(1986: 79–80) argues that in simpler societies the culture and the economy are in relative harmony, with the economy reflecting the cultural classifications and the latter quite effectively satisfying the individual need for discrimination. Complex societies, by contrast, are characterised by functional specialisation at the social level and by cultural pluralism and relativism. Here one finds not only enormous momentum in the value-homogenising drive of the exchange system but also publicly recognised classifications underwriting commodification and operating side by side with innumerable schemes of valuation and singularisation that have been devised by individuals, social categories and groups and may conflict inexorably not only with public commodification but also with one another.

An examination of local food from this perspective opens new possibilities of interpretations and new avenues for research. It might, for example, be interesting to trace the social and cultural trajectory of local food through the successive transmutations of commodification and de-commodification, with social relations of production and exchange both playing an important role as mechanisms of material and immaterial value production. It is perhaps worth noting from a theoretical viewpoint that a rigid interpretation of commodity fetishism could have the effect of obscuring the differences that lie behind different types of commodity exchange, especially in the case of place-specific food products (Gibson-Graham 2006).

In the individual and collective fight to redefine cultural, symbolic and also social values, local food is simultaneously subjected to contradictory forces of commodification, de-commodification, homogenisation and singularisation (Kopytoff 1986: 76). Transactions in farmers' markets may furthermore, by shortening the food chain and establishing direct links between producers and consumers, help to unveil rather than obscure the economic, social and environmental conditions of production, thus making possible a re-composition of the specialised, segmented knowledge of the long-distance commodity trade. Exchange of 'meaningful commodities' (Guthman 2002) may not only serve to redistribute value and rent but may also contribute to affirming common intangible ethical and political values and, in so doing, creating or strengthening social bonds and/or social networks in the context of a moral economy.

Transcending the traditional Marxist emphasis on the social relations of production, local food points to the importance of the innovative organisation of social relations at the point of exchange, between producers and consumers, as potential driver for the construction of new food communities.

The relation between 'de-commodification' and 'singularisation' on the one hand and 'resistance' on the other is a topic requiring further research investigation. In their origins the former are cultural concepts; the last-mentioned social and political. Appadurai argues that it is politics (in the broad sense of relations, assumptions and contests over power) that links value and exchange in the social life of commodities. The constant tension between the existing frameworks (of prices, bargaining, etc.) and the tendency of commodities to breach these frameworks in search of a re-definition is political, that is to say, pertaining to power (Appadurai 1986: 57).

Van der Ploeg (2007: 1) recently made the point that 'a more comprehensive concept of resistance can play a more prominent role in sociology, especially when it comes to sustainability in rural areas and food production'. He distinguishes three forms of resistance: overt struggle (typical of the working class fighting for better terms and conditions of work), sabotage (typical of traditional peasants, see Scott 1985) and, last but not least, direct intervention in the organisation of labour and production.

Sabotage is a form of passive resistance but resistance of the third kind represents production and action, based on innovativeness and autonomous co-operation between producers and consumers:

> One important feature of these new forms of resistance, especially relevant to sustainability, is that they entail searches for, and constructions of, *local* solutions to global problems. ... Individually these expressions are innocent and harmless: considered together they become powerful and change the panorama. (van der Ploeg 2007: 3–4; emphasis in original)

Returning to innovativeness in the organisation of production and consumption, the concept of 'resistance' in its various forms by its nature entails the cultural concept of de-commodification. Both ideas bring to the fore the role of new subjectivities and of the social movements that fight consciously to win self-determination and autonomy from the global forces of the economy.

From neo-liberalism to new 'food communities'?   In an effort to make intelligible the multiplicity of local food initiatives, which often become prescriptions about what to eat and how to consume food (Guthman 2007a), Maye, Holloway and Kneafsey (2007) distinguish between a 'product and place' and a 'process and place' approach. 'Process and place' initiatives (farmers' markets, CSA, etc.) are seen as radical, politically oppositional, alternatives, offensive strategies because they draw into question the social and ethical values of the dominant food system, seeking to create a direct relationship between producers and consumers.

The 'product and place' approach, by contrast, has the appearance of a defensive strategy. Its aim is to produce geographically specific food products that can be sold outside the production region as niche market commodities (Maye et al. 2007: p. 5). Supported by policy schemes such as the Protected Designation of origin (PDO) and Protected Geographical Indication (PGI) labels that were introduced by the EU in the 1990s, this strategy comes over as a weak alternative: it focuses on rural development objectives for marginal areas and offers no truly radical alternative to the conventional food supply chain. For some authors such schemes are not alternative at all. Labelling is an instrument of neo-liberal politics, part of a shift from government to governance, limiting the right of access and creating scarcity through enclosure (Guthman 2007b).

These oppositional authors put forward a simplistic reading of the PDO and PGI policy. It is not just some final characteristics of the product that the PDO/

PGI labels certify but also the entire process by means of which it was grown and transformed. The link between 'product and place' derives precisely from such a specific process of production, from local knowledge and a local culture embodying knowledge of how such a food might be produced and consumed. It is also interesting to note that the GI labels do not bestow exclusive rights on techniques, animal breeds or plant varieties, but simply protect geographical names. Anyone can copy the production techniques for parmesan, feta, cheddar or oscypek cheese and commercialise their products without any authorisation being required, as long as they do it under a different name. It is the identity and the reputation of the name that is protected, and protected as a collective, not a private good. It is protected, that is to say, as a good that belongs to a community of producers in a specific geographical area. Any producer in that region (even outsiders who operate there) can use it as long as he/she observes the rules that have been negotiated. It is a collective good which justifies community rights. It is particularistic and exclusionary, as the domestic convention (Boltanski and Thevenot 1991) and the concept of community imply. It is regarded as a defensive strategy in the sense that it aims to protect 'what the market leaves after it has filtered out everything else' (Pratt 2007). But it implies a quest for place-based differences rather than a drive towards homogenisation. This means that GI labels may open up a possibility for preserving and valorising local identities and ways of life, as against the global appropriation of local resources. In that sense they may be seen as a form of cultural resistance to commodification (Kopytoff 1986). To the extent that they offer and elaborate political and institutional instruments making possible the management of collective goods they may be considered not only a defensive, but also an offensive strategy against the neo-liberal rush to individualisation and homogenisation.

Guthman (2007b) underlines the contradictions and paradoxes in the use of neo-liberal tools to protect community and collective goods. But the stories of biodiversity, bio-piracy and free software convey the message that in order to avoid a commons, a *res universitatis* (i.e. a thing belonging to everybody in a community) becoming *res nullius* (a thing belonging to nobody), it is necessary to devise protective institutions and new regimes of regulation[4] (Rose 1986).

Although it is generally understood that management of the commons and of common-pool resources is affected by the increasing scale of social interaction, new theories and concepts are needed if there is to be firstly recognition and then

---

4 If today we have something called 'free software', we are indebted for this to Richard Stallman, a former researcher at IBM, who in the 1980s, amidst the extension and strengthening of intellectual property rights legislation, was able to use the copyright law to protect free software by means of *the GNU General Public Licence* (GNU GPL). The GNU GPL allows everybody to use, study, modify, and redistribute free software on the proviso that he/she does it under the same conditions of the GNU GPL (see http://www.gnu.org/philosophy/free-sw.html). It excludes those who do not share the values of 'free software'– the outsiders to the (place-less) community of 'free software'.

analysis of the transformation that is taking place in the new millennium in the commons and in the 'community' (Dolsak and Ostrom 2003). In many disciplines calls are being issued for renewed attention to the concept of community. In economics, Bowles and Gintis (2002) speak of community governance being likely to acquire greater rather than less importance in the future as a complementary form to the state and the market.

> Far from representing holdovers from a premodern era, the small-scale local interactions that characterize communities are likely to increase in importance as the economic problems that community governance handles relatively well become more important. (Bowles and Gintis 2002: 422)

Territory,[5] the institutions of microfinance, the production of free software through voluntary participation are all seen as being underwritten by some kind of community governance. Gibson-Graham (2006) try to develop a sociological discourse of the 'community economy' which articulates a set of concepts and practices able to provide potential co-ordinates for counter-hegemonic projects. Finally, Etzioni (2006a) calls for a *new* approach in the form of a responsive or democratic communitarian social philosophy counteracting liberalism. While the latter focuses on the individualistic conception of self-interest, the former favours a balance between liberty and social order and between particularistic (communal) and society-wide values and bonds. Unlike the old neo-communitarianism, it takes as its starting point the assumption that both the universalistic demands for human rights and the particularistic demands of communities have strong moral standing. It recognises also that the two may be reconciled through compromises that are both morally defensible and sound (Etzioni 2006b).

In the global movements around local food we should recognise that there are new food communities emerging with quite specific features, that they are trying to link together and reconcile universal and particularistic/collective claims when they propose what may seem to be a paradox: empowerment of local food communities as the best strategy for asserting and implementing the universal right to food.

Science and knowledge in the post-positivist era. What place for local knowledge? In the debate on local food there has been an enrichment in meaning in the concept of the 'local', which has come to be associated not only with geographical locations but also with particular communities, particular histories, particular institutions. One other important constitutive element of locality is its specific, collective way of being in particular places, producing situated knowledge and elaborating a particular method for knowing things (Jasanoff and Martello 2004: 13) which is often labelled 'local knowledge' as opposed to scientific or expert knowledge.

---

5   See the literature on industrial districts and local production systems (Becattini 1989, Garofoli 2003).

The CORASON project was conceived as an attempt to analyse the interrelations among different forms of knowledge – scientific, managerial, local (Box I.1) – in the process of constructing sustainability in rural areas, starting from the observation that 'changing society in a sustainable direction means both changing knowledge processes and relationships, and using knowledge to manage resources for rural development in a sustainable way' (Tovey, Bruckmeier, Mooney 2009: 265).

Box I.1   Knowledge forms and knowledge producers

---

1 Scientific knowledge generated by researchers in clearly defined research roles. Criteria: specialised, discipline-bound or interdisciplinary, methodologically guided, may be experimental, documented/written, public and published, learned in public and controlled/certified education and training, neutral with regard to persons, age, gender, social organisation, produced by researchers.

2 Managerial knowledge generated or used in resource management, programme and project management, political, administrative and economic decision-making, including planning. Criteria: shares many criteria with scientific knowledge and its specialisation; is mainly learned in public and controlled/certified education and training; is more clearly and explicitly bound to use of power and decision-making and normative criteria; not always public and published; often about routines and procedures; can also be informal, person-bound and based on individual experience.

3 Local knowledge as locally specific, context-and actor specific. Criteria: locally and culturally specific/particularistic – context-bound or situated, often orally transmitted, person-bound, experience-bound or more experiential than scientific knowledge; not neutral with regard to person, age, gender, social organisation, status; bound to production and resource use in agriculture; learned in informal and private contexts of family and face-to-face interaction, in neighbourhoods, from local cultural traditions and practices; intergenerational transmission; print and other media may be of increasing importance in local knowledge use and transmission.

Source: Bruckmeier 2004.

Knowledge is today considered the most important resource for economic development, but there still persists an *urban bias* in the conception of technological progress, which is thought to be linked only to the scientific knowledge produced in the urban *milieux* of the universities, government and industrial laboratories, especially in the fields of informatics, telecommunications, biotechnology and nanotechnology. Rural areas are by contrast often characterised as 'lacking in human capital' and rural societies are said to be suffering a 'knowledge deficit' (Bruckmeier and Tovey 2009c: 276–277) which hinders the spread of global technologies.

Recent social science studies have criticised the triumphalism of technical progress based on modern science, disputing its capacity to capture the full complexity of natural phenomena. Modern science and technologies have engendered a risk society, in which people perceive themselves to be constantly endangered by scientific and technological projects and products, whether through economic crisis or through ecological destruction (Beck, Giddens and Lash 1994). Modern science has in fact never really attained the status of being a superior form of objective, universal knowledge, as the rhetoric may perhaps have suggested. Latour (1987, 2004) argues that the ontological separation between nature and society and facts and values that is often represented as being part of the 'constitution of modernity' has in fact never been realised. The results of scientific research are always socially constructed. Their success is judged by their capacity to build social networks 'acting at a distance' as a means of creating the social conditions for their own diffusion.

Nowotny, Scott and Gibbons (2001) describe the innovation of post-positivist science as being a shift from *mode 1* to *mode 2*. In *mode 1* the context of discovery was considered to be the domain of scientific creativity. Scientific methods of justification were portrayed as de-contextualised, 'that is to say, detached as much as possible from social aspects of the worlds from which they had arisen and in which they were practised' (Harding 2008: 81). In *mode 2* the *loci* of knowledge production have shifted from universities to industry and government laboratories. Science is now always mission- directed and by consequence is even more contextualised. Focused as it has been since its emergence on solving practical problems, scientific research is organised in such a way that it transcends disciplinary boundaries, involving multi- and trans-disciplinary teams, in which not only researchers and scientists are called to participate, but also target groups, users and other lay persons.

The growth of uncertainty and complexity in society, as illustrated among much else by the proliferation of controversies between experts, further underlines science's inability to deliver a socially binding definition of truth. The sites of problem formulation and negotiation move away from the institutional domain of industry, government and universities – dubbed 'the triple helix' by Leydesdorff and Etkowitz 1998 – into the public space, the *agora*, where 'science meets the public' and the public speaks back to science. People are forced to enter the scientific debate in an experimental frame of mind, with experience rather than data and procedures becoming the decisive factors in handling ambivalence and uncertainty (Beck 1992 and 1997, Harding 2008). Participation by the public, bringing all its experiential baggage with it into the debate, makes science more socially robust and the knowledge system more open and more democratic.

The trans-disciplinary approach opens the stage of knowledge production to lay persons, stakeholders, ordinary citizens. But the rural context and local knowledge are not specifically discussed in the new post-positivist scenarios. They rather constitute the object of debate in development and ethnographic studies of local cultures in Third World countries (Sillitoe et al. 2002, Bruckmeier and Tovey 2009c).

The originality of the CORASON project lies in the way it highlights the relevance of local knowledge as a necessary component in the sustainable management of local resources in Europe. It could of course be argued that in Europe the prevalence of industrial agriculture and the fact that a large part of the rural population has undergone a process of formal education and professional training have both undermined the foundations of local knowledge. The CORASON project set out to explore the role of local knowledge in contemporary rural development and it indicates that the best way of interpreting the present situation is to posit an interaction between, and a blending of, knowledge forms as processes trough which local knowledge and its related practices are updated, not finally eroded (Bruckmeier and Tovey 2009c: 270).

Local food is a privileged domain for exploration of knowledge dynamics, not least because food production and preparation are among the oldest of human activities: activities within which different forms and processes of knowledge have found optimum expression and become consolidated. A second reason for interest derives from the fact that in the second half of the twentieth century agriculture became a special field for application of scientific knowledge, through the spread of what is today known as the 'green revolution', a development admittedly disastrous for the patrimony of knowledge accumulated by generations of farmers.

Revitalisation of local food economies necessarily implies a renewed attention to local conditions of production and consumption. Local food networks may not only represent resistance to the globalised, placeless reorganisation of food chains but may also serve to challenge a continuous trend towards simplification and homogenisation of agricultural techniques and agro-ecosystems, leading to a revaluation of traditional/local forms of knowledge and techniques and their recognition as a specific and important resource in the management of agricultural and natural ecosystems.

A critical analysis of the notion of 'local knowledge', with discussion of its ambiguity, is developed in CORASON project publications (Bruckmeier and Tovey 2009c, Fonte 2008). The concept is also explored by the various authors who have written chapters for this volume. Tovey and Bruckmeier (2009b) and Fonte (2008) stress the importance of drawing distinctions between *tacit* knowledge and *lay* knowledge.

Tacit knowledge is understood as being 'the sort of knowledge which we use, more or less unconsciously, to manage our interactions with other people' (Bruckmeier and Tovey 2009c: 273). Created through normal processes of socialisation, this is a form of knowledge transmitted pre-discursively in a community through its social norms and habits. It is important in rural development because it helps to strengthen informal social networks and social relations, promoting trust and social cohesion. Lay knowledge, by contrast, is 'about "objective reality", practical causal connections or "how things work"'

(Bruckmeier and Tovey 2009c: 273). It is a technical form of knowledge acquired through particular experiential circumstances and transmitted by specific 'local experts' in informal learning situations. It differs from 'scientific' knowledge in that it is neither standardised nor formal. Its variability (linked as it is to specific places and cultures) has earned it a status that is inferior to that of 'scientific' knowledge.

The authors of this volume agree on the importance of local knowledge in the organisation of alternative, sustainable food economies, while offering different perspectives on how it should be defined or characterised. Kvam, for example, stresses the importance of the tacit components in lay local knowledge, to the extent even of making it difficult for the two to be distinguished; Adamski and Gorlach introduce a category of 'adaptive local knowledge' to denote a modification, indeed a 'misuse', of traditional forms of lay knowledge under the pressure of economic opportunities opened up by mass tourism. Papadopoulos focuses on a specifically ecological variant of local knowledge, analysing the practical skills and the intelligence that are acquired through interaction with a constantly changing environment. Bruckmeier, but also Dargan and Harris, stress how the boundaries between 'lay' and 'expert' knowledge become uncertain in the construction of local food, an expert being, in that context, a person with expertise in specific traditional and artisanal practices of food production.

Notwithstanding all these differences, common understandings and findings do emerge from the case studies: local knowledge is not only an important resource for local development but is also a constitutive element in the identity of rural communities and in construction of their sense of place. The analysis of local food projects does not presuppose an opposition between local/lay and global/scientific forms of knowledge. It is rather an analysis of institutional processes, social mechanisms and networks through which ideas and ways of acquiring knowledge are empowered and legitimated.

Local food and the dynamics of knowledge

Fonte (2008) has elsewhere identified two types of conceptual reference for the case studies reported in this volume: a re-connection perspective and an origin-of food perspective (see Boxes I.2 and I.3).

The re-connection perspective supports strengthening of the social relations between producers and consumers at the exchange site as a way of strengthening rural community and augmenting the sustainability of food production systems. In the origin-of-food perspective, 'local' acquires a temporal dimension, denoting a 'place' where common history and a common belonging have consolidated into collective norms, traditional forms of knowledge and 'typical' products.

Box I.2   Case studies in the reconnection perspective

Ireland
The C— Farmer Market in Tipperary, south-east Ireland, was established by the C— Development Association with the aim of attracting people to the village of C— on Saturdays and promoting the sale of a wide range of local products.

Germany
Netzwerk Vorpommern is a voluntary association started in 1995 by a group of active organic food consumers. The initiative gradually grew, with various activities supporting new projects for a sustainable local and regional development.

Scotland
The Skye and Lochalsh Horticultural Development Association, in Scotland, was set up in 1995. It is a network of actors committed to support horticulture on Skye and teach local farmers horticultural skills that have gradually become lost.

Sweden
The Eldrimner initiative is a rural network for the small-scale refinement of agricultural products centred in Rösta, in the municipality of Ås in Jämtland.

Box I.3   Case studies in the origin of food perspective

Portugal
The construction of the Barrancos cured ham PDO (Protected Designation of Origin).

Italy
The construction of the 'Aspromonte National Park Product' certification.

Poland
The valorisation of oscypek cheese.

Norway
Valdres Rakfisk brand (traditional fermented fish).
Kurv frå Valdres BA (traditional salami).

Spain
Utiel-Requena PDO wine.
Requena sausages Protected Geographic Indication.

Greece
Mavro Messenikola wine production 'Quality Wine Produced in Specific Region' (VQPRD).
Nemea wine production (VQPRD).

From the CORASON case studies it clearly emerges that the understanding of localness from the two different perspectives reflects their differing agro-food contexts. The context for the first perspective is what is called a 'food desert',

Table I.1   From the 'cold' negotiation of the market to the face-to-face relations of local food production

Consumers/ Markets	Rural development strategy	Social relation consumers – producers
Local/Local	Territorial development through valorisation of regional food; integrated rural development strategies	Face-to-face routine relations
Local/Distant	Migrants markets	Face-to-face regular but spatially discontinuous relations
Distant/Local	Rural tourism	Face-to-face discontinuous relations
Distant/Distant	Product/sector strategies of rural development Certification for access to differentiated markets	Market connection through information/certification

Source: Own elaboration.

i.e. a place where there is no potential for local provisioning of food and where supermarkets are the only place to shop for food. The context for the second is an environment of socio-economic marginalisation, persistence of small farms and traditional food production and consumption practices.

Local lay knowledge is important in both contexts. In the marginalised areas by-passed by modernising programmes of agricultural industrialisation, local lay knowledge may take the form of 'traditional' knowledge, associated with pre-industrial practices of production and transmitted from generation to generation of farmers. In the context of 'food deserts' we see how the efforts to re-localise food necessarily implies an effort to re-create or create *ex novo* (as in eastern Germany) a local lay knowledge of growing and preparing food. Thus, side by side with efforts to valorise and mobilise traditional local knowledge, we find in European rural areas efforts to re-create the conditions for development of 'non-traditional' forms of local lay knowledge, from a variety of sources, formal and informal, oral and written, and with prominent involvement of social movements such as the movement supporting organic agriculture.

Consumers have a special role to play in these processes: elaboration of a new definition of quality demands their involvement in the food system. Direct relation with producers at the level of exchange is one way of strengthening trust and reciprocity in the rural community, establishing and sustaining a common sense of place, fostering tacit and lay knowledge through operation of the local food system (see Dargan and Harris's contribution to this volume).

In marginal rural areas depopulated through emigration, local markets have declined and lost the ability to provide a sustainable livelihood for the local population. Very often rural development strategies aim at inverting this trend and revitalising local economies, both by producing local food that will be traded at a distance and by attracting tourists. At the same time local markets are also expanding as, in distant places, migrants maintain their local food culture and, through it, a link to their local community.

'Local markets', bearing in mind these processes, could be subdivided into two components: consumers and place of exchange (Table I.3). A whole spectrum of variations is possible from the 'cold' negotiations of distant markets to the 'warm' sociality of direct exchange in local markets (Callon 1998), implying different types of relations between producers and consumers.

Re-skilling farmers and consumers in the new local food economies

In the food desert created by trade- and export-oriented industrialisation of the food production, processing and retailing industries, most uncodified lay knowledge about how to produce food crops and how to prepare them for consumption has been expropriated from farmers and consumers:

> [A]rtisan production and processing of food has existed before, but the modernisation of agriculture during the past century led to an 'intellectual expropriation' of the local producers and farmers and their tacit knowledge about agriculture and food production. (Bruckmeier et al. 2006, 12)

> Local knowledge and skills in food production have largely vanished, even among rural populations. (Bruckmeier, in this volume: 118-119)

> With produce readily available in the supermarkets, fewer and fewer people grew their own food, and the pool of tacit knowledge around this type of food production was gradually lost. (Dargan and Harris, in this volume: 85)

The knowledge needed by the small artisan producers in these networks has to do firstly with learning how to grow food in accordance with non-conventional agricultural practices that take into account local conditions and resources. Initiatives to relocalise food systems include attempts to educate both food growers and consumers in matters of food quality and consumption practices. In the Scottish and Swedish cases a key objective of the project organisers was to re-skill farmers in agro-food practices that had been lost in their area and to re-educate consumers in the characteristics of local foods and methods for preparing and cooking them.

Scientific knowledge is not always considered appropriate by local farmers, given the scale of their production and the specific difficulties of the growing conditions they face. In their daily routine they need to be able to avail themselves of the local expertise of other farmers and of the residents of their area generally.

This makes it possible for new combinations of lay and expert knowledge to be generated, and local growers come over time to be recognised as 'experts'. The newly created knowledge is then shared with other local growers by word of mouth, through mentoring schemes, and through printed materials (see Dargan and Harris, in this volume).

The Eldrimner initiative in Sweden included setting up a resource centre to convey local knowledge of small-scale food production and food processing (cheese-making, pork butchering and jam-making) to wider groups of local actors. It provides courses on how to improve product quality and assists with the procedures involved in starting and managing small enterprises. For the Eldrimner initiative the 'expert' is not a scientific specialist but someone with experience, 'somebody who has already done it'. To revitalise local knowledge various methods have been followed: knowledge is compiled from elderly people in local communities as well as from many other sources, through contacts with local producers in other countries, in literature as well in archives, and through the information and networking in the project, which often resulted in new members with special knowledge joining the project (see Bruckmeier, in this volume).

In our Irish case study many stallholders at the farmers' market are, or have been, members of the organic movement, which to them is an important source of knowledge about how to produce using small-scale, environmentally friendly techniques. Also important are other informal and formal sources of knowledge, including older farmers, experience and common sense, books, courses, networks and contacts with 'experts'. Consumers are involved in exchanging knowledge about food quality and ways to prepare food, especially at the point of purchase through interaction with the grower/seller, but also in other events like food tasting, exhibitions and school programmes (see Tovey, in this volume).

The East German case draws attention to the fact that the concept of 're-localisation' implies local production as a historical starting point (Siebert and Laschewski, in this volume). But, as the authors argue, in many peripheral rural areas of Central and Eastern Europe agriculture has always been export-oriented and characterised by a history of expulsion and mass emigration. In such circumstances it is difficult for local actors to find a common tradition from which to initiate a process of re-creating a locality. Locality therefore has to be built again from scratch; the ecological paradigm and the de-contextualised concept of organic farming offers a useful framework of reference for the regionalisation of food production. Consumers' knowledge of food appears to be the most significant impulse behind the creation of a new tradition of local food.

These examples suggest that scientists from universities or bureaucratic-managerial experts from governmental development agencies are not the best experts, and scientific knowledge not the most relevant form of knowledge, for local food initiatives. Sometimes scientific knowledge may be an appropriate starting point, but it needs to be integrated, adapted and mediated by those with expertise and trained in specific traditional and artisan modes of food production.

Local knowledge is rebuilt through experience, including experience of exchanges with other growers, with farmers in other regions (nationally and internationally), experience of relationships with the consumers at farmers' markets, or through formal and informal contacts with experts. In this process of creation, re-invention and consolidation of local knowledge, new social networks are created and rural communities strengthened. The definition of 'expert' is broadened to include non-scientists; knowledge production becomes more inclusive.

Recovering and valorising traditional knowledge

On the peripheries of modernity and agro-industrial development local knowledge has been conserved firstly in the form of traditional knowledge, as part of the local culture of growing, producing and preparing food in a specific socio-agro-ecosystem. Cultivars adapted to specific locations are the outcome of centuries of collective communal work on domesticating and adapting plants and animals to the geographical micro-habitat. They embody characteristics both of geographical places and of the empirical knowledge of generations of farmers.

> Preparing semi-fermented trout has been a food tradition since the sixteenth century or earlier, with the producers sourcing trout from local lakes. (Kvam, in this volume: 203)

> Oscypek is a smoked cheese made of sheep's milk or a mixture of cow's and sheep's milk. It is an important part of the shepherding tradition, with a history going back to the fifteenth century ... For hundreds of years it was produced in the mountains by local shepherds. (Adamski and Gorlach, in this volume: 174)

> [The Utiel-Requena area] is a traditional wine-producing region, with one of largest, but at the same time most compact, vineyard areas in Spain. The production of wine here dates back to prehistoric times... (Buciega et al., in this volume: 219)

> The Alentejano-breed pig (*Sus ibericus*) has constituted the basis of the local diet over the centuries due to the range of products it supplies and its ease of preservation, using simple techniques that make possible year-round consumption. (Rodrigo and Veiga, in this volume: 135)

Undervalued and dismissed under the technocratic assumptions of national and local development agencies during the agro-industrial era (van der Ploeg 1986; Benvenuti et al. 1988), traditional lay knowledge attracts new interest today. Markets and policies articulate a demand for quality and for regional diversification of food, necessitating a step back from the homogenisation of industrial agriculture. New technological and institutional developments, such as biotechnologies and the strengthening of intellectual property rights on seeds, have intensified the preoccupation with conservation and valorisation of biodiversity.

Traditional indigenous knowledge accompanying the practice and the conservation of biodiversity, especially in developing countries, has become a valuable asset, to be defended from appropriation by private interests.

A wide debate has developed in international fora (the United Nations Conference on Trade and Development (UNCTAD), the World Intellectual Property Organization (WIPO), the World Trade Organization (WTO) and numerous non-governmental organisations such as the Action Group on Erosion, Technology and Concentration (ETC Group) or GRAIN[6]) on the value of traditional knowledge and the necessity for it to be protected. In these contexts traditional knowledge is seen as knowledge that is generally, 'not produced systematically, but in accordance with the individual or collective creators' responses to and interaction with their cultural environment' (WIPO 2002: 1). It does not perform a specialised function in society, but rather embodies cultural values as an element integrated into a vast and mostly coherent complex of beliefs and knowledge that is for the most part held collectively and transmitted both orally and through common practices, from generation to generation. In this context the term 'traditional' can be used of a form of knowledge

> only to the extent that its creation and use are part of the cultural traditions of communities. 'Traditional', therefore, does not necessarily mean that the knowledge is ancient. 'Traditional' knowledge is being created every day, it is evolving as a response of individuals and communities to the challenges posed by their social environment. In its use, traditional knowledge is also contemporary knowledge. (WIPO 2002: 1)

Arguably, then, it is not so much the contents or forms of knowledge that distinguish the 'traditional' and 'local' from the 'scientific' and 'managerial'. It is more the specific way in which they are created and transmitted. The CORASON research makes it clear that local and traditional lay knowledge persists in many European rural areas, not only in the southern, Mediterranean, countries, but also in the marginalised areas of northern European countries such as Norway or eastern European countries like Poland. A marginalisation process lasting for decades blocked the co-evolution of traditional knowledge in response to changes in the functions of food and new consumption habits. We accordingly find in our case studies that for certain types of production traditional lay knowledge may have the reputation of being static or outdated. Relocalisation of food, though, sets in motion processes of recovery and valorisation of traditional lay knowledge that result not only in interaction and dialogue, but also in confrontation, with other forms of knowledge and other actors, experts and managers. In the next sections

---

6   GRAIN is not an acronym but is the name of 'an international non-governmental organisation (NGO) promoting sustainable management and agricultural biodiversity on the foundation of people's control over genetic resources and local knowledge' (www.grain.org).

we propose to examine two special instances of this dynamic that are accorded particular emphasis in our case studies: the elaboration of origin (or provenance) certification and the nexus between experts and lay knowledge in the wine sector.

Provenance certification: opportunities and risks for local knowledge

A discussion of local food and local knowledge cannot avoid taking into account certification. Certification has become the dominant route for recovering, codifying and valorising the lay knowledge embodied in local products. It is a contested process, in which local lay knowledge comes up against other forms of managerial and scientific knowledge. In her presentation of the Italian case study, Fonte (in this volume) draws attention to the many approaches to be found in the literature on certification. Certification can be seen as a tool of governance in a system of civic agriculture (Oosterveer 2007), a neo-liberal tool in a new food regime based on quality governance, an information tool or a hybrid forum for the development of a dialogue among different forms of knowledge.

Certification stands as an opportunity or looms as a risk in all our case studies. It introduces local networks to an adjustment process whose economic, social and cognitive results are not defined *a priori* and are dependent upon the power relationships inherent in the process of its construction: local actors may lose or gain significant bargaining power and win or forfeit representation in the development of certification (see Rodrigo and Veiga, in this volume). In the reconnection perspective, certification is mostly perceived as the risk of de-linking consumption from production (see Tovey, in this volume). But it is not regarded as a priori incompatible with a local food economy. The Eldrimner project, for example, sees the development of certification for small-scale products as a way for them to become more independent of national and EU funding (see Bruckmeier, in this volume).

It is, of course, first and foremost from within the origin-of-food perspective that certification is considered and discussed, being presented as an inclusionary or exclusionary economic process leading either to expropriation or to an improvement and updating of local knowledge.

The case studies of Portugal and Italy (the development of Barrancos cured ham and the Aspromonte National Park certification) provide deep insights into the evolution of the certification process. In both cases the interest in certification first emerged among groups external to the producers, the Department of Zootechny at the University of Evora in the Portuguese case and the managers of the Aspromonte National Park in the Italian case. The certification process is initiated through selection of one or more exemplary farmers. Their production practices are observed, some improvements or modifications are suggested in production (most commonly in relation to hygiene) and production protocols are compiled (i.e. codification is carried out). Certification of origin for a local product thus involves compilation and selection of the available stock of local traditional knowledge as well as interaction with expert, managerial or scientific knowledge (see Rodrigo and Vega, in this volume and also Fonte, in this volume).

The two chapters by Rodrigo and Vega and by Fonte illustrate the opposite results to which such processes may lead. The 'ethnographic' description of the Barrancos cured ham certification process in Portugal is very impressive. The image of the university researcher in charge of the certification process who spends a year 'recording the various stages of manufacture (of the local producers) and listing the unforeseen occurrences, without involvement in the technological matrix' conveys a powerful impression of the top-down process that will lead to economic restructuring of the sector and exclusion of the local actors, both from participation in the cognitive process and from the economic benefits of certification.

Local product certification in the Aspromonte National Park (Italy) is by contrast envisaged as a civic action, aimed at improving the image of the locality and strengthening both the economy and the community. Certification becomes a process of negotiation among local actors, a cognitive process with aspects of participative intervention in development, serving to increase the local actors' awareness of the importance of local knowledge and the value of local resources (also see Kvam, in this volume).

The chapter on the valorisation of oscypek cheese, produced in the Podhale region of southern Poland, introduces additional dimensions into the analysis of the effects of certification (see Adamski and Gorlach, in this volume). Traditional knowledge may be appropriated not only by experts but also by other local producers. The complex relationship between a traditional product, local knowledge, rural development and certification is here well illustrated. The cheese is part of the shepherding tradition of the Tatra Mountains and is produced in summer in mountain sheds, using non-pasteurised sheep milk. It was made by shepherds for their own consumption and for sale in the local villages. The social and political changes of the 1990s, in particular the fall of the communist regime, the decrease in sheep stock and the development of mass ski tourism in the Tatra Mountains, all contribute to the great economic transformation opening up new opportunities for oscypek cheese, which now becomes a valuable commercial asset.

The proliferation of economic opportunities favours the emergence of new production networks, further developing and transforming traditional knowledge. Three separate social and economic networks develop around the re-elaboration of the traditional knowledge, each offering economic opportunities to different local actors, each differently embedded in both the conventional and the alternative food systems.

Will the institutional process for PDO certification of the oscypek cheese limit rather than expand opportunities for the economic development of the Tatra mountains? Is local knowledge of the production procedure for oscypek cheese a collective property of the traditional shepherds or does it belong to the whole community? Will 'misuse' of the traditional knowledge end with its final erosion and homogenisation of production? These are some of the new questions opened up by the certification process for oscypek cheese.

The risk of appropriation of local knowledge by experts and big manufacturers is greater when products possess the potential to become 'global' products, that is to say, when production reaches a minimum quantity sufficient for industrial production and the link to local consumers and the local food culture is weakened (as in the Portuguese and the Polish case studies). When small niche products are the object of valorisation, certification may constitute an important element for activation of an integrated rural development strategy by local actors. Local producers and citizens often promote participative certification schemes as part of a more comprehensive initiative for valorisation of the local cuisine through fairs and festivals to attract tourists into the area, particularly former residents who have migrated to other places. We here see deployment of a multiplicity of post-industrial rural development strategies, with tourism as their common element. Rural tourism has the potential to create complementarities, synergies, cohesion as between the different rural activities of a territory. The traditional or local lay knowledge that is mobilised interacts, by contrast, mostly with the managerial knowledge that is necessary for setting up and administering rural development projects (see especially the valorisation of cold meat in the village of Requena, Spain and the Norwegian case studies).

The nexus between traditional and expert knowledge: the case of winemaking

In many food processes, such as the production of oscypek cheese in the Podhale region of Poland, Barrancos cured ham in Portugal, the fermented fish and salami of Valdres in Norway, traditional artisan knowledge is the key element from which the product's excellence is derived.

In other food sectors, such as olive oil production and even more so wine production, the contribution of expert technical knowledge to the production process is of the utmost importance for attainment of what are today considered high standards of quality. It is in the initiatives to valorise the origin of wine and olive oil (in Greece, Spain and Italy particularly) that the limits of traditional lay knowledge start to become evident, as may be seen from the felt need for a nexus to be established with technical and scientific knowledge (Buciega et al. in this volume, Papadopoulos in this volume). The environment for knowledge production is highly institutionalised, through specialised technical schools, co-operatives and PDO institutions. Further elaborations for the development of these issues emerge in the Italian, Greek and in the Spanish case studies.

Local varieties of wines (Mavro Messenikola in the Lake Plastiras area and Ayiorghitiko in the Nemea area in Greece; Bobal in Utiel-Requena, Spain; Nerello Mascalese and Nerello Cappuccio in Palizzi, Aspromonte, Italy) have adapted over centuries to their specific agro-natural habitat, thanks to the work and the empirical knowledge of generations of farmers, who were also winemakers. Since the mid-twentieth century, however, wine has ceased to be a subsistence product consumed by the farmers' families. It has been transformed into a commercial good. Vineyard cultivation is segregated from winemaking and the sector goes through a process of commercialisation and specialisation culminating in the co-existence of separate economic and social structures. The wine industry in Europe today is evidently something complex, comprising family wine cellars (where winemaking remains linked to the farm), specialised commercial enterprises and social co-operatives.

Along with this differentiation process, in the process of which vineyard cultivation has become something more and more separate from winemaking, specialised public schools have been established, with corresponding professionalisation of the technical knowledge required for producing high-quality wines. Professionalisation and reliance on formal knowledge has been strengthened since the 1970s, when there was a turn to quality wines and the 'oenologist' emerged as the 'expert' who understands the chemical process of wine fermentation. The travelling oenologist, who sells his knowledge to many different winemaking companies – the 'flying winemaker' – has become a powerful international actor in the global industry (Lagendijk 2004 quoted by Papadopoulos, in this volume).

Isolated from the evolution of the markets and the product's new roles, traditional lay knowledge of winemaking has come to be seen as outdated:

> Traditionally wine was produced for self-consumption and for the local market and responded to different functions and tastes compared to today. It was an energetic drink, targeted for consumption within the year, rather than ageing. Only new techniques can create the conditions to keep and even improve wine characteristics during ageing. (The president of Qualiter Co-operative, in Fonte, Agostino and Acampora 2006: 21)

In the case of the Utiel-Requena PDO wine (Spain) the limitations of traditional lay knowledge in winemaking may be attributed to the fact that the area in the past was associated with a different specialisation: production of *doble pasta*, which was used to add colour to other wines. But the establishment of oenology schools in the 1960s and subsequently has led to technical and expert knowledge taking the lead in the process of winemaking, marginalising local knowledge (see Buciega et al. in this volume).

In Greece the diffusion of an agro-industrial and productivist logic, with its stress on high yields and increased quantities, has meant the loss of local lay knowledge of vineyard cultivation and winemaking:

> In the past vineyard yields were smaller but the wine was of much higher quality. And other products, for example *tsipouro,* were also made from the remains of grapes. The new tacit knowledge based on agro-industrial logic has displaced the former repertoire of practices and of experiential knowledge. There was a break with former knowledge repertoires, justified on the basis of the higher incomes and guaranteed prices... (Papadopoulos, in this volume: 256)

This separation between lay and expert knowledge in the evolution of the wine industry has produced a gap between 'industrial wines' and 'terroir wines'. The quality of the former is associated primarily with the brand and with winemaking techniques, while for the latter it is linked to the ensemble of properties conveyed by the concept of 'terroir', i.e. a conjunction of human (history, cultural, technical) and natural characteristics (local variety of grapes, soil and micro-climate). There is thus a perennial tension between the two concepts of quality, also implying a different dynamics of knowledge. According to Buciega et al. (in this volume), the mode of incorporation of new knowledge, primarily codified technical and managerial knowledge, into the wine production process in the Utiel-Requena region (Valencia, Spain) was such as to preclude interaction and communication between traditional/lay and codified/technical knowledge. Nevertheless, the development of Labels of Origin with their emphasis on the ecology and culture of specific places has the potential to re-embed wine 'in the natural processes and social context of its territory' in a system that is 'nested with multiple levels of co-ordination from the local to the global' (see Buciega et al. in this volume: 224).

The chapter by Papadopoulos (in this volume) makes the point that there is a certain convergence in the quality and knowledge trajectory of 'industrial' and 'terroir' wines. An illustration is provided to corroborate what may seem a paradoxical finding: the traditional farmer is not always able to participate in the construction of wine quality based on the territorial identity of the product. He may remain locked in the agro-industrial logic of high quantity, supported by the productivist policy of local institutions such as the Union of Wine Co-operatives in the Lake Plastiras area. By contrast, the success of the Nemea area in constructing the 'terroir' for a quality strategy in winemaking is attributed to a capacity to generate interactions and exchanges between different forms of knowledge within the area and with other areas. The local winegrowers possess a stock of tacit and lay knowledge linked to the Ayiorghitiko variety of grapes, while the new wineries that have relocated their activity in Nemea bring the scientific knowledge and the dynamism that is necessary for reconstruction of the locality as a quality wine area. The rhetoric of traditional local knowledge and the local/expert nexus play a vital role in construction of the quality narrative, issuing a challenge to the conventional, industrial wine sector (Papadopoulos, in this volume).

Concluding remarks

Local food can be seen as a political project pursuing the construction of new food communities among producers and consumers, centred around shared civic values of equity, justice and holistic sustainability. No model is more 'alternative' than any other for the accomplishment of this objective: community supported agriculture, the farmers' market, certification schemes – all are equal contenders. Every form of local food is susceptible to appropriation and commodification by the dominant global economy. But cultural anthropology teaches us that commodification is a specific process that may be counteracted by an opposite process of de-commodification where products are attributed values other than their exchangeability (e.g. local food value). One paradox of local food is its capacity to embody de-commodification in the same market place, re-embedding the exchange act in sociality and (in some cases) in a project, common to producers and consumers, of building an alternative food economy. The great contribution of local food literature is precisely its identifying and stressing the importance of exchange relations in the local market (as opposed to the global market) in the construction of new models of food production and in promoting a 'moral economy', as against the commodifying push of the global economy.

But our aim, both in the CORASON project and in this volume, goes beyond this. Placing at the centre of our analysis the dynamic existing between different forms of knowledge (scientific, managerial, local) and the role assigned to local knowledge in the development of local food, we would like to stress that no new food economy is possible without a reform of the dominant scientific and knowledge-production processes.

The case studies considered in this volume suggest that local knowledge in the European countryside cannot be dismissed as useless or totally eroded. That established, the fact remains that efforts to rebuild new food communities will face problems of recovering, valorising, re-inventing or even re-building local ecological knowledge of the context in which food is grown, prepared and consumed. The new food communities must be constituted not only as reflexive political subjects but also as learning communities where democracy is predicated on the capacity to recognise importance, status and dignity in the different knowledge forms and their bearers, not the least being local knowledge and those possessing it.

References

Appadurai, A. 1986. Introduction: commodities and the politics of value, in *The Social Life of Things: Commodities in Cultural Perspective*, edited by Appadurai, A. Cambridge: Cambridge University Press, 3–63.

Becattini G. 1989. Riflessioni sul distretto industriale marshalliano come concetto socio-economico. *Stato e Mercato*, 25, 111–128.

Beck, U. 1992. *Risk Society: Towards a New Modernity*. London: Sage.

Beck, U. 1997. *The Reinvention of Politics: Rethinking Modernity in the Global Social Order*. Cambridge: Polity Press.

Beck, U., Giddens, A. and Lash, S. 1994. *Reflexive Modernization: Politics, Tradition and Aesthetics in the Modern Social Order*. Cambridge: Polity Press.

Benvenuti, B., Antonello, S., De Roest, C., Sauda E. and van der Ploeg J.D. 1988. *Produttore agricolo e potere*. Roma: Consiglio Nazionale delle Ricerche – Istituto Per la Ricerca nell'Agroalimentare.

Bicker, A., Sillitoe, P. and Pottier, J. 2004. *Investigating Local Knowledge: New Directions, New Approaches*. Aldershot: Ashgate.

Blythman, J. 2005. Organic food is not necessarily the automatic choice for the ethical consumer. *The Ecologist*, 17 June 2005.

Boltanski, L. and Thévenot, L. 1991. *De la Justification. Les économies de la grandeur*. Paris: Gallimard.

Bowles, S. and Gintis, H. 2002. Social capital and community governance. *Economic Journal*, 112(November), 419–436.

Brown, L.R. 1970. *Seeds of Change: The Green Revolution and Development in the 1970s*. New York: Praeger.

Bruckmeier, K. 2004. Theoretical and conceptual framework (final version). *CORASON-project WP 2 Input paper*. Available at: http://www.corason.hu [accessed: 16 June 2009].

Bruckmeier, K. and Tovey, H. 2008. Knowledge in sustainable rural development: from forms of knowledge to knowledge processes. *Sociologia Ruralis*, 48(3), 313–329.

Bruckmeier, K. and Tovey, H. (eds.) 2009a. *Rural Sustainable Development in the Knowledge Society*. Aldershot: Ashgate.

Bruckmeier, K. and Tovey, H. 2009b. Natural resource management for rural sustainable development, in *Rural Sustainable Development in the Knowledge Society*, edited by Bruckmeier, K. and Tovey, H. Aldershot: Ashgate, 1–19.

Bruckmeier, K. and Tovey, H. 2009c. Beyond the policy process: conditions for rural sustainable development in European countries, in *Rural Sustainable Development in the Knowledge Society*, edited by Bruckmeier, K. and Tovey, H. Aldershot: Ashgate, 267–288.

Bruckmeier, K., Engwall, Y. and Höj-Larsen, C. 2006. *Local Food Production and Knowledge Dynamics in Rural Sustainable Development*. Manuscript: CORASON-project, report work package 6; Gothenburg University, Human Ecology Section. Available at: http://www.corason.hu [accessed: 25 November 2009].

Buck, D., Getz, C. and Guthman J. 1997. From farm to table: the organic vegetable commodity chain of northern California. *Sociologia Ruralis*, 37(1), 1–20.

Burawoy, M. 2005a. For public sociology. *American Sociological Review*, 70, 4–28.

Burawoy, M. 2005b Third-wave sociology and the end of pure science. *The American Sociologist*, 36, 152–165.

Buttel, F.H. 2000. Ecological modernisation as social theory. *Geoforum*, 31(1), 57–65.

Callon, M. (ed.) 1998. *The Laws of the Markets*. Malden, MA: Blackwell.

DeWeerdt, S. 2008. Is local food better? Yes, probably – but not in the way many people think. *World Watch Magazine*, 21(3). Available at: http://www.worldwatch.org/node/6064 [accessed: 5 October 2009].

Dolsak, N. and Ostrom, E. 2003. *The Commons in the New Millennium: Challenges and Adaptation*. Cambridge, MA: MIT Press.

DuPuis, E.M. and Goodman, D. 2005. Should we go 'home' to eat? Toward a reflexive politics of localism. *Journal of Rural Studies*, 21(3), 359–371.

DuPuis, E.M., Goodman, D. and Harrison, J. 2006. Between the local and the global: confronting complexity in the contemporary agri-food sector research. *Rural Sociology and Development*, 12, 241–268.

Etzioni, A. 2006a. A neo-communitarian approach to international relations. *Human Rights Review*, 7(1), 69–80. Available at: www.gwu.edu/~ccps/etzioni/articles1.html [accessed: 15 October 2009].

Etzioni, A. 2006b. Communitarianism, in *The Cambridge Dictionary of Sociology*, edited by Turner, B.S. Cambridge: Cambridge University Press, 81–83.

Evenson, R.E. and Gollin, D. 2003. Assessing the impact of the Green Revolution 1960 to 2000. *Science*, 300(5620), 758–762.

Fonte, M. 2008. Knowledge, food and place: a way of producing, a way of knowing. *Sociologia Ruralis*, 48(3), 200–222.

Fonte, M. and Grando, S. 2006. *Local Food Production and Knowledge Dynamics in the Rural Sustainable Development*. Input paper for the Working Package 6 (Local food production) of the CORASON Project. Unpublished manuscript.

Fonte, M., Agostino, M. and Acampora, T. 2006. Italy south WP6 country report for the CORASON project. Available at: http://corason.hu/download/wp6/wp6_sitaly.pdf [accessed: 5 October 2009].

Garnett, T. 2007. *Animal Feed, Livestock and Greenhouse Gas Emissions: What are the Issues?* Paper presented to the Society of Animal Feed Technologists, Coventry, 25 January 2007.

Garofoli, G. (ed.) 2003. *Impresa e territorio*. Bologna: il Mulino.

Gereffi, G., Humprey, J. and Sturgeon, T. 2004. The governance of global value chains. *Review of International Political Economy*.

Gibson-Graham, J.K. 2006. *Postcapitalist Politics*. Minneapolis, MN and London: University of Minnesota Press.

Granberg, L., Kovách, I. and Tovey, H. (eds.) 2001. *Europe's Green Ring*. Aldershot: Ashgate.

Guthman, J. 2002. Commodified meanings, meaningful commodities: re-thinking production-consumption links through the organic system of provision. *Sociologia Ruralis*, 42(4), 295–311.

Guthman, J. 2003. The trouble with 'organic lite' in California: a rejoinder to the 'conventionalisation' debate. *Sociologia Ruralis*, 44(3), 301–316.

Guthman, J. 2007a. Why I am fed up with Michael Pollan et al. *Agriculture and Human Values*, 24, 261–264.

Guthman, J. 2007b. The polanyan way? Voluntary food labels as neoliberal governance. *Antipode*, 39(3), 456–478.

Harding, S. 2008. *Science from Below: Feminism, Postcolonialism and Modernities*. Durham, NC and London: Duke University Press.

Harris, E. 2008. *Exploring Localism in Alternative Food Networks*, Presentation at the Scottish Colloquium on Food and Feeding, December 12. Available at: http://localfoods.files.wordpress.com/2008/11/local-agendas-diagram.png [accessed: 5 October 2009].

Hendrickson, M.K. and Heffernan, W.D. 2002. Opening spaces through relocalization: locating potential resistance in the weaknesses of the global food system. *Sociologia Ruralisi*, 42(4), 347–369.

Hinrichs, C.C. 2003. The practice and politics of food system localization. *Journal of Rural Studies*, 19(1), 33–45.

IAASTD 2009. *Agriculture at a Crossroads*. Washington DC: Island Press. Available at: http://www.agassessment.org/docs/ [accessed: 16 October 2009].

Jasanoff, S. and Martello, M.L. 2004. Introduction: globalization and environmental governance, in *Earthly Politics: Local and Global in Environmental Governance*, edited by Jasanoff, S. and Martello, M.L. Cambridge, MA: MIT Press, 1–29.

Kloppenburg, J. Jr. 1991. Social theory and the de/reconstruction of agricultural science: local knowledge for an alternative agriculture. *Rural Sociology*, 56(4), 519–548.

Kopytoff, I. 1986. The cultural biography of things: commoditization as process, in *The Social Life of Things: Commodities in Cultural Perspective*, edited by Appadurai, A. Cambridge: Cambridge University Press, 64–91.

Latour, B. 1987. *Science in Action*. Cambridge, MA: Harvard University Press.

Latour, B. 1998. From the World of Science to the World of Research? *Science*, 280(5361), 208–209.

Latour, B. 2004. *Politics of Nature: How to Bring the Science into Democracy*. Cambridge, MA: Harvard University Press.

Leydesdorff, L. and Etzkowitz, H. 1998. The triple helix as a model for innovation studies. *Science and Public Policy*, 25(3), 195–203. Available at http://users.fmg.uva.nl/lleydesdorff/th2/spp.htm.

Lockie, S. 2009. Responsibility and agency within alternative food networks: assembling the 'citizen consumer'. *Agriculture and Human Values*, 26(3), 193–201.

Lyson, T.A. 2004. *Civic Agriculture: Reconnecting Farm, Food, and Community*. Medford, MA: Tufts University Press.

Marsden, T., Banks, J. and Bristow, G. 2000. Food supply chain approaches: exploring their role in rural development. *Sociologia Ruralis*, 40(4), 424–438.

Maye, D., Holloway, L. and Kneafsey, M. 2007. Introducing alternative food geographies, in *Alternative Food Geographies: Representation and Practices*, edited by Maye, D., Holloway, L. and Kneafsey, M. Oxford: Elsevier, 1–20.

McMichael, P. 2008. Multi-functionality vs. food sovereignty? *Sociologia Urbana e Rurale*, 30(87), 80–99.

Moore, O. 2006. Farmers' markets, and what they say about the perpetual post-organic movement in Ireland, in *Sociological Perspectives of Organic Agriculture: From Pioneer to Policy*, edited by Holt, G. and Reed, M. Wallingford: CABI Publishing, 18–36.

Morgan, K., Marsden, T. and Murdoch, J. 2006. *Worlds of Food Place, Power and Provenance in the Food Chain*. Oxford: Oxford University Press.

Mosher, A.T. 1966. *Getting Agriculture Moving: Essentials for Development and Modernization*. New York: Praeger.

Nowotny, H., Scott, P. and Gibbons, M. 2001. *Re-Thinking Science. Knowledge and the Public in an Age of Uncertainty*. Cambridge: Polity Press.

Oosterveer, P. 2007. *Global Governance of Food Production and Consumption: Issues and Challenges*. Cheltenham: Edward Elgar Publishing Ltd.

Patel, R. 2007. *Stuffed and Starved: From Farm to Fork, The Hidden Battle for the World Food System*. London: Portobello Book.

Pirog, R. and Benjamin, A. 2003. *Checking the Food Odometer: Comparing Food Miles for Local Versus Conventional Produce Sales to Iowa Institutions*, Leopold Center for Sustainable Agriculture. Available at: www.leopold.iastate.edu [accessed: 23 October 2009].

Pirog, R., Van Pelt, T., Enshayan, K. and Cook, E. 2001. *Food, Fuel, And Freeways: An Iowa Perspective on How Far Food Travels, Fuel Usage, and Greenhouse Gas Emissions*. Leopold Center for Sustainable Agriculture. Available at: www.leopold.iastate.edu [accessed: 23 October 2009].

Polanyi, K. 1957. *The Great Transformation: The Political and Economic Origin of our Time*. Boston: Beacon Press.

Pratt, J. 2007. Food values: the local and the authentic. *Critique of Anthropology*, 27(3), 285–300.

Pretty, J.N., Ball, A.S., Lang, T., and Morison, J.L. 2005. Farm costs and food miles: an assessment of the full cost of the UK weekly food basket. *Food Policy*, 30(1), 1–19.

Renard, M.-C. 1999. *Los Intersticios del la Globalización: Un Label (Max Havelaar) para los Pequeños Productores de Café*. Mexico City: CEMCA.

Rose, C. 1986. The comedy of the commons: Custom, commerce and inherently public property. *University of Chicago Law Review*, 53, 711–774.

Scott, J.C. 1985. *Weapons of the Weak: Everyday Forms of Peasant Resistance*. New Haven, CT and London: Yale University Press.

Shultz, T.W. 1964. *Transforming Traditional Agriculture*. New Haven, CT: Yale University Press.

Sillitoe, P., Bicker, A. and Pottier, J. (eds.) 2002. *Participating in Development: Approaches to Indigenous Knowledge*. London and New York: Routledge.

Strassen, S. 2003. *Commodifying Everything: Relationships of the Market*. London: Routledge.

Tovey, H., Bruckmeier, K. and Mooney, R. 2009. Innovation in rural development and rural sustainable development, in *Rural Sustainable Development in the Knowledge Society*, edited by Bruckmeier, K. and Tovey, H. Aldershot: Ashgate, 243–266.

van der Ploeg, J.D. 1986. *La ristrutturazione del lavoro agricolo*. Roma: Edizioni per l'Agricoltura.

van der Ploeg, J.D. 2007. Resistance of the third kind and the construction of sustainability. Paper presented to the XXII ESRS conference plenary session on sustainabilities, Wageningen, 23 August.

Weber, C.L. and Matthews, H.S. 2008. Food-miles and the relative climate impacts of food choices in the United States. *Environ. Sci. Technol*, 42(10), 3508–3513.

WIPO 2002. *Information note on traditional knowledge. WIPO international forum on intellectual property and traditional knowledge: our identity, our future.* Muscat, 21–22 January.

PART I
Reinventing Local Food and Local Knowledge

Chapter 1

'Local Food' as a Contested Concept: Networks, Knowledges, Nature and Power in Food-Based Strategies for Rural Development[1]

Hilary Tovey

Introduction

This chapter reports on a contestation around the meaning of 'local food' which is ongoing in a single place, south County Tipperary in Ireland. The contest is over whether 'local food' means reforming the food system so as to allow those increasingly marginalised by it, local farmers, to receive better returns on their food products, or radically alternativising it by constructing a local food exchange system, viz. a Farmers' Market.

Within global capitalism, tensions between radicalism and reformism are of course longtime features of left wing politics (socialism versus social democracy); they are also extensively discussed, specifically in relation to food, in Raynolds, Murray and Wilkinson's (2007) collection of papers on Fair Trade. These note that in recent years the Fair Trade movement has developed an increasing interest in positioning itself in mass markets and in organising business partnerships which involve large-scale traders, distributors and retailers – 'a strategy that is causing considerable concern and debate within the movement' (Murray and Raynolds 2007: 9) and that increasingly differentiates Fair Trade from the Alternative Trade Organisations which were important to its beginnings. The assumption of Alternative Trade Organisations – that the best way to promote global equality in the food trade is to try to set up different channels of commodity exchange and that the credibility and integrity claimed for these can be upheld through face-to-face relations of trust – are undermined by a move towards expanding exchange networks and the use of increasingly bureaucratised, rationalised certification systems and of independent certification bodies using 'increasingly formalised rules, standards and product labelling procedures' (Murray and Raynolds 2007:

---

1 This chapter has been previously published as an article in *International Journal of Sociology of Agriculture and Food*, 16(2), 21–35. [Online]. Available at: http://www.ijsaf.org/. Thanks to the editors of the journal for permission to reprint it.

18), which come between producers and consumers. Barrientos, Conroy and Jones (2007: 53) summarise the disagreement as between trying 'to advance *an alternative* to mainstream trade', and trying 'to advance Fair Trade *within* the mainstream of commercial retailing' [emphases in original]. On one side the argument is that generating greater benefits for small producers in developing countries requires a move beyond selling only to 'politically correct' outlets and consumers, on the other that developing partnerships with large food retail corporations and large-scale food producers will compromise Fair Trade's roots in cooperative principles and support for small farmers.

The case I discuss here is local rather than global but reveals similar tensions and pressures. Whereas Raynolds and colleagues largely take a developmentalist approach to Fair Trade, tracing this as a process of contested change over *time*, my study is rather one of contested action in *space*, where the two visions of 'advancing' the local food project co-exist and are found within separate but overlapping networks. I will suggest that the struggle between them is not only over strategies to improve economic returns to such producers, but also over the social forms and relations of production which are seen as appropriate for 'rural development' in the Irish context. The network engaged in 'alternativising' food exchanges in County Tipperary builds on a vision of co-operative knowledge- and opportunity-sharing as the basis for development, while that engaged in 'reform' of the conventional system envisions the formation of individual entrepreneurs and a top-down provision of commercial and production knowledges, emphasising 'innovation' in food products as the way to gain access to the global system.

A recent paper on production-consumption food networks (Holloway et al. 2007) argues strongly against starting the analysis of any given 'food project' with dichotomising categories of the sort used above – 'alternative' versus 'conventional', 'alternativising' versus 'reformist'. In the authors' view, the focus should be on finding ways to analyse the diversity and heterogeneity which characterise such projects, which tend to be obscured when they are labelled into one or other category in advance: 'Rather than categorising heterogeneous modes of food provisioning as alternative, we explore how particular food projects can be understood as arranged across a series of inter-related analytical fields in ways which make their operation possible' (Holloway et al. 2007: 3). They suggest that:

> there should be other ways of thinking about food networks which retain a sense of the diversity and particularity of different food networks, but which also allow us to say something useful about them in terms of relations of power and struggles over how food production and consumption should be arranged in society. (Holloway et al. 2007: 5)

They also provide a methodological framework for capturing the diversity in arrangements exhibited by individual food projects, focusing on the site of production, methods of production, supply chains and arenas of exchange, producer-consumer interactions, motivations for participation, and the construction of identities for the different actors and groups involved. This does prove useful in showing how a large number of projects which might be loosely grouped together as 'alternative' are actually arranged in fairly distinctive ways. Whether it leads to a useful analysis of 'power and struggles' might be questioned, and I return to this in the last part of my chapter.

While I agree that we need to recognise heterogeneity in examining food networks and projects, I am not convinced that we can dispense with the notion of 'alternativity'. It is still useful to distinguish those food projects which rest on oppositional attitudes to the capitalist food economy from those which attempt to join that economy on more favourable terms. The arrangements which most interest Holloway et al. (2007: 7) are the ways in which producer-consumer relations are organised, and they freely admit that they have taken 'limited account of the important roles played by government agencies and other organisations'. Yet it may be – and the case study reported below does, I believe, suggest this – that it is precisely in relation to state and other development agencies that the 'alternative-conventional' distinction is most useful. My study focuses on relations between local producers and state development policy agencies rather than on producer-consumer relations, unabashed by the recent consensus in the food studies literature that focusing on producer-consumer relationships is the way to develop 'integrated' analyses of food systems as totalities. If this places my chapter rather clearly in the category of the 'European school' of research on alternative food networks which, according to Goodman (2003), is interested primarily in how these shape or contribute to European rural development, it also addresses themes which the same author allocates to the North American literature, characterised as interested more in possible challenges to global economic and environmental injustices than in rural development per se. An alternative classification might distinguish between those studies which are primarily interested in analysing ideas of 'food quality' in the context of producer-consumer relations (Murdoch et al. 2000, Goodman and DuPuis 2002, Sage 2003), and those which have a more political agenda in relation to the place of agriculture and food production within economic, trade and political systems, whatever their geographic-intellectual location.

I first outline the economic and social context in which my case study is located, and then elaborate the current situation through a focus on three significant dimensions: networks, knowledges and orientations to nature. This provides the basis for a discussion of power resources and mechanisms in this rural site, and their impacts on an alternative food project.

Context: a 'food desert'?

County Tipperary, in the southern centre of Ireland, is historically renowned for its good land and agricultural productivity. In the flat lands of the centre, farms are larger than the Irish or European average and the soil is deep and rich; there is some grain production, some well-known horse studs, and in the past, some apple orchards for the cider industry, but farming is predominantly specialised in beef production and dairying. Around the county boundaries the terrain becomes mountainous, and here the soil is thinner, wetter and more acid, farm sizes are smaller, and particularly with the ending of the sugar beet industry, farmers are turning to coniferous afforestation and off-farm work. Our research was carried out in the southern part of County Tipperary, which possesses both sets of characteristics. The county has been a significant node in national and international agro-food systems for centuries and is characterised by the presence of many small and larger market towns, although the proportion of the population living in rural areas is still higher than the national average.

Historical processes, both local and global, have transformed this rich and fertile farming area into a peripheral site within the global food industry. In the colonial period, beef and butter production were already organised as export industries. In the past century, three institutions have continued and intensified this structure: the dairy cooperatives, the large-scale beef processors and Teagasc, the semi-state Agriculture and Food Authority, which operates extension services for farmers, advises on and implements national and EU policy, and conducts 'research and development' for commodity production and marketing.

The Cooperative movement (1890s–1920s), a movement based primarily around dairy farmers which sought to increase their market strength and upgrade and modernise their production techniques (Tovey 2001), found an early base in County Tipperary. Today nearly all milk producers still sell their milk to a dairy cooperative (a very few have started on-farm milk processing for farmhouse cheese production or other consumer products), but since the late 1980s the co-operatives have amalgamated, centralised and transformed themselves into corporate actors with global reach. They operate under a policy regime which increasingly favours traders over producers (McMichael 2005). In County Tipperary many small independent co-operatives were still found in the 1960s, but by the 1980s most had amalgamated into one large organisation, Golden Vale co-op, which today is part of the multinational food company Glanbia which produces a range of milk-based products and is a global importer and exporter of fractionated milk elements to use in consumer-ready foods.

Meat production largely remained outside the co-operative movement, and meat processing (slaughtering, de-boning, packaging) and distribution remained in private ownership. The dominant actor in the south Tipperary meat industry is Anglo-Irish Meats Ltd., a large commercial company owned in Ireland but with plants both there and in Britain. Ninety per cent of beef produced in Ireland is exported, primarily to Britain and other EU countries, and there is also still a live cattle trade, controlled by a small number of big traders.

South Tipperary commodity farmers are thus tied into a food processing and exporting system in which they hold little power, over either their own production practices or the prices they receive. Tight integration of farmers into the food processing system has been a goal of state agricultural policy since the 1980s and has been encouraged by Teagasc through its research, advisory and training personnel; the 'agricultural system' in which they were first incorporated as suppliers of raw commodities has been replaced by an 'agro-industrial system' dominated by international corporate actors, increasingly detached from the locality and sourcing their raw inputs wherever conditions are best for them. While the effects are not as dramatic in terms of concentration of production and dislocation of landholders as in US agriculture, for instance, the agro-food economy in south Tipperary shows how global changes in 'the temperate grain-livestock complex' (Weis 2007) have restructured the farming landscape. Over the past two decades, even farmers on good soil and relatively large and modernised farms have struggled to secure a livelihood, while smaller, poorer farmers have abandoned the occupation of farming in substantial numbers.

Food distribution and consumption practices mirror the changes in the production system. Fairs and markets largely disappeared from local towns after the establishment of cattle marts by the national farmers' organisation (IFA) in the 1970s. The marts were intended to achieve better prices for farmers by auctioning all cattle in the same ring rather than through a series of private one-to-one encounters between farmer and cattle buyer; but they also suited the growing concentration of the meat processing industry, met emerging urban concerns about hygiene and safety, and effectively ended one key site for local food exchanges. More recently, corner shops in towns and villages have been declining, displaced by large Irish, British and German-owned retail chains and by the growing trend of siting shops at petrol stations along motorways; these 'mini-supermarkets' are usually part of an international retail chain such as Spar and carry a range of imported or non-local mass production food stocks, with sometimes also a small number of locally-sourced items, mainly bread and cakes. McDonalds and other fast food chains have made inroads into the local towns. A rich food producing region is being transformed into a 'food desert' (Fonte 2008) where consumption is spatially disconnected from production, diets are shaped by global retail and consumer-ready food processing corporations and food exchanges have become routinised and depersonalised within mass retail outlets.

Irish people generally do not spend much on food (the average, at around 8 per cent of household expenditure, is one of the lowest in the EU). Convenience and a need for 'fuel' (Miele 2001) have dominated Irish consumer perceptions of food. However, in the last five years or so as the economy has boomed, food has become 'fashion' (Miele 2001), especially among more affluent consumers. Food journalists, television programs ('celebrity chefs') and competition among supermarket chains have helped to create an interest in 'good food', whose meanings range from the 'exotic' to the 'healthy/safe' to the 'alternative'. Primarily centred on the larger cities (Dublin, Cork, Galway), this is now beginning to influence consumption habits throughout the country; a Slow Food Convivium has recently been established in south Tipperary, Farmers' Markets are proliferating in the area, and the organic movement, which has had a presence since the early 1980s, is also increasingly recognised and institutionalised.

'Local food': contested versions

Very little remains in this part of Ireland of what might be described as 'traditional' food, either in what is grown, how it is processed, or in the cuisine of the area. Nevertheless, the concept of 'local food' has become a focal point for local rural development efforts in the past five years. Two main groups of actors are involved. One, a loose network of actors connected with a particular Farmer's Market, understand what they are doing as an attempt to re-localise the food distribution system, embracing both local producers and local consumers, in an area which has been incorporated into a global food system for a long period of time. This is expected to increase livelihood opportunities in local agriculture and food processing, and provide a local social space for civil society from which further developments may come. Some of these actors are embedded within larger social movements around both food and environmental issues found in different parts of rural Ireland over several decades (Tovey 2007), while others draw on community development discourses. The idea of local food is also gradually being picked up by some key development institutions dealing with rural Ireland, such as Teagasc, Bord Bia (the National Food Promotion Board) and LEADER;[2] the main carrier of the institutional understandings of the concept in South Tipperary is the local LEADER Action Group. But in this case, the meaning of 'local food' is much closer to the European understanding of it, as foodstuffs 'traditional' to the locality, or capable of being represented as such, which can use their local origin to reach into and colonise global niche or 'quality food' markets. Overlapping networks and relationships between the two groups of actors help to conceal the presence in the county of a struggle for discursive ownership of the 'local food project', and behind that, of conflicting visions of 'rural development'.

The project to relocalise the food system is associated with a Farmers Market established in a small town in South Tipperary (referred to here as C—) about 18 miles from the county seat, Clonmel. It was set up in 2001, the first to be established in the county although several more have followed since. Farmers Markets are a good illustration of Holloway et al. (2007) call for a recognition of heterogeity and difference in food projects often grouped together under a single classification. Moore (2006) distinguishes three different types in the development of Farmers Markets in Ireland. The 'Pioneers' were the first to be set up, usually in areas (such as West Cork) which have had a strong 'alternative' food culture since the 1970s. They are generally run by a committee of the stallholders, who establish more or less formal rules about what can be sold, who can sell, and from what distance (usually around 30 miles) products can be brought for sale; these rules attempt to establish both 'localness' of products and producers, and some degree of 'natural embeddedness' of the product (the producer must have

---

2   LEADER is a European Union Initiative aiming at promoting rural development. The LAG (acronym for 'local action group') is the subject intended to implement the program at local level.

a direct connection with the production process, either as grower or as craft food processor). The pioneering markets often faced hostility in establishing themselves: local authorities, responding to pressure from town food businesses, tried to close them down under legislation forbidding 'casual trading', and health and safety authorities put obstacles in the way of selling home-produced foods such as raw milk cheeses, cooked meats or smoked fish. Next to be established were the 'Privately Run Farmers' Markets', mainly found in the east of the country and particularly around Dublin. They are run by 'benevolent dictators' (Moore 2006) – individual entrepreneurs with a passion for food who determine the organisation of trading and what products can be sold. In these markets, all fresh fruit and vegetables (although not other produce) must be certified organic, but need not be 'local', and stallholders are often traders rather than producers: they can import much of what they sell, and sell at a number of different markets around the country. These markets expand the selling of food as a cultural object (Jordan 2007) and a signifier of taste and social status; they attract customers through their promise to provide 'exotic', 'craft' and 'heritage' foods. Finally, Moore (2006) distinguishes the 'Participatory' type of Farmers Market; generally the most recently established, these have an organisational structure similar to that of the Pioneers but have worked with rather than been confronted by local authorities and rural development groups such as LEADER in setting themselves up. Under the guidance of those 'outside' institutions, the Author suggests, the committees in charge of Participatory Markets usually try to ensure that the participating farmers are low-income farmers, and rules about both 'local' and 'natural' embeddedness are quite closely policed.

The Farmers' Market at C—

The Market discussed here falls into Moore's third type, but is perhaps unusually autonomous. It came into existence before the LEADER LAG became interested in 'local food' or in Farmers' Markets, and was established by a local civil society grouping which called itself the C— Development Association, who saw a market as a way of 'bringing people into town on Saturdays because the town of C— every Saturday was practically deserted' (Association member interview). The Development Association wanted the Market to sell a fairly wide spectrum of produce, particularly products which were 'indigenous to the area'. To minimise conflict, they brought local shopkeepers into the Association at an early stage of planning. They found a suitable site, in the car park of an old grain store which had been converted into a craft shop, persuaded the county council to provide water, electricity, and a waste collection service to it, and advertised for stallholders. Once the market was set up, the Development Association left it to be run by the stallholders themselves as a committee; there is some limited oversight by the county council which charges each stallholder 150 euros a year for their stall (this is considerably cheaper than in the Privately-Run type of markets). Stallholders also pay insurance against accidents around their stalls, around another 250 euros a year, which was negotiated collectively with an insurance company set up in the 1980s by the Irish Farmers' Association to provide insurance to the agricultural sector; and pay seven euros a week into a market development fund, used for occasional advertising in local newspapers or to pay entertainers (face-painters, musicians, a juggler) for special occasions such as the Christmas, Easter and St. Valentine's Day markets.

Despite the existence of a stallholders' committee, stallholders' own accounts of how the market is organised suggest a strong element of informal self-regulation. Decisions about what to sell, or with whom to go into or out of partnership, are made by individual stallholders themselves, following more or less tacit understandings of what the market is about (to sell 'local' food, 'quality' food, to support the livelihoods of local producers, to restore personal relations between producers and consumers, to contribute to local 'development'). Nevertheless, they do act collectively on occasion; in one instance a local company which is a large distributor of organic meats wanted to take a stall in the market and the committee refused.

> We felt that [company name] was importing an awful lot of stuff from Germany and places like that and we did feel it was commercialising the market a little too much and we would rather give local suppliers the possibility of marketing their produce. (Stallholder interview)

The C— market is quite small; starting in 2001 with six stallholders, the committee hoped to operate with 12. In 2005/6 there were 11: two bakers, a 'fish man', a savouries confectioner and a sweets confectioner, an 'apple man', a pate stall, two meat and poultry stalls, and two vegetable stalls. All sell their own produce, although they sometimes carry products for other local producers as well, usually without charge – none is only a 'middleman'. Most cannot get a full livelihood out of selling at the market, but it offers a better, more convenient and more interesting return on production than other outlets such as selling to commodity processors or to retail shops. They rely on other sources of income – a pension from previous employment, cattle or other commodity farming, off-farm work, farm support payments and REPS (agri-environmental) payments, small LEADER grants – to survive. Nevertheless they regard the market as a success: this is measured in terms of the volume of customers coming through and the sociability of the atmosphere.

> We have brought people back into town, we have done what the traders [town shopkeepers] wanted and we are doing something for ourselves as well. We are doing a service for ourselves ... Now there was something there I just wanted to say to you about the market itself. What I found during the summer was that people were coming there and stopping and talking in little groups, and that's something that's totally missing when you go to supermarkets or anything like that ... Nobody is in a hurry out there, they are all chatting away, you know

that's something in a rural area that you need, and it's there that development begins, I think. (Stallholder interview)

The market aims to supply basic items of household consumption, such as meat, bread and vegetables. They hope to convert their 'neighbours' to the idea that local food is better food. Apart from the pate stall, the market does not engage in selling the 'exotics' found in the Dublin markets, for example. While all the meat sold in it is certified organic, the vegetables are organically grown but not certified; their 'quality' rests on the personal knowledge and 'relations of regard' (Sage 2003) between seller and consumer. Nevertheless stallholders enjoy introducing their customers to new food experiences and exchanging cooking and cuisine tips with them; part of their objective is to encourage locals to develop a wider diet, based mainly on foods once eaten in the area and since forgotten. A vegetable stallholder described how:

> When I introduced spinach first it wasn't really known in C—, would you believe, and one woman came up and she said 'What's that?' 'That's spinach', I said. 'Oh, that's the stuff that keeps flies out of the kitchen, isn't it?'. Needless to say she bought none of it! I was trying out a few crops to see how they would go, I put in a lot of spinach and I got very worried because nearly three or four markets and nobody bought any, and then suddenly it took off and people began to buy, and then came back looking for more. (Interview to a vegetable stallholder at the Farmers' Market)

Like most parts of Ireland, South Tipperary has experienced a rapid growth in the number of immigrants living and working in the area, which is almost certainly linked to the capacity of the farmers' markets in the area to flourish. Nevertheless most of the market's regular customers appear to be Irish people from the town and from its rural hinterland, covering distances of 10 to 20 miles and including quite elderly rural people with limited incomes. Among the customers we interviewed, the committee's philosophy of direct selling of household staples seemed to be understood and reciprocated. Over half were 'regular' buyers (weekly or fortnightly) while the rest were either first timer locals or non-residents returning to visit relations; very few were tourists or others passing through the town. Their main reasons for coming were that food from the market had the same sort of taste as their own home-grown produce in the past or that they remembered from childhood, and a strong dislike of supermarket shopping as 'claustrophobic', aggressive, and untrustworthy. Nevertheless most customers only bought a small part of their weekly shop at the market.

The LEADER Local Action Group

The second group of actors who have embraced the concept of 'local food' as a vehicle for local development is the LEADER LAG for South Tipperary. This

LAG has been in existence since the start of the LEADER Programmes in the early 1990s, but it was not until 2004 that it realised that 'food was an area that had to be developed, particularly with Tipperary's strong food heritage (…) Tipperary would have a rich local food heritage with apples, seeds and grain' (interview with Food Project Development Officer). A Tipperary Micro-food Strategy Group was set up in May 2004 to develop the idea; chaired by an academic from University College Cork's Department of Food Business and Development, it included about 12 'local stakeholders' drawn from the private sector (representatives from an organic meats distributor, a speciality food retailer, a national supermarket chain, a prestigious local restaurant), and the public agro-food sectors (representatives from Teagasc, from the Agricultural Co-operatives in south and north Tipperary, from the South Tipperary County Enterprise Board) among others. Its report set out three strategies for promoting small food production in the area: development of outlets and capacity; provision of support and services; and provision of education and awareness for consumers as well as producers.

Initially the LAG wanted to set up 'some kind of branded initiative, similar to the West Cork Fuschia brand' (a designation of origin label shared by a number of West Cork producers and aimed at tourists, national and international markets as much as local ones). This now awaits funding in the next LEADER programme, and at the time of our study they were concentrating instead on encouraging farmers into direct selling of produce, and establishing new Farmers' Markets as outlets for that.

LEADER defines its main role in promoting rural development as *networking*. In the local food promotion case, this meant first, bringing together a lot of different agencies who 'all have a finger in the food pie', and 'facilitating' them to work together. Second, it means linking up 'interested producers' with these different bodies, by holding open days and seminars and through direct contacts. LEADER can also help producers with grants or mentoring, or if they cannot provide these, they pass the person on to other sources such as the County Enterprise Boards. They have made a limited direct input into local food promotion, for example by holding food fairs, producing a guide (the *Savour Tipperary* Food Guide) to local foods for tourists, and organising competitions around food in local schools.

Tipperary LEADER's food strategy builds on the idea that food has recently become 'very popular' in Ireland; with the increased media interest in food as entertainment and the sophistication of consumer tastes, boosted by the Tipperary Slow Food Convivium and new consumer health and environmental concerns,

> people are looking at eating out experiences and home entertaining, they are not going to put up with the traditional bacon and cabbage if they are home entertaining, they are going to have something more exotic, and they will want to have some local talking point – cheese, for example. (Food Project Development Officer interview)

'Local food' is thus assimilated into a fashionable diet which also includes imported products. In this changing consumption context, LEADER's strategy is to 'try to differentiate Tipperary foods'. 'Local food' means giving local produce a clear designation of origin.

> The local people realise there is good food in Tipperary but for any tourist coming through, at the moment we have people driving from say Wexford to Donegal, they generally eat the same stuff no matter where they eat, you don't realise you are in a different part of the country, you go in and order your bacon and cabbage but they are not saying this is Limerick or South Tipperary bacon, or cabbage grown in Clonmel, you are not giving them that local feel. (Food Project Development Officer interview)

LEADER's interest in this form of local food promotion is shared by Bord Bia, the national food marketing agency, which first began to mention 'local food' in its Annual Report of 2003. Bord Bia cooperated with the county council, the Tipperary Heritage Society and other local groups in establishing a 'heritage food fair' in C— in September 2006, to reflect its status as a 'heritage town'. Validation of 'local' food appears to require that it can be recast as 'heritage', or 'reinvented tradition'.

How does supporting Farmers' Markets come into this strategy? The LEADER Development Officer is keenly aware of the changing funding regime for European agriculture, and works closely with Teagasc, through its 'Options Programme' for the Single Farm Payment regime, to find opportunities in this for local food producers. Expanding direct selling of farm produce is seen as a significant 'option'. If farmers realise that they can't stay in commodity farming cost effectively, the Food Project Development Officer says, they should look at direct selling to 'add value' to their product:

> An ideal way to work is to try your product through a very cost effective basis going through a Farmers' Market, you get a very immediate feedback, and obviously if it looks quite good and there is a bit of demand for it you could look at distribution beyond the Farmers' Market, maybe deliver locally to shops or get into one of the distribution companies who are doing artisan food products … We would view the Farmers' Markets as an ideal kind of a test ground or incubation area for a basic product to get out, and how to modify the product. (Food Project Development Officer interview)

Direct selling is linked to innovation in food production. In deciding to grant-fund producers, for example, LEADER's practice is to

> focus on the innovative aspect, if there is already 50 organic egg producers in the country we are not going near number 51 … Generally projects at the beginning are not funded, say at a national level, and say then after a while they become

accepted and taken up by the department as part of a mainstream programme, so obviously the trailblazers if you want to call them that, might come knocking at our door. (ibid)

'Local food' and its role in promoting local development, then, are understood very differently by these two groups of actors. The LEADER LAG has largely adopted a 'Mediterranean' concept of it, emphasising a revalorisation of 'traditional' cuisine for global marketing, while the C— Farmers' Market group might be seen as closer to the American and northern European concept of 'community supported agriculture'. How the two 'local food' projects overlap and contest each other in the same space is the subject of the next section of the chapter.

Networks, knowledges and nature

Our study of the C— Farmers' Market was undertaken less for its own sake than as a route into understanding some dynamic features of the 'local food' contestation in South Tipperary. A key dynamic to capture power relations and potential future developments are the relationships in which participants are involved, and we tried to trace network relationships for both LEADER and Farmers' Market actors. We were interested in the sorts of knowledges which were shared and transmitted through network relationships, and in seeing whether tracing knowledges could help to establish network nodes and boundaries. A third issue of interest was the extent to which, and ways in which, participants were not just 'socially embedded' but also 'ecologically embedded', which seemed likely to provide insight into the extent of their 'alternativeness' or otherwise.

Network relationships

Given the presence of two competing versions of 'local food' as a strategy for development, one oriented towards an individualist style of interventionism which we might call 'picking winners' and the other to a more collectivist, cooperative or egalitarian style of development, it was not unexpected to find that these are circulated within and supported by two relatively distinct social networks. As I will suggest further below, however, the boundaries between these networks are more fuzzy and permeable than a first description allows for.

Sellers at the C— Farmers' Market exhibit a high degree of 'local embeddedness' (Winter 2003) and mutual cooperation in their relations with each other and with other small producers and food artisans in the area. People move in and out of partnerships on stalls, and stallholders take trips together to visit other Farmers' Markets to see what they can learn to develop their own market. As the first to establish in Tipperary, they are also approached for advice by groups thinking of setting up another Farmers' Market in the county, and although this may ultimately cut into their own sales they have been willing to share their own experience and knowledge. They occasionally sell the produce of other local foodworkers on their own stalls; this tends to be without charge or is repaid through some sort of barter system.³ A vegetable stallholder gets free manure for his crops from a local farmer who takes it from stud farms in the area and stacks it for rotting down on his farm; in return he sells some of the farmer's vegetables for him at the market. The 'apple man' who produces apple juice from his own orchard also processes apples for other small producers who do not have their own juicing plant; he knows five or six other small apple growers in the region and if they are distributing juice over any distance they often share the costs of lorry transport between them. He also runs a Farm Shop which is his primary mode of direct selling, and which brings him into contact with other small producers in the region whose products he sells there. Thus actors affiliated to the C— market have between them wide-ranging contacts with other artisan food producers in the region, strengthened through an ethos of 'neighbourliness'.

While primarily focussed on 'the local', this network does stretch across larger social spaces. Involvement in the organic movement brings some into contact with other food producers and food activists on a national and sometimes international level. One of the stallholders, who has a Masters degree in horticulture, teaches part-time in the University of Limerick, about an hour's drive away, and another worked for years in the meat processing industry and with the Department of Agriculture before taking over the family farm, and retains contacts from those years. And nearly all the C— market stallholders have some contacts with LEADER, collectively or individually.

LEADER also operates through building cooperative relationships, but this is understood as 'networking' rather than 'neighbourliness', that is, networks are created to function for some instrumental aim rather than as webs of sociability. Networking brings together potentially successful local food innovators with a range of state and other authorities (Teagasc, Bord Bia, the Health Boards, Food Science and rural development departments in local universities and institutes, the local authorities, County Enterprise Boards, Tourism committees, The Irish Farmers' Association, the Irish Countrywomen's Association and so on) who can help them to develop their enterprise. It also draws in expertise from regional and national level food industries, particularly at the processing and retail end of the food chain. Expertise from these companies is used to assess the potential viability of projects submitted to the LEADER LAG for support. Inclusion in such networks leads to contacts which stretch well beyond the local. The 35 LEADER groups in Ireland share a National Small Food Co-ordinator who works closely with the LEADER governing body for Ireland and with the Ministry for Rural Development; in 2005 she organised the attendance of LEADER-supported small food firms at the National Food Exhibition in Dublin, seven of whom came from County Tipperary.

---

3   See Tovey (2006) for similar findings about the use of barter among organic food producers in West Cork.

LEADER's extended networks encourage an orientation which values local direct selling more for what it can eventually contribute to the national economy, through exports to niche markets, than to local economies and local livelihoods. An example of the sort of project which gets support through this networking is a relatively new venture in farmhouse sheep's milk cheese, flagged to us by the LEADER Food Project Development Officer as a project which 'ticks all their boxes'. It is part of an extended family enterprise which has already produced a cow's milk cheese of international renown. The sheep's milk cheese has recently received EU recognition as an Irish local speciality product; it is sold through personal contacts with up-market food retail outlets and a few of the 'higher-quality' supermarket chains, through hotels and restaurants and through wholesalers primarily in the UK, Australia, US and Japan who sell it on to similar outlets in those countries. Like the cow's milk cheese on whose reputation it has to a degree 'piggy-backed', it is marketed as a 'local' product, named after a medieval religious monument which is a national tourist attraction. In interview, its producers recounted how cheese-making was a feature of life in local monasteries and after these were closed down in the fifteenth and sixteenth centuries, the knowledge of cheesemaking passed to local producers 'who produced new cheeses but all based on the traditional product'. But the key selling point of the cheese is 'differentiation ... it is distinctly different in that it's a sheep's milk cheese and costs twice as much to produce' than its famous sibling, and 'it is therefore twice as expensive and aimed at the top delis and restaurants, not the multiples'. It positions itself as a cheese of 'high quality', meaning that it is best appreciated by connoisseurs or those with money to spend on food.

The producers of this cheese have little contact with other artisan food producers in the area. The cheese may be found at some of the 'Privately-Run' Farmers' Markets around Dublin but it is not sold at the C— or other local markets. Their significant networks appear to be the national and regional networks of experts facilitated by LEADER, and, in particular, a network of 'good food promoters' stretching across both Ireland and the UK which links together food journalists, restaurateurs, media chefs, Slow Food Consortium members, speciality food shop owners and a few select food producers. Within this network it is the aesthetics of food which dominates judgements of food quality; 'localness' is primarily a reinvention of 'heritage' to market difference.

But the LEADER LAG is also, given its constitution, 'locally embedded' to a degree. At least two members of the C— Stallholders Committee have close contacts with LEADER networks, having served on their Micro-Food Strategy Group in 2004 or acting as 'experts' in various consultations organised by them; several of the stallholders have taken LEADER-sponsored courses or have submitted projects to LEADER for support. The crossover between networks becomes evident when we look at the knowledges and knowledge discourses used by local food activists in the two network nodes.

Knowledge dynamics

The LEADER 'micro-food strategy' aims to provide knowledge to those who want to innovate with food products and to get involved in direct selling of food. 'Facilitating' the development of local food means bringing producers into networks of experts where they can acquire the necessary knowledge to succeed. The approach prioritises 'information shortage' as the main problem facing those starting up new food processing businesses: this includes information about scientific management of food safety, but particularly 'practical hands-on expertise' in introducing new foods into markets. LEADER tries to bring new food producers into contact with successful entrepreneurs in the food industry, who possess this practical business experience. The relevant food knowledges for LEADER, then, are *technical* (hygiene and safety) and *commercial* knowledges.

Asked about the knowledges they need, on the other hand, the artisan producers connected to the C— market primarily talk about food *production* knowledges. Their trading and selling skills are seen as part of their larger repertoire of skills in social interaction, not needing to be codified or formally taught. But rebuilding local knowledge about how to grow or process artisan foods is understood as a significant part of a project to relocalise food exchanges, and in the absence of much inherited local tradition to draw on, knowledge is acquired in various ways. Some are informal, as knowledges are circulated through everyday social interaction with other small producers, and some are more formal. Many opportunities for knowledge building come through the organic movement. During the early years of establishing himself, a vegetable stallholder attended 'five or six' weekend courses at the National Organic Centre in County Leitrim, which is independent of the state and run by organic movement enthusiasts. He also learns from older farmers in his area, having a sense that in returning to organic production methods he is returning to how farming was done in Ireland half a century ago; his production knowledges are shaped by 'traditional' practices as well as the more codified knowledges about organic farming in circulation today. Similarly organic meat producers attend courses run by IOFGA (one of the three organic certification bodies in Ireland), go on organic farm visits, and meet and learn from other producers. Others rely more extensively on accredited 'expert' knowledges. The 'apple man', for example, has used his horticultural training to modernise his family's orchards, which have been replanted in a 'non-traditional' way (in his father's time, the apple trees were large and were grazed underneath by sheep and even cattle, today his orchards are planted with low-growing trees 'and there's no livestock in them') and grow only varieties specialised for eating or juicing. His knowledge about how to produce apple juice was mainly got from the UK company which sold him the juicing equipment, and from 'a few books'.

One of the meat producers at the market takes courses on marketing and selling and subscribes to international meat magazines: 'I would be looking at product developments and how to present your product'. He also takes a keen interest in craft knowledges, particularly about butchering: 'I try to visit the Craft Butchers'

Association shows and that, I attend some of their courses just to be up to date on what's happening'. Two years ago he went to Scotland to learn another craft – 'how to smoke product and to be able to retail it (…) I take in cured bacon from other people and I smoke it for them, or [when he has pigs on the farm] I smoke and sell it myself'. But he also takes advice and knowledge informally, from the organic movement and from older cattle farmers:

> I think if we share knowledge we learn a lot more together ... We've produced beef on the hoof for generations at home on the farm, and we use our local butcher to slaughter our cattle, you have generations of families working in the abattoir, in the butcher's stall, in their family businesses, we're using old knowledge that's there in the beef and sheep industries. (Meat producers at the Farmers' Market)

The knowledge dynamics around 'local food' in this area emerged as rich and complex. It is not possible to say that the LEADER network disseminates 'expert' knowledges while the Farmers' Market network shares 'lay' or 'local' knowledges: rather, two 'knowledge cultures' (Morris 2006) co-exist in the area, both of which blend scientific, or formally codified, knowledges acquired in educational settings or from books with relatively uncodified knowledges learnt more tacitly, from experience, conversation and observation. However, one is a production-oriented knowledge culture which makes more use of informal, local or tacit knowledges, and the other a marketing-oriented knowledge culture which makes more use of codified and 'expert' knowledges. The artisan producers around the C— market participate more in the first, LEADER, the national food institutions, and local producers caught up in these networks such as the sheep milk cheese makers described earlier, participate more in the second; but most actors in the artisan producer network move between the two with relative ease. The impact of the different knowledge cultures is shaped by the higher status of the marketing-oriented knowledges, which can be seen in some instances to interfere with and disorient the reproduction of the 'localised food' philosophy.

Nature

Not all the producers in the Farmers' Market network are certified organic producers. Those who produce and sell meat products value the certification, but many others, particularly the vegetable growers, are 'post-organic' or 'movement organic': they are deeply influenced by organic ideas and use organic production methods but rely in selling on face-to-face contact with customers rather than on certification. Different discourses and judgements of 'food quality' are in circulation in the network, reflecting tensions around selling a 'certified' or 'branded' product. The 'movement organic' vegetable growers emphasise 'freshness', 'flavour' and locally grown' as the strengths of their produce, and the 'apple man' says that his experience in 'how to grow a good tasting apple is … a big part of the quality thing'.

His apple juice is a branded product (although sold only in local outlets), but it is called not after the locality but after the apple variety used in it. One of the meat stallholders, however, who values direct selling for the feedback it gives him on the 'quality' of his product and for its contribution to the welfare of the community, is nevertheless planning to brand his produce (as 'Tipperary Organics') in order to reach national and international markets. His discourse of food quality blends aesthetic, commercial and ecological discourses, linking quality to hygiene and traceability as well as to breed variety and to production methods (for example, the use of well rather than mains water for his cattle, because this allows 'the nutrients that are leached through the soils to be recycled, trace elements and minerals, back into the animal') which influence the taste of the meat.

Whether certified or not, the producers in this network can be described as 'ecologically embedded'. They connect their own interest in localising food exchanges to wider environmental concerns about 'food miles', and talk at length about the practices they have introduced to ensure environmental sustainability in food production, from building up soil fertility to increasing biodiversity on their land, to only using recyclable packaging materials and treating and recycling waste. Producers supported by and oriented to LEADER networks also talk about the need to set 'very high standards' for quality and environmental management in their enterprise. The sheep's cheese producers, for example, have installed a HACCP system of surveillance to monitor health and safety regulations and systems to reduce waste in energy and water in production. However, their commitment to environmental protection emerges as more a concern to manage risk than an interest in ecologising food. Their environmental concern is articulated as a concern for the quality of the food *enterprise* as a manager of risk, whereas for the C— market network actors, the way nature is managed is central to the goodness of the *food* itself.

Power in rural development

It is difficult to draw clear boundaries between networks around food and development in this local space; particular actors are more deeply involved in one set of social relationships rather than another, but few if any completely lack contact across the margins. This blurring of boundaries between actor clusters is important for an understanding of how power works in the local space.

Much of the recent literature on 'alternative' food movements and projects argues against views of capitalism as hegemonic and as unchangingly capable of repressing, subverting or appropriating practices of resistance towards it. Raynolds and Wilkinson (2007: 42) discuss 'the cyclical process of corporate appropriation and social movement outflanking' currently reshaping Fair Trade, where social movement initiatives are appropriated into conventional circuits and alternative products transformed into new consumer foods, but in order to emphasise that this process *is* cyclical and in turn stimulates new social movement initiatives.

Massey (2000) similarly rejects the view that globalisation and economic neoliberalism are inescapable processes which necessarily subsume alternative economic projects, with 'every attempt at radical otherness being so quickly commercialised and sold or used to sell' (Massey 2000: 281). Leyshon and Lee (2003: 16) represent capitalism as fragile and open to challenge, evidenced by the proliferation of examples of 'performing the economy otherwise'. Holloway et al. (2007: 6) question conventional views of power and spatiality, challenging representations of space as a fixed and static order of dominance and arguing for a 'more processural' understanding of space as an arena of resistances and oppositionalities, 'part of an entangled and continually remade web of relations' many of which are 'ambiguous' for the exercise of power or resistance to power. Recognising the 'ambiguous' spaces between domination and resistance and treating power as relational means that 'there are always possibilities available for reimagining and restructuring those relationships towards different ways of doing things' (Holloway et al. 2007: 6). Recognising the 'diverse geographies of power' enables us to see it as 'a process, as productive (rather than simply repressive), central to all social relations and crucially, "not something to be overthrown, but rather to be used and transformed" (Cresswell 2000: 264) by actors opposing social power relations which oppress or restrict them'. (Holloway et al. 2007: 5)

My analysis of contestation around 'local food' as a vehicle for sustainable rural development in South County Tipperary offers a less sanguine conclusion about the operation of power within a local space and the strength of capitalism to subsume and divert 'alternative' food actors. Capitalism, to be sure, cannot prevent 'reimagining and restructuring' food relationships and economies, but it may still be able to undermine and block effective implementation of alternatives. The Tipperary case suggests that capitalism has indeed a formidable capacity to subsume and appropriate – even, or perhaps particularly, when it is operating through apparently non-capitalist agencies such as a LEADER LAG. Interactions within a relatively small-scale rural area often appear unstructured by hierarchies of power; nevertheless, power resources are unequal in their local distribution, knowledges possess unequal symbolic capital, and network boundaries are blurred or disregarded by those for whom other ways of doing things have no legitimacy. Power acts through the uncertainty and deference of small rural producers, many of whom are fearful of what the future holds for their livelihoods and profession. As Guthman (2004) observes, in her study of the co-existence of large-scale agribusiness and a counter-cultural organic movement in California in the 1980s and 1990s, a movement for a radically alternative food system may be subverted by some of the counter-cultural actors themselves as they come to adopt conventional business ambitions. In the local space of south County Tipperary, the power of conventional economic thinking operates economically, by structuring the provision of development aids, but much more important is how it operates discursively, supported by one-way processes of network co-optation.

Market-led expansion of 'local food' projects has impacts on the social forms and relations of production (Barrientos et al. 2007). In the Tipperary case the 'reformist' vision of local food as 'innovative', reconstructing local heritage to brand and differentiate itself in order to move into exchange networks which include international markets, international tourists, and some national upper class or fashionable retail outlets, encourages the individualisation of producers and competitive relations between them. It reproduces state perceptions of agriculture as an economic sector which is significant primarily for its contribution to export earnings, much less as a source of rural livelihoods. The 'alternative' vision of local food as local in both production and consumption regards the practices of food-based livelihoods as in themselves a contribution to local development as they create and reproduce local interaction, social relationships and civil society. Although most of the C— market stallholders do not secure a full livelihood from this, their activities help to sustain in production a much wider set of small farmers and artisan producers. One stallholder also provides levels of direct employment in the locality from his Farm Shop and juicing enterprise which are much higher than those provided by the LEADER-backed food entrepreneurs we interviewed. Potential to provide local employment or to contribute to the maintenance of a local economic network appear lower on the list of criteria used by the LEADER LAG to determine to which local food projects funding and other resources should be devoted than marketing buzzwords do. In the words of the Food Project Development Officer, 'Local, artisan and organic would be the three kind of catchwords, and you would be hoping to get two of them at least when a new product comes in'.

As the holder of economic and discursive development resources, the LEADER network is more able to extend its boundaries and to co-opt new members than the alternative network is. Although some alternative producers seem able to move between the two networks without compromising their own alternative goals, such as the apple juice producer who is part of LEADER's 'expert' advisory network yet who persists in selling his own very successful product locally and has no interest in expanding into larger markets, many others add on the discourses picked up in LEADER contexts to their own alternative discourses, fracturing the knowledge culture and the understandings of an 'alternative' within the C— market network. 'Relocalising' food exchanges is associated with circulating more money in the local economy and supporting other local producers, but is also associated by some network members with attracting and retaining tourism in the area. Opposition to branding and certification derives in part from a residual agrarian populist antipathy to 'big business', shared by both some 'movement organic' producers and some of the certified organic producers who prefer to sell at the Farmers Market than through wholesale or retail chains. But some of those who express such opposition also embrace the commercial and technical knowledges they are exposed to from their contacts with LEADER courses and networks, leading to industrial discourses of 'quality food' as hygienic, safe, traceable and differentiated. There is a constant pressure on alternative network actors to become less 'locally embedded' which arises, ironically, from the co-existence of reformist and radical critiques of the capitalist food system within the same small local space. The 'ambiguities' of the relations of power across space, while not affecting the capitalist market discourse of local food, work to disorganise and destablise relations within the alternative network and to make the project of an alternative local food system vulnerable to transition into forms more compatible with capitalist development policy.

References

Barrientos, S., Conroy, M.E., Jones, E. 2007. Northern social movements and Fair Trade, in *Fair Trade: the Challenges of Transforming Globalisation*, edited by Raynolds, L.T., Murray, D. and Wilkinson, J. Abingdon: Routledge, 51–62.
Cresswell, T. 2000. Falling down: resistance as diagnostic, in *Entanglements of Power: Geographies of Domination/Resistance*, edited by Sharp, J., Routledge, P., Philo, C. and Paddison, R. London: Routledge, 256–268.
Fonte, M. 2008. Knowledge, food and place: a way of producing, a way of knowing. *Sociologia Ruralis*, 48(3), 200–222.
Goodman, D. 2003. The quality 'turn' and alternative food practices: reflections and agenda. *Journal of Rural Studies*, 19(1), 1–7.
Goodman, D. and DuPuis, E.M. 2002. Knowing food and growing food – beyond the production-consumption debate in the sociology of agriculture. *Sociologia Ruralis*, 42(1), 5–22.
Guthman, J. 2004. *Agrarian Dreams: The Paradox of Organic Farming in California*. Berkeley, CA/London: University of California Press.
Holloway, L., Kneafsey, M., Venn, L., Cox, R., Dowler, E. and Tuomainen, H. 2007. Possible food economies: a methodological framework for exploring food production-consumption relationships. *Sociologia Ruralis*, 47(1), 1–19.
Jordan, J.A. 2007. The heirloom tomato as cultural object: investigating taste and space. *Sociologia Ruralis*, 47(1), 20–41.
Leyshon, A. and Lee, R. 2003. Introduction: alternative economic geographies, in *Alternative Economic Spaces*, edited by Leyshon, A., Lee, R. and Williams, C. London: Sage, 1–26.
Massey, D. 2000. Entanglements of power: reflections, in *Entanglements of Power: Geographies of Domination/Resistance*, edited by Sharp, J. Routledge, J., Philo, C. and Paddison, R. London: Routledge, 279–286.
McMichael, P. 2005. Global development and the corporate food regime, in *New Directions in the Sociology of Global Development*, edited by Buttel, F.H. and McMichael, P. London: Elsevier, 265–299.
Miele, M. 2001. Changing passions for food in Europe, in *Agricultural Transformation, Food and Environment*, edited by Buller, H. and Hoggart, K. Aldershot: Ashgate, 29–50.
Moore, O. 2006. Farmers' markets, in *Uncertain Ireland – a Sociological Chronicle 2003–2004*, edited by Corcoran, M. and Peillon, M. Dublin: Institute of Public Administration, 129–142.

Morris, C. 2006. Negotiating the boundary between state-led and farmer approaches to knowing nature – an analysis of UK agri-environmental schemes. *Geoforum*, 37(1), 113–127.

Murdoch, J., Marsden, T. and Banks, J. 2000. Quality, nature and embeddedness: some theoretical considerations in the context of the food sector. *Economic Geography*, 76(2), 107–125.

Murray, D.L. and Raynolds, L.T. 2007. Globalisation and its antinomies: negotiating a fair trade movement, in *Fair Trade: the Challenges of Transforming Globalisation*, edited by Raynolds, L., Murray, T.D. and Wilkinson, J. Abingdon: Routledge, 3–14.

Raynolds, L.T., Murray, T.D. and Wilkinson, J. (eds) 2007. *Fair Trade: The Challenges of Transforming Globalisation*. Abingdon: Routledge.

Raynolds, L.T. and Long, M.A. 2007. Fair/Alternative Trade: historical and empirical dimensions, in *Fair Trade: the Challenges of Transforming Globalisation*, edited by Raynolds, L., Murray, T.D. and Wilkinson, J. Abingdon: Routledge, 15–32.

Raynolds, L.T. and Wilkinson, J. 2007. Fair Trade in the agriculture and food sector: analytical dimensions, in *Fair Trade: the Challenges of Transforming Globalisation*, edited by Raynolds, L., Murray, T.D. and Wilkinson, J. Abingdon, Oxon: Routledge, 33–47.

Sage, C. 2003. Social embeddedness and relations of regard: alternative 'good food' networks in south-west Ireland. *Journal of Rural Studies*, 19(3), 47–60.

Tovey, H. 2001. The cooperative movement in Ireland: reconstructing civil society, in *Food, Nature and Society*, edited by Tovey, H. and Blanc, M. Aldershot: Ashgate, 321–338.

Tovey, H. 2006. New movements in old places? The alternative food movement in rural Ireland, in *Social Movements in Ireland*, edited by Connolly, L. and Hourigan, N. Manchester: Manchester University Press, 168–189.

Tovey, H. 2007. *Environmentalism in Ireland: Movement and Activists*. Dublin: IPA.

Weis, T. 2007. *The Global Food Economy: The Battle for the Future of Farming*. London: Zed Books.

Winter, M. 2003. Embeddedness, the new food economy and defensive localism. *Journal of Rural Studies*, 19(1), 23–32.

Chapter 2
Creating a Tradition That We Never Had: Local Food and Local Knowledge in the Northeast of Germany

Rosemarie Siebert and Lutz Laschewski

Introduction

Relocalisation of food systems has become an important paradigm for rural movements. It is connected with a demand for better quality in food, the quest for sustainability and threatened rural livelihoods. Despite the fact that this debate has a bearing on some common trajectories of modern food systems, local food movements have adopted a slightly different focus. For instance, some authors have distinguished between a North American and a European perspective on local food (Goodman 2003, Holloway et al. 2007). Fonte (2008: 202) has distinguished between a reconnecting and an origin-of-food perspective. According to her analysis, the model of reconnecting food producers and consumers 'often develops in a context of longstanding export-oriented agriculture and the loss of food culture, where food is provided only by big supermarkets and there is no outlet for local agricultural production'. By contrast, the 'origin-of-food' perspective has evolved in territories such as the Mediterranean countries, which 'were latecomers to industrial development and never fully completed their "great transitions". Rather, they have passed through a process of economic and social marginalisation (marked by depopulation) that they are now trying to reverse' (Fonte 2008: 202–203).

Relocalisation of food cannot be identified with a single approach but need to be interpreted in a number of different ways, providing a rationale for some rather different social practices (Winter 2003, Feagan 2007). Our argument is that the different meanings of relocalisation also pertain to different kinds of social relationships, and that the diversity has to be acknowledged if the relationship between local food and local knowledge is to be addressed. Morris (2006: 115) has argued that rather than defining attributes of knowledge, it is more important to examine 'social, historical and institutional relations in which knowledge develops and is represented'. As a consequence, some authors have referred to the concept of 'knowledge cultures' (Tsouvalis et al. 2000, Morris 2006). In this view, knowledge is a social achievement, something that is produced through interaction in social situations. To some extent, knowledge itself is a collective resource in the form of patterns of experiences stored by organisations, institutions or networks.

In this sense, knowledge formation and (organisational) learning are the outcome of institutional routines and regulations to collect process and document personal knowledge (Levitt and March 1988, Willke 2001).

From a sociological point of view, a particular type of knowledge is defined not by its content but by the way it is embedded in social relations. *Scientific* knowledge is described as highly de-contextualised, specialised, and 'standardised' (Kloppenburg 1991, van der Ploeg 1993, Wynne 1996) and deemed valuable in many contexts. *Local* (or tacit) knowledge, in contrast, is highly variable and non-universal. These two types of knowledge generate different types of practices. Scientific knowledge encourages practices aimed at controlling and standardising local conditions; local knowledge also aims at control, but in such a way that adaptive flexibility towards the uncontrolled is still recognised as a necessary attribute (van der Ploeg 1993, Wynne 1996).

Starting from this premise, a study of relocalisation of food and knowledge has to address the reconfiguration of social relations and the creation and transformation of knowledge that accompanies it. The term 'local food production' is ambiguous. Clearly all food is produced locally. In the rural sociology literature, the terms 'local food production' is used in reference to the effort to 'relocalise' and to 'resocialise' food production and consumption (Marsden et al. 2000). However, when we speak about 'relocalisation' of food, there is often an implicit assumption that localised food has been the historical starting point, and that the modern world is departing from this rather 'natural' condition. Hence, in many rural contexts this means that local actors seek to revitalise and re-invent local food traditions. Yet, most peripheral rural areas have been export-oriented, and agricultural production has had to adapt to changing demand, political conditions and market forces. Also, some parts of Europe have a history of expulsion, mass migration and re-location. In such circumstances, it is difficult for local actors, who often come from different backgrounds, to find a common tradition that can be easily related to existing natural conditions.

With this wider debate in mind, the present chapter is concerned with knowledge dynamics and localisation of food in a post-socialist setting in the former German Democratic Republic (GDR). While local food systems have found considerable recognition in research in West European and North American contexts, comparatively little research has been done on local food systems in a post-socialist context. In the following we present a case study of an emerging local organic farming network in the northeast of Germany. Some of the issues are typical of a number of post-socialist ruralities in Central and Eastern Europe (CEE). However, CEE countries have followed rather different historical paths before, during and after socialism; this is true of the degree of collectivisation as well as of the development strategies for post-socialist rural restructuring (Hann et al. 2003). With regard to post-socialist developments in agriculture, the trajectory in East Germany appears to be more akin to that in the Czech Republic or Hungary than to agricultural change in Poland, in Romania or in Bulgaria. Yet, even within

East Germany cultural and geographical situations vary – in particular between the northern and the southern states.

To a large extent the northeast of Germany – by comparison with many western European countries – has a comparatively recent history of rural settlements, made possible through draining and land reclamation in the course of the straightening of watercourses during the last two centuries. It subsequently developed into a 'granary' for the highly populated urban centres that grew rapidly during the period of industrialisation. Aside from household and subsistence production, these areas never had any localised food production. During the last centuries, this part of Europe has also had a record of state-planned rural settlement as well as of expulsion and mass migration. Under such circumstances it is difficult for local actors, who often come from different cultural backgrounds, to find a common tradition and to create a common local identity that can be easily related to existing natural conditions.

Further, many examples of relocalisation of food, especially in the European debate, seem to involve rural areas where farm structures are dominated by family holdings for which the processing of farm products and direct trading may offer an opportunity to increase household income. The present chapter, by contrast, examines a specific rural region in the northeast of Germany in which farm structures have undergone successive radical transformations in recent history. Before the Second World War much of the land belonged to large (and often feudal) estates. In 1945, through a land reform agreed upon as part of the Potsdam Treaty, such holdings were expropriated and distributed among farm workers and small holders, with priority given to the thousands of refugees who had been expelled from the former German territories. In 1945 the river Oder also became the new border between Germany and Poland, turning the region, which used to be part of the hinterland of the former German city of Stettin (since then Szczecin), into a peripheral area.

Under socialism rural society underwent fundamental restructuring. Family-based farming and fisheries were collectivised. From the 1970s onwards rural society was based on industry-like farm estates that also played a central role in local social and cultural development. As far as social order is concerned, productivism in the former GDR entailed a much more substantial institutional restructuring than in western societies. After 1989 and unification for the third time in less than 50 years rural eastern Germany underwent a major process of restructuring which once again imposed a new social order on rural society, at the same time ushering in a radical decline in production and employment. Moreover during that period new farmers from western Germany and other western European countries moved into the area and set up new farm businesses. The speed and scale of the changes that hit eastern German rural economies are almost without historical precedent, and have shaped rural development in the region to the present day (Laschewski and Siebert 2001, 2004).

The area today serves as a 'worst case' example for economically depressed countryside in Germany and has been studied intensively by 'experts' from all academic fields. It faces huge demographic changes, and its rural economic outlook is perceived as almost hopeless. For some years now, regions such as this have been the focus for a debate on rural decline, in which technocratic ideas of taking a proactive approach to sparsely populated areas have attracted considerable public attention (Berlin-Institut für Bevölkerung und Entwicklung 2006). However, agriculture and food processing remain central to the region's economy, and local protagonists have exerted great efforts to rebuild local food systems.

In regions such as this one, the food system has historically been export-oriented and dominated by large-scale agriculture. Prior to 1945 virtually no tradition existed of small-scale local food production and handicrafts. What little there was almost completely disappeared after the collectivisation programs of the 1960s. For all these reasons, there were few or no vital elements of local food production existed in this area. However, in a structurally disadvantaged region such as the northeast of Germany that is dominated by agriculture, small-scale local food production that would take advantage of local opportunities could make an important contribution to rural development. A growing demand for locally produced food is also being generated by tourism, which has been the fastest growing economic sector over the past two decades.

One significant obstacle to establishing a local food system is the absence of local knowledge and local identities. But notwithstanding this handicap a number of local food networks succeeded in emerging from the remnants of former, or recently-formed, food clusters, incorporating new actors and inventing new 'local' products. Being able to connect with local knowledge in only limited ways, these start-ups depend on experiences from other regions and/or have to experiment with new approaches.

The questions we ask in this chapter are: what does localisation of food mean in a context such as this? And what are the knowledge dynamics entailed in it? To answer these questions, we examine the case of an organic food network.[1] The objective of the case study is to analyse the differences in dynamics of the interrelations between local and expert knowledge in the selected example. To that end, we will identify the different actors and the different uses of knowledge and analyse the processes of codification, transfer and learning of knowledge.

We have chosen this case because much of the local food debate is related to organic farming. In the eastern German context 'organic' adds further complexity to local practices, given that organic agriculture was an almost unknown cultural concept that found broader recognition only after unification, although quite similar models of environmentally-friendly farming were also debated in the

---

1 The analysis of the case study reported in this chapter is based on the Germany Report for the Work Package 6 of the EU Project CORASON (A cognitive approach to rural sustainable development: the dynamics of expert and lay knowledge) which analyses the dynamics of knowledge in the valorisation of local food (Siebert et al. 2006). The research was conducted through analyses of documentary materials, official data and in-depth interviews with key actors in local development projects and regional administrations.

GDR. 'Organic farming' is predominantly a West German/European concept. Increasing support for organic farming has made it possible for there to be a rapid increase in the amount of land under cultivation for organically grown crops, but most of the organic produce is exported from the region in question to western German urban centres (including Berlin), while organic food consumption in eastern Germany remains at a very low level (Hagedorn et al. 2003). The most important protagonists in the eastern German organic farming and food movement come from western Germany. Our argument is that while on the surface what is involved has the appearance of a 're-connecting' perspective, to use Fonte's (2008) typology, a closer look reveals that there is no 're'-connection between producers and consumers. Localisation of food in this particular geographical context in no way implies the existence of a common, shared historical model. In this particular instance a local food system must first be introduced. Organic farming here is merely an imported model that has yet to be adapted to the local context.

The study region

The region of the case study is the Odermündung in northeastern Germany. It includes the two counties (Kreise) of Uecker-Randow (UER) and Ostvorpommern (OVP) in the federal state of Mecklenburg-Vorpommern (MV) (see Figure 2.1). The region is bordered to the north by the Baltic Sea and the island of Rügen, and to the East by the Polish region of western Pomerania. Historically, until 1945 the two counties comprised the hinterland of what is now the Polish city of Szczecin that was the German Stettin.

The region is predominantly rural in character and lacks industrial centres. Only a handful of smaller towns exist there: the university city of *Greifswald* and a few flourishing sea resorts. The region in general is comparatively distant from industrial and service centres and its economic structure is weak, reflecting the absence of a strong industrial history. Under socialism shipbuilding and fisheries were located in the harbour cities of *Rostock* and *Wismar*. Besides those two industries, agriculture and food production had always been the economic base of the region. After German unification, all these economic activities came under tremendous pressure. Since then the region has struggled with high unemployment rates ranging between 20 per cent and 30 per cent. Its population has declined by around 18 per cent since 1989 due to steady outmigration, especially of the younger generation, resulting in a substantial ageing of the population.

Today, agriculture and food production remain important economic activities. In 2005 there were 446 agricultural enterprises in OVP and 118,000 ha under cultivation. Employing 1,431 people (OVP) and accounting for 3.9 per cent of total gross value added, agriculture is of much greater relative importance for economic development in this region than elsewhere. In the UER district 258 agricultural enterprises manage about 79,000 ha, with an average farm size of about 280 ha. The farm structure in eastern Germany is very different from the farm structure

Figure 2.1   Case study region

in the western part of Germany: in the east farming is dominated by large-scale agriculture, only a few of the farms are family operated, and agricultural production is highly mechanised.

Following unification, the transformation process resulted in a farm structure dominated by international companies prioritising production for the market, notably of dairy products. This made for an effective production system with low labour costs that generated competitive advantages for the companies involved but had minimal effect on the local economy. The majority of agricultural outputs are finding outlets as raw products on national or European markets without added value from processing. This sectoral specialisation has impeded adequate integration of agriculture into the regional economy. A number of surveys (Landkreis Vorpommern 2004, Stiftung Odermündung 2004, Die Region Odermündung e.V. 2002, Regionaler Planungsverband Vorpommern 2001, 2002,

2004) have mentioned the lack of manufacturing facilities and the fact that there is so little local distribution of agricultural products. The mono-crop structure and the absence of a diversified agricultural sector are the most conspicuous negative aspects of the new farm structure.

One activity with strong potential in the region is organic farming: more than 15 per cent of agricultural firms in the county of Ostvorpommern produce organic products, a share far exceeding the national average. But 'organic farming' is a predominantly West German/European concept developed in this region only after unification. Increases in the level of funding available for organic farming have made it possible for there to be a rapid increase in the amount of land being employed for organic farming purposes.

Organic farming in this region is different from that in the western part of Germany. In the west organic farming is small-scale and consumer-oriented, selling products directly to the consumer. In the study region, by contrast, organic farming is predominantly large-scale. Most of the products are exported unprocessed, shipped raw to other regions due to the lack of processing facilities. Much of organic farming consists of extensive cattle husbandry on marginal land, creating little in the way of added value.

Despite its weaknesses, the food industry sector is an important component in production, accounting for a large part of the region's exports. Proximity to the metropolitan markets of Berlin and Hamburg has served to boost regional food exports to the point where they now comprise about 45 per cent of total regional exports. The quality of regional food products, meat and bread is often celebrated through awards from the German Agricultural Society (DLG). The food industry is clearly outwardly oriented, with food exports twice as high as imports.

Despite – or perhaps because of – its impressive performance, the food sector has not experienced any significant increase in the amount of labour it employs. Production gains have tended to be achieved through greater mechanisation and automation. Nevertheless, in an environment of declining overall employment, the food sector has at least kept a relatively stable number of workers in employment.

The main items processed by the food industry in these regions have been the predominating products of agriculture and fisheries (fish, cereals, meat, oil seeds, milk), usually for export purposes. International companies such as Nestle, Hipp, Nordzucker, Danesco, Rubave, with processing facilities in north Germany, are the market leaders. State and private institutes of hygiene with modern facilities ensure efficient and safe quality controls for industrialised food production. Small-scale local food production is of only minor importance. The post-unification economic transformation has contributed to the up-scaling of food-processing units. On the other hand, socialism had already largely destroyed handicraft culture.

There is no significant level of direct contact with consumers; only a few producers in the region have close connections with local processors and retailers. Supply chains are quite long and anonymous, the origin of products more or less meaningless, and quality aspects are often in conflict with efficiency and competitiveness of pricing.

Opportunities now exist for some of these structural deficits to be overcome as a number of local stakeholders have it on their agenda to establish local supply chains. They are aware of the narrowness of the product spectrum currently available, which includes just the main arable goods (grain, sugar, beets and potatoes) and they are working toward extending the range of locally-produced agricultural goods.

The case study: Netzwerk Vorpommern

In 1995 local consumers from the academic milieu in the Greifswald founded a food cooperative, which they organised as an association. Their idea was to establish a small market for environmentally conscious consumers, meeting the increasing demand for organic food. No local knowledge was available to assist with the launching of this initiative. The initiators had only lay knowledge of agriculture, but their life style was in accordance with ecological paradigms. The original idea was later broadened to include attempts to become active in local and regional development, initiating or supporting new projects for sustainable development. The aim was that local farmers and vegetable gardeners should be able to benefit from local demand. An association was therefore formed to establish a regional network, called 'Netzwerk Vorpommern'. The network now exists, with its members closely interlinked in an informal organic food network that has set itself the task of improving organic food production and food marketing in the region of Vorpommern. Most of its members are newcomers who established their businesses after unification despite a certain lack of requisite knowledge. They were obliged to draw on experience and knowledge from outside the region, first and foremost from groups concerned with ecology and organic food. But their orientation was and is clear: they favour sustainable rural development and they prioritise local consumers.

Part of their philosophy is a strong orientation toward high standards of quality. Among the organic associations, they prefer DEMETER and BIOLAND because of their stronger production guidelines. The first project of the association was to centralise the supply of regional and organic food to the public. The revenues of the cooperative and the farmers' market showed that there was enough demand for these products. The food cooperative subsequently evolved into the organic food-shop Keimblatt. A green market was organised on the town square in some years in collaboration with other producers. The range of services was later supplemented by a direct distribution system to consumer households. The declared objective of the organic food-shop was to strengthen the relationship between local producers and consumers. The emphasis of the food-shop is on freshness, local origin and biological quality of its products (fruit and vegetables, frozen food, bread, meats and cold cuts/sausages). About 40 per cent of the products come from the region of Vorpommern or other nearby areas (Mecklenburg and northern Brandenburg). More than 50 per cent of the cheese is produced in the region. Farmers themselves manage the sales of their vegetables inside the store. The organic food shop had nine employees, four of them in full-time jobs. It was meant to become a meeting point for residents of Greifswald concerned about health and the environment; the hope was also that it would strengthen the connection between producers and consumers. The store also bridged the gap between the large number of organic producers and the small number of locally sold products. With its enlarged sales space it enriched the retailing structure and service sector by providing an attractive point of reference for quality- and health-oriented consumers. The university city of Greifswald with its purchasing power has obvious qualifications as a source of potential expenditure on premium products. Focusing on local products, the organic food shop applies a communication strategy that helps foster regional identity on both the demand and the supply side. The opportunity to meet consumer demand directly provides an evident incentive for farmers to diversify their production, for example, in vegetables and fruits. Producers selling directly to the end consumer also enjoy increased profit margins.

In 1998 the organic food shop was separated from the association and launched as an economically independent enterprise which, however, returns its profits to the cooperative to finance the association's non-profit-making activities. The company is now called Naturkost Vorpommern GmbH (Biofood West Pomerania Ltd). The store's net profits, as indicated, go to the association, which uses the funds to further its mission (for example, environmental protection through organic cultivation; regional development initiatives). The primary aim of Naturkost Vorpommern GmbH is to market organic products of improved quality. The association is organised in such a manner that substantive control of its initiatives is exercised jointly by employees, consumers, and producers.

To summarise, the 'Netzwerk Vorpommern' association makes a significant contribution to sustainable development in the region. It started by concentrating on the regional supply of organic food and went on to exploit the market potential both of university staff and of tourist visitors to the region. The organic food shop has been conceptualised as a promoter of increased regional organic food production, an initiator of joint marketing initiatives and a supply organisation. The revenues of the shop, as a component of the non-profit association, are therefore reinvested for the purpose of developing new projects. The association's primary objectives are environmental protection through organic food production, improvement of the local economy and support for self-initiated and self-organised projects.

Apart from these activities, the association is also active in local and regional politics. At the municipal level, there are intensive contacts with other retailers to arrange activities in support of municipal development generally and, in particular, development of the municipality as a tourist attraction. However, this collaboration with the municipality has produced mixed results. Sometimes the cooperation is very good, but over the years some surprising impediments have emerged. Often they are attributable to personal antipathies and distrust between members of the association and the municipality.

The association lobbies the government to enhance governmental support for organic farmers and organic food supply chains. The association's leading role in the regional organic farming network is explicable simply by the fact that its members possess economic expertise (for example, in marketing and business management) as well as being highly competent political managers. They are thus motivated to encourage and nurture activities favouring sustainable rural development.

A phase of intensive negotiation and coordination was initiated with a view to creating an alternative producer network, but it was unsuccessful. Agreement had been reached on investment capital, management, logistics, and arrangements for price formation and quality control had been worked out. In the end, however, the stronger partners felt that the balance between costs and benefits was not favourable enough to justify their signing the contracts. Asymmetrical power relations in the network, distrust among its members and unwillingness to delegate decision-making were the main reasons for the failure of this initiative. It was not the 'hard' factors such as contracting, quality criteria or prices that proved to be the obstacles to collective action. It was rather the 'soft' factors of power and trust that made key actors reluctant to join the network and sign the agreements.

The first opportunity to extend and strengthen the organic food sector has thus failed. Nevertheless, the rate of growth of the shop's turnover seems likely to provide the kind of economic power that will make it possible for them to look for new projects, such as an organic supermarket or a scheme for collaboration with seaside hotels in resort areas. One key problem will be that of securing more qualified employees, a difficulty already encountered during the establishment of the shop-in-shop butchery.

Expert knowledge of food production and processing, especially in relation to quality characteristics, is a crucial prerequisite for the marketing of organic food, even if the main priority is successful retailing. Consumers of organic food pay more attention to product information (Rämisch 2001, Recke et al. 2004). They tend to demand traceability, so that retailers have to be well informed about the farms or food processors that are supplying them. For the purposes of this case study that meant that managers were emphasising that a high proportion of products should be of regional origin. Short supply chains make it possible for information to be detailed, thereby making the chains transparent, something especially necessary with meat products, about which consumers have become quite sensitive.

Knowledge of marketing is an essential resource for successful business management even if there is no direct competitor in the local market. High-intensity communication is helpful in convincing even occasional buyers.

Most training to do with production and marketing is either self-training or on-the-job training. There are two main reasons why the initiative is steadily continuing to grow. One is the ongoing process of individual and collective learning. The other is the strong regional identity that that provides an incentive for collective voluntary action. Increased managerial skills go hand in hand with better organisational capacities and greater professional know-how by employees. The manager becomes responsible for the food market's quality control on the procurement side of the food market and for the training of employees. He has therefore had to learn to gather information on his own initiative because he has not been educated in these subjects. A continuous quality management system in the shop has been developed in collaboration with the employees. Along with professional magazines, direct contacts to suppliers, producers and wholesalers are used as information sources. The professional know-how in business management and food quality is partly implicit and partly codified. To evaluate product quality it is necessary to have extensive information about the production process and the qualifications of the producers. This knowledge depends on local conditions and personal experiences and could hardly be codified.

On the other hand the experiences gained in the attempt to set up a supply chain, not to mention some other actions serving to strengthen the regional market for organic food, could possibly be codified. Such managerial knowledge could be generalised and transferred to other regions. The range of the transfer, however, is restricted by the scope of the regional network, which has strong endogenous ties but few links to external organisations.

As the organisation leaders indicated to us, the university is an especially favourable context for organic food marketing; it can also counteract the brain drain by integrating well-educated graduates. Local politics could provide assistance for the establishment of new business or projects, which motivated young academics can complement in endogenous capacities.

Missing from the history of this project are any links to the faculties for food processing and marketing at the University of Applied Science in Neubrandenburg. The main sources for knowledge transfer are the training programmes run by wholesalers or information from trade and producer associations.

The Netzwerk Vorpommern started with the idea of establishing a small market for environmentally-conscious consumers, meeting the increasing demand for health food. Most of its members are newcomers who established their businesses after reunification but initially did not have the requisite knowledge for those businesses. For that reason, they used experiences and knowledge from outside the region, above all from groups concerned with ecology and organic food. They possess a clear orientation to sustainable rural development and to local consumers.

Conclusions

In this chapter we have described a case in the *Odermündung* region of northeast Germany in order to understand the relocalisation of food in a context which, for various historical reasons, cannot easily draw upon old and well established local food practices. In such a context relocalisation means that local actors have to (re-)invent local food traditions by attempting to establish direct relations between producers and consumers.

The Netzwerk aims to establish conditions for the local distribution of organic food. It consists of small farmers, manufacturers and retailers. Localisation in this case is seen as establishing local producer networks and producer-consumer relationships. 'Local' is primarily understood in a geographical sense (short distance), a notion directly related to sustainability. There is no conception of specific local food practices or a regional identity linked to food. Most of the actors involved are newcomers who established their businesses some time after unification. Some of them show great innovative capacity, discovering market niches and new distribution channels. The knowledge dynamics entailed by this appear to be more complex, since they connect forms of expert as well as lay knowledge, scientific as well as local knowledge, and technical as well as managerial knowledge.

The network has an orientation toward sustainable rural development and local consumers. But the effects on employment remain marginal. The network has so far been more successful in fostering capacities for self-learning capacities and generating social capital. But a great potential exists for closer cooperation, particularly in optimising costs for marketing and distribution.

The Netzwerk Vorpommern came into being after German unification. It could not fall back on local knowledge. The region has hardly any tradition of small-scale or ecological food production, so that it is impossible really to talk about a relocalisation of production. In the absence of prescribed local models, the de-contextualised concept of 'organic farming' has offered a useful framework for the local actors to establish new local food relations. The actors of the Netzwerk are mostly newcomers and belong to a group of ecology-minded people that has gradually been establishing itself. Their lifestyle follows an ecological paradigm that has great potential for encouraging sustainable rural development. Their mission was to contextualise the relatively abstract concept of 'organic farming' by connecting producers and consumers. These efforts promote regionalisation of food production. The key impediments to attaining a stronger position in the market are lack of expertise and deficits in local, scientific and managerial knowledge.

The approach chosen by the actors suffers from a major weakness: the absence of food practices that are vital elements of local cultures. The *Netzwerk* appears to be fixated on a naturalistic conception of food. It targets a niche market comprised of highly environmentally-concerned consumers. The knowledge dynamics addressed by the actors, accordingly, have to do with technological processes, nature, managerial and marketing issues. The concerns and knowledge of the consumers are taken as given. The relocalisation of food appears to suffer from the lack of locally-oriented food practices by local actors. In addition, the high proportion of large-scale organic farming in eastern Germany makes it possible for such products to be obtained from producers outside the region. The knowledge of consumers about food and their ability to cook seem to have been until now the main driving forces behind the development of local food circles.

References

Berlin-Institut für Bevölkerung und Entwicklung. 2006. *Die demografische Lage der Nation*. München: Verlag d+v.
Die Region Odermündung e.V. 2002. *Regionales Entwicklungskonzept für den ländlichen Raum. Bewerbungsbeitrag im Wettbewerb „REGIONEN AKTIV – Land gestaltet Zukunft" des Bundesministeriums für Verbraucherschutz*, Ernährung und Landwirtschaft.
Feagan, R. 2007. The place of food: mapping out the 'local' in local food systems. *Progress in Human Geography* [Online], 31(1), 23–42. Available at: http://phg.sagepub.com/cgi/reprint/31/1/23 [Accessed 26 January 2009].
Fonte, M. 2008. Knowledge, food and place. a way of producing, a way of knowing. *Sociologia Ruralis*, 48(3), 200–222.
Goodman, D. 2003. The quality 'turn' and alternative food practices: reflections and agenda. *Journal of Rural Studies*, 19(1), 1–7.
Hagedorn K., Laschewski, L. and Steller, O. 2003. *Institutionelle Erfolgsfaktoren des ökologischen Landbaus*. Organic e-prints. Available at: http://orgprints.org/7321/ [accessed 3 November 2009]
Hann, C. and the 'Property Relation' Group (eds.) 2003. *The Postsocialist Agrarian Question*. Münster: LIT Verlag.
Holloway, L., Kneafsey, M., Venn, L., Cox, R., Dowler, E. and Tuomainen, H. 2007. Possible food economies: a methodological framework for exploring food production-consumption relationships. *Sociologia Ruralis*, 47(1), 1–19.
Kloppenburg, J.R. Jr. 1991. Social theory and the de/reconstruction of agricultural science: local knowledge for an alternative agriculture. *Rural Sociology*, 56(4), 519–548.
Landkreis Vorpommern 2004. *Regionales Entwicklungskonzept* (REK) Landkreis Ostvorpommern.
Laschewski, L. and Siebert, R. 2001. Effiziente Agrarwirtschaft und arme ländliche Ökonomie? Über gesellschaftliche und wirtschaftliche Folgen des Agrarstrukturwandels. *Berliner Debatte Initial – Zeitschrift für sozialwissenschaftlichen Diskurs* 12(6), 31–32.
Laschewski, L. and Siebert, R. 2004. Power and rural development: social capital formation in rural East Germany, in *Power and Gender in European Rural Development*, edited by Goverde, H., de Haan, H. and Baylına, M. Aldershot: Ashgate, 20–31.
Levitt, B. and March, J.G. 1988. Organizational Learning. *Annual Review of Sociology*, 14(August), 319–340. Available at: http://www.itu.dk/

~kristianskriver/b9/Organizational%20Learning.pdf [Accessed: 26 January 2009].
Marsden, T., Flynn, A. and Harrison, M. 2000. *Consuming Interest: The Social Provision of Food.* London: UCL Press.
Morris, C. 2006. Negotiating the boundary between state-led and farmer approaches to knowing nature: an analysis of UK agri-environment schemes. *Geoforum*, 37(1), 113–127.
Rämisch, G. 2000. *Regionale Marktchancen für Produkte des Ökologischen Landbaus – dargestellt am Fallbeispiel Klostergut Scheyern und Großraum Pfaffenhofen an der Ilm.* FAM-Bericht 41. Aachen: Shaker-Verlag.
Rämisch, G. 2001. Regionale Marktchancen für Produkte des ökologischen Landbaus. *Berichte über Landwirtschaft*, 79(2), 212–233.
Recke, G., Zenner, S. and Wirthgen, B. 2004. Situation und Perspektiven der Direktvermarktung in der Bundesrepublik Deutschland – eine Analyse der Angebots und Nachfrageseite, *Angewandte Wissenschaft* Heft 501. Münster-Hiltrup: Landwirtschaftsverlag.
Regionaler Planungsverband Vorpommern (Hrsg.) RPV 2002. *REK Vorpommern*, Greifswald.
Regionaler Planungsverband Vorpommern (Hrsg.) RPV 2004. *Möglichkeiten zur nachhaltigen Entwicklung der Odermündung im Bereich des Peenestroms, des Achterwassers und des Stettiner Haffs unter besonderer Berücksichtigung maritimer touristischer Nutzungen.* Greifswald.
Regionaler Planungsverband Vorpommern RPV (Hrsg.) 2001. *Entwicklungspotenziale der Landwirtschaft in der Planungsregion Vorpommern.* Greifswald.
Siebert, R., Dosch, A. and Laschewski, L. 2006. CORASON WP6 Local food production. *Germany Country Report.* Available at: http://corason.hu/download/wp6/wp6_germany.pdf [accessed: 3 November 2009].
Statistisches Amt Mecklenburg-Vorpommern 2008. *Statistisches Jahrbuch* Mecklenburg-Vorpommern, Schwerin.
Stiftung Odermündung 2004. *Jahresbericht der Lokalen Aktionsgruppe LEADER+* Anklam, Unpublished manuscript.
Tsouvalis, J., Seymour, S. and Watkins, C. 2000. Exploring knowledge cultures: precision farming, yield mapping, and the expert-farmer interface. *Environment and Planning A*, 32, 908–924.
van der Ploeg, J.D. 1993. Potatoes and knowledge, in *An Anthropological Critique of Development: The Growth of Ignorance*, edited by Hobart, M. London: Routledge, 209–227.
Venn, L., Kneafsey, M., Holloway, L., Cox, R., Dowler, E. and Toumainen, H. 2006. Researching European 'alternative' food networks: some methodological considerations. *Area*, 38(3), 248–258.
Willke, H. 2001. *Systemtheorie III: Steuerungstheorie.* Stuttgart: Lucius und Lucius.

Winter, M. 2003. Embeddedness, the new food economy and defensive localism. *Journal of Rural Studies*, 19(1), 23–32.
Wynne, B. 1996. May the sheep safely graze? A reflexive view of the expert-lay knowledge divide, in *Risk, Environment and Modernity: Towards a New Ecology*, edited by Lash, S., Szersynski, B. and Wynne, B. London: Sage Publications, 44–83.

Chapter 3
The Reconstruction of Local Food Knowledge in the Isle of Skye, Scotland

Lorna Dargan and Edmund Harris

Introduction

In this chapter, the role played by knowledge in the reconstruction of local food systems is explored using the Isle of Skye in Scotland as a case study. Attempts to rebuild local food systems rely on the valorisation of local foods by both producers and consumers, and knowledge is central to the discovery or construction of the value of local foods (Fonte 2008, Kirwan 2004). By exploring the production, negotiation and application of different forms of knowledge in an emergent local foods system on Skye, this case study examines the ways in which producers and consumers have reconstructed their knowledge of local food. As the local food sector has developed on Skye, local actors have combined the scientific and codified knowledge of national-level agricultural advisors with lay knowledge of local places and growing conditions, and, in doing so, have started to rebuild the community's tacit knowledge relating to food production and consumption at the local scale. In absence of strong support for local food systems from the national government, islanders have taken the lead in developing a food system that emphasises face-to-face contact between producers and consumers, short supply chains, and quality-assurance based on social relationships of trust. The burgeoning social intercourse around products and varieties through which such relationships are built contributes to the reconstruction of the community's tacit knowledge of local food.

Using data gathered as part of the CORASON[1] project, this chapter explores the emergence of a local food system on the Isle of Skye, using the Skye and Lochalsh Horticultural Development Association (SLHDA) as a case study. The chapter explores the roles of different forms of knowledge in the development of the SLHDA, and also introduces some of the key elements of the emerging local food system on the island. While the emergence of a local food system on Skye has taken place under the leadership of local residents, its contribution to the

---

1   CORASON stands for the 'Cognitive approach to rural sustainable development'. For more information, see the Introduction to this volume and the project website at http://www.corason.hu/. This chapter draws on research carried out for Work Package 6 which addressed the dynamics of knowledge in the valorisation of local food.

sustainable development and rural regeneration agendas is important, particularly since Skye's economy is heavily reliant, first, on the export of high-value food and drink products and, second, on the tourism industry. The enhanced value of local foods and the link between the valorisation of local foods and the branding of Skye as a pure, unspoilt, natural environment tie the emergent local food system to the island's economic future, and to those industries that rely on non-local perceptions of Skye and the food produced there.

The following section introduces some current approaches to the study of local food systems, and positions food system relocalisation within a broader landscape of alternative food politics. Recent work on different forms of knowledge is also summarised, forming the foundation for an exploration of the knowledge dynamics in Skye's emergent local food system. Qazi and Selfa (2005: 47) have observed that alternative food network strategies are 'regionally contingent due to particular agro-industrial histories and consumer and producer politics', highlighting the importance of understanding the historical and political context for case studies. The third section, therefore, outlines the history of agricultural production in Scotland and on Skye, and introduces the policy context surrounding Scottish food and agriculture. The final sections then focus on Skye, detailing the key agencies involved in efforts to rebuild a local food system, and exploring the movement and mobilisation of different forms of knowledge through these emerging networks.

Contextualising alternative food politics and knowledge

Theorising alternative food networks

During the past 30 years a wide range of 'alternative' approaches to food systems have emerged in reaction to 'conventional' or agro-industrial food systems, which are characterised by high-input, large-scale and corporate agriculture that feeds a globally integrated distribution and retail network dominated by supermarkets. Initially these alternatives focused on organic production (Vos 2000), but as criticism has mounted of the 'conventionalisation' of organic agriculture (Guthman 2004, Holt and Amilien 2007), alternative food politics have shifted to address the spatiality of food systems and the potential for relocalising networks of food production and consumption. Kimbrell (2002: 1) describes how the 'great psychological and physical distance between consumer and food production creates a tragic disconnect between the general public and the social and environmental consequences of the food being grown and eaten'. Concern for the social and environmental impacts of conventional food systems now drives popular interest in the relocalisation of food production and consumption.

In attempts to relocalise food systems, place is of central importance. Feagan (2007: 38) describes how, as food systems are stretched further across space, we experience a range of 'disconnections and *disembedding* ... loss of agricultural resilience and diversity, degradation of the environment, dislocation of community, loss of identity and place' [emphasis in original]. In response to the resulting 'systemic placelessness' (DuPuis and Goodman 2005: 360) of conventional food systems, relocalisation aims to embed food production and consumption in local places and to build direct relationships between producers and consumers.

As interest in localisation has expanded and started to assume the 'features of a new orthodoxy or paradigm' (Fonte 2008: 200) in alternative food politics, it has attracted increased critical attention from sociologists and geographers, who are keen to point out that 'the local is not everywhere the same' (Allen et al. 2003: 63). Recent reviews of alternative food projects have identified two approaches to embedding food production in place: 'product and place' and 'process and place' (Maye et al. 2007, Watts et al. 2005). The first refers to the embedding of food *products* in specific places, thus providing information about provenance to consumers. In this case, the food products are not necessarily retailed to 'local' markets, and through the adding of economic value, this approach is often used to stimulate rural development (Goodman 2004: 7, Ilbery and Maye 2005). The second form refers to attempts to embed food production and consumption processes in 'local' places, and involves emphasising 'the social and ethical values associated with particular supply chains' (Maye et al. 2007: 7). Goodman (2003) has suggested that this distinction is reflected in a difference in alternative food politics between Europe and North America, with the reformist 'product and place' approach prevalent in Europe, and the more radical and politically-motivated 'process and place' approach dominant in the US and Canada. Goodman (2003) attributes these differences to the degree of involvement of agro-food scholars in policy processes – involvement which is greater in the European context – and to other contextual aspects such as the slow process of Common Agricultural Policy reform in the EU. The case study examined in this chapter, however, better fits the 'process and place' category and alongside other case studies collected as part of the CORASON research, renders problematic this geographical distinction between approaches to alternative food politics.

Building on the CORASON research case studies, Fonte (2008) proposes an alternative 'dual perspective' on local food networks – one that reflects the distinctions described above, but that acknowledges that both variants are present within Europe. First, projects initiated by grassroots social movements aiming to rebuild local producer-consumer links characterise the 'reconnection perspective' (Fonte 2008: 202). The second perspective 'repositions local food production in relation to values associated with territory, tradition and pre-industrial production practices' and is termed the 'origin-of-food perspective' (Fonte 2008: 202). Here, the origin of food products is emphasised (as in the 'product and place' category discussed above) and the strategy is often explicitly linked to rural development goals. The rebuilding of relationships between local producers *and consumers* is not so central to this latter perspective. As is discussed later in this chapter, the emergent local food network on Skye better fits the first perspective, aiming to reconnect producers with consumers on the island and to market local food to local people. The remaining part of this section focuses on the role played by knowledge in local food networks viewed through the 'reconnection perspective'.

The role of knowledge in local food networks

Central to local food projects seeking to reconnect producers and consumers is a desire to resist the perceived disempowering effects of conventional globalised food systems (Feagan 2007, Holloway et al. 2007). The knowledge needed to rebuild functioning local food networks, however, has often been lost in areas characterised as 'food deserts' – where agriculture is large-scale and export-oriented, and where food distribution and retail are dominated by supermarkets. As Fonte (2008: 212) observes, 'initiatives to relocalise food systems include both attempts to educate or self-educate growers, and conscious attempts by growers to re-educate consumers about food and food consumption'. An important question in considering the politics of food system relocalisation concerns the sources of knowledge drawn upon in these processes of re-education, and the dynamics through which they are transferred between groups.

In their comparison of the role of knowledge in conventional and organic farming networks, Morgan and Murdoch (2000: 171) detail how as farming was industrialised in the UK since the 1950s, 'farmers' local knowledge was displaced by standardised knowledge emanating from up-stream supply industries'. They describe how the resulting 'conventional [food] chain is biased towards standardised knowledge with the effect that tacit knowledge is debased so that it cannot easily be drawn upon once this chain moves into crisis'. Morgan and Murdoch draw here on a distinction between a 'scientific' knowledge type, which tends to standardise and codify information, and a 'tacit' knowledge type that is 'personal and context-dependent' and difficult to transfer except through shared practice or experience (Morgan and Murdoch 2000: 161). Tacit knowledge is that created through 'normal processes of socialisation' and can be described as pre-discursive – captured by Polanyi's (1966) aphorism 'we know more than we can tell'. In social networks, shared tacit knowledge helps build and reinforce relationships, trust and social cohesion (Fonte 2008: 210). Alongside 'lay' knowledge – that gathered through practical experience – 'tacit' and 'lay' knowledges are sometimes referred to as 'local' or 'traditional' knowledge.

Farmers' loss of knowledge about food production for local markets is paralleled by the 'deskilling' of consumers in the food system. As processed and packaged foods from distant sources become more prevalent, consumers have become increasingly distanced from the raw ingredients of their diets, and no longer know how their food has reached the supermarket shelves. Jaffe and Gertler (2006: 157) suggest that as a result of this 'deskilling', 'most consumers lack the scientific and practical knowledge to make choices that reflect their fundamental interests in health, longevity, and obtaining value for money'. The re-skilling of consumers in a local food network requires knowledge both of the food itself – how to prepare meals from fresh ingredients and how to preserve seasonally-available products

– and knowledge of how that food has been produced, and by whom. Knowledge constructed and mobilised to value local foods over industrially-produced foods often has to operate in the face of concerted marketing and publicity campaigns from the agro-industrial food complex. As Jaffe and Gertler (2006: 158) conclude, 'the struggle over relevant knowledge and know-how is key. Skilled consumers will be vital to the positive transformation of food systems'.

It is clear that developing new ways of producing and consuming food locally requires new ways of knowing that must be constructed by individuals and communities from a range of knowledge sources. Through the transfer of knowledge and through the practice and experience of new knowledge–based activities *in the local context*, a local food system can emerge through the situated construction of a new knowledge culture. This chapter examines the knowledge production and circulation processes at work in the emergence of a local food system on Skye, and in particular, demonstrates how non-local 'expert' knowledge has been contextualised in this particular local setting.

Agricultural production and policy in Scotland

Agricultural production in Scotland

Farming is a significant sector in Scotland's rural economy, contributing £2bn to Scotland's GDP (1.4 per cent) and constituting the dominant land use (Scottish Executive 2001: 21).[2] In 2007, 79.5 per cent of Scotland's land was used for agriculture (Scottish Government 2008a: 30). There were approximately 51,300 agricultural holdings in 2006 employing some 67,100 people (Scottish Executive 2007: 64) which, while small compared to other sectors, is important in rural areas and plays a significant role in maintaining the viability of some more remote communities. The percentage of farmers working part-time, however, is relatively high with 77 per cent of farmers and spouses having non-farm sources of income in 1998/99 (Macaulay Institute/University of Aberdeen 2001: 28), a trend which is set to increase.

Despite representing a sizable proportion of Scotland's land use, much agricultural land is of poor quality, with 86 per cent designated as Less Favoured Areas (LFAs)[3] (Scottish Executive 2007: 68). Approximately 11 per cent of grants and subsidies to Scottish agriculture in 2006 were made through LFA payments. In all, grants and subsidies to farmers totalled £527.4m in 2006, forming a large proportion of Scotland's total income from farming of £577.5m (Scottish

---

2 The Scottish Executive was rebranded as the Scottish Government in 2007.

3 Less Favoured Areas are those with poor climate, soils or other characteristics resulting in low yields from agriculture designated by national governments within the European Union, and eligible for specific subsidies under the Common Agricultural Policy. In Scotland, this support is through the Less Favoured Areas Support Scheme (LFASS).

Executive 2007: 6). Our case study region of the Highlands and Islands suffers acutely from these problems. The region is mountainous, troubled by poor soils and an extremely difficult climate which is very wet and windy, and which receives less sunlight than lowland regions. Traditionally, therefore, agriculture in northern and western Scotland has focused predominantly on hill cattle and sheep farming, and crofting (a small-scale, low-input form of agriculture, whose characteristics will be illustrated below), while agricultural production in the lowland areas has concentrated on livestock, dairy production and arable cropping.

A significant proportion of Scotland's agricultural production is generated by small-holdings, with almost 50 per cent of total holdings being less than 10 ha in 2006 (Scottish Executive 2007: 71). Many of these are crofts, which represent the key agricultural tradition in the Scottish Highlands where this case study is located. Crofting is a small-scale, low-input form of agriculture. Land is less intensively farmed, and is treated with fewer pesticides and herbicides (Hunter 1991, RSPB and Scottish Crofters' Union 1992). Crofting emerged in the eighteenth century from the notorious Highland Clearances, a historical event that dramatically altered the landscape and population structure of Scotland. During the Clearances, tenant farmers were forcibly removed from their holdings by landowners and either transported overseas or moved to very small, infertile plots known as crofts. Land was converted from mixed farming to animal husbandry, which remains the dominant form of agriculture in the Highlands and Islands.

The crofts onto which many tenants were moved during the Clearances were characterised by lower quality soils and a lack of common grazing land, and were often situated in marginal areas such as hillsides or coasts. The landlords (or Lairds) who had previously managed estates according to a feudal system were looking for a cheap supply of labour, and so deliberately constructed crofts to be of an insufficient size to support subsistence farming, forcing crofters to work off the croft in order to survive. As a result, the region has a strong culture of pluri-activity, as crofters and their families continue to combine small-scale agriculture with work off the croft in order to supplement their incomes.

The policy context for agricultural production in Scotland

Despite the prominence of agriculture in rural Scotland and the Scottish economy, a nationally distinctive food culture has not developed. The Scottish culinary tradition has, until modern times, been based around simple food cooked plainly. Oats, potatoes, turnips, meat, game and fish have been staple ingredients. Government agencies acknowledge a lack of imagination in Scottish cooking, but often highlight Scotland's 'quality' exports such as meats, fish and whisky. Speaking about 'Scotland's Farming Future', for example, Scottish Agriculture Minister Richard Lochhead stated that 'Scotland has an international reputation as a food-producing nation' and said 'It makes me proud to see our successful, market driven food companies producing world-renowned Scottish specialities like whisky and Scotch beef' (Lochhead 2009: 28).

Responsibility for food and agriculture policy is at present divided between a number of government and regulatory agencies, including the Department for Environment and Rural Affairs, the Food Standards Agency, and the Scottish Food Certification Agency. In order to produce a more coordinated policy approach to all food-related issues, however, the Scottish Government (with the help of private sector companies, such as Scottish Food and Drink) has started to develop a National Food Policy, seeking to address dietary health, economic growth through agriculture, food processing and food exports, sustainable development, and climate change (Scottish Government 2008b).

Despite current attempts to integrate Scottish agricultural policy with sustainable development objectives, the agro-industrial model remains dominant, focusing on large-scale, technologically intensive farming, which demands that farms should increase the scale of production to meet the demands of the supermarket-controlled retail sector. The 'Forward Strategy for Scottish Agriculture' (Scottish Executive 2001) acknowledges the need for continued subsidies, but calls for farmers to increase their incomes and profitability by becoming more competitive and better exploiting market opportunities.

The fact that many small-farms will struggle to survive in this market–oriented and competition-driven environment is recognised, but smaller operations are encouraged to diversify. Acknowledging that they cannot compete on cost, they are directed to produce for higher-value export-orientated niche markets. Despite the large number of small producers in Scotland, the Executive seems to perceive them as an obstacle to the competitiveness of Scottish agriculture. The Executive's message to farmers is clear: 'if people are not prepared to change, little can be done to help' (Scottish Executive 2001: 41).

The national agro-industrial model has clear implications for agricultural production on Skye, which is dominated by small-scale farming. While the Forward Strategy does not *directly* challenge the mode of production in the Highlands, it clearly states that larger farms are more competitive and, therefore, more likely to survive without recourse to dwindling subsidies. However, this push towards expansion is being resisted locally. The regional agencies responsible for developing food production on Skye continue to offer support to small-holdings, which have been the backbone of agricultural production in the Highlands and Islands. The crofting model has long been supported at the regional level, despite national opposition (Hunter 1991), as a means of halting depopulation by keeping people on the land. This more localised model sees farms and crofts as more than just businesses, recognising their crucial role in the economic, social and cultural landscape. As a consequence, support for small-scale agricultural production is perceived to be critical for sustainable development in the region. It is in this policy context of regional support, but relative national indifference, that this case study operates.

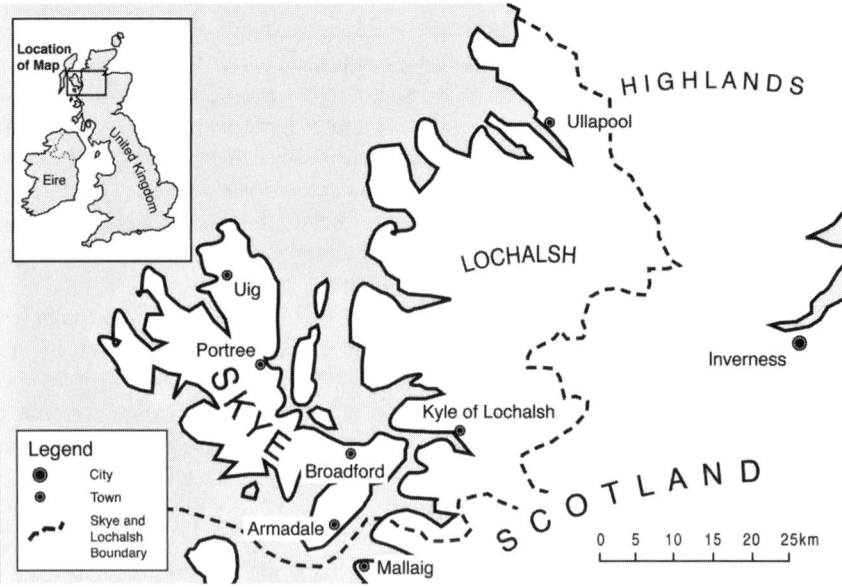

Figure 3.1 Map showing location of Skye and Lochalsh

The case study area

Skye and Lochalsh

Skye and Lochalsh is situated within the Highlands and Islands region of Scotland, in the Highland Council local authority district. It is one of the most remote parts of the UK, two hours drive from the nearest administrative centre of Inverness. It has a population of just under 12,000 people, 76 per cent of whom live on the Isle of Skye, and 24 per cent of whom live on the adjacent mainland area of Lochalsh, and has a population density of just 4.4 people per km^2. For many, Skye represents the typical highland landscape: sparsely populated and rugged, and shaped by generations of crofting. Skye's mountainous and inhospitable landscape is a strong draw for a tourist industry that has been developed around the attractions of this 'wild' and 'unspoilt' terrain.

There are 23,300 agricultural holdings in the Highlands and Islands Enterprise (HIE) area, representing 53 per cent (HIE) area, representing 53 per cent of Scotland's total agricultural area, of which 17,800 are crofts (HIE 2007a). Agriculture focuses on livestock rearing, and crofts are characterised by 'more rough grazing, less intensive use of improved grass, less cash cropping, more fodder crops, less winter cereals, much lower overall stocking rates, more sheep, less cattle finishing and a much smaller dairy sector' (Scottish Agricultural College 2003: 3).

HIE states that one of the key weaknesses of the agricultural sector in the Highlands is its dependence on direct subsidies, and that farmers are increasingly reliant on income from their spouses or from non-agricultural activities. While this does mean that farmers are multi-skilled workers, it also means that farming is not perceived to offer good employment prospects by younger people, and the harsh nature of the climate and the work involved discourages many from entering the agricultural sector. The implications of this are twofold. First, outmigration of young people leads to population stagnation and loss in some of the more remote areas. Second, the region is increasingly reliant on imports in many sectors, and particularly in horticulture. Despite these difficulties, food and drink production is an important part of the region's economy, with almost 300 largely small-scale businesses engaged in processing, of which 61 per cent employ fewer than 10 people (HIE 2007b). Food and drink have significant export values, and led by whisky and fish, they are the top exporting sectors in the Highlands and Islands region (HIE 2007b).

Interest in local foods on Skye

Skye has a long history of resistance to the national agro-industrial model of agricultural development. The lack of national support for an appropriate model for farming on Skye has led local actors – both producers on Skye and Highland Council Officers – to develop their own approaches to local food development. The SLHDA is an example of such a locally-led initiative, aiming to redevelop horticulture on Skye. Following a sustained period of national support for export-oriented agriculture for national and international markets, horticulture on Skye had almost completely disappeared and had been replaced by imported food retailed through supermarkets. Prior to the arrival of supermarkets on Skye in the 1960s, many residents had produced food on the available land, a process that was demanding but necessary due to a lack of alternative food provision. The arrival of supermarkets on the island changed people's attitudes towards the value of their own produce, leading to a growing perception that supermarket goods were better than their own produce. With produce readily available in the supermarkets, fewer and fewer people grew their own food, and the pool of tacit knowledge around this type of food production was gradually lost.

During the 1980s and 1990s, interest in local food started to grow. Skye was becoming an increasingly popular tourist destination, and the island landscape was marketed as pure and unspoilt, with food produced on the island an important part of this image. Tourists and in-migrants seeking a rural idyll expected food that was 'fresh' and 'natural', leading locally produced goods to become more highly valued in the community (Árnason and Lee 2003).

The Skye and Lochalsh Horticultural Development Association

The SLHDA was established in 1994 by a Highland Council Economic Development Officer based on Skye. The officer, who had a long-standing personal interest in horticulture and wanted to promote horticulture on the island, met with a local grower who shared her views, and then arranged a public meeting to make contact with other interested residents. The result was the SLHDA. Initial funding came from the Highland Council and SALE (the local enterprise company), which now also provides some training to members of the network. At first, the project struggled to persuade local people that horticulture was a worthwhile or viable activity on Skye. Since the island's agricultural activities had traditionally focussed on lamb and cattle, there was some scepticism from crofters and other local people that the area could support horticultural activities, especially on such poor soils and in the harsh climate. To tackle this scepticism, the SLHDA founders led by example and piloted a trial with four growers on 0.25 acre plots. The Project Coordinator argues that the increasing success of the network is because growers demonstrated that Skye could support horticulture, a message which resonated at a time when Common Agricultural Policy (CAP) reform was forcing farmers and crofters to rethink their business strategies.

The network has grown from 12 to over 60 members, who pay an annual subscription to remain part of the network. The growers lead the network and serve on the committee alongside growing and selling their own produce. Growers using organic methods are in a minority, and participate in the SLHDA alongside those using non-organic methods. There is one full-time development officer who is responsible for overseeing the project, in collaboration with a committee of 19 people including a Treasurer, Secretary, and growers. This committee is responsible for generating new ideas for project funding applications, producing the monthly newsletter, and organising the Annual General Meeting (AGM) and the associated annual report. By 2001, the network was generating sufficient sales to warrant the creation of a trading company, which was formally established in 2002. The company handles bulk order buying and the weekly Farmers' Market in Portree, through which growers sell their produce to the public.

The SLHDA has no permanent funding other than membership subscriptions, and receives funding for individual projects from other agencies. As a result, while the overall aim remains the same – 'to promote the use of local produce and products and to replace "imported" products and skills' (SLHDA 2008a) – the network undertakes many different activities in order to access funding. Recent projects have focussed on developing connections between SLHDA growers and local schools through education and growing food on school grounds (SLHDA 2008b).

The SLHDA places no constraints on who may join the network. Many 'incomers' (migrants to Skye) have joined, including those who have arrived with the explicit intention of engaging in horticultural activities and those who have arrived looking for a 'rural idyll', and have decided that growing their own food is a part of that vision. For novice farmers such as these, horticulture seems far less intimidating than livestock rearing. Many SLHDA growers are only engaged in horticulture part-time, but engage in network activities alongside full-time farmers.

The SLHDA Coordinator reports having found it harder to engage with crofters, perhaps due to their traditional emphasis on livestock rearing rather then horticulture, and observes that horticulture can be very time-intensive and thus is less well suited to part-time crofting lifestyles. The network maintains that crofters are welcome and is currently trying to include more local crofters. However, the rising cost of production is a significant barrier to widening participation. Many growers on Skye operate on an extremely small-scale and at the edge of viability, and have little capacity to respond to rising or unexpected costs. Many growers use polytunnels to expand the growing season in Skye's difficult climate. Vulnerability to severe weather remains, however. When Skye experienced severe gales in January 2005, many polytunnels were lost or destroyed, in addition to causing widespread crop damage. Horticulture does not have the same safety net of grants and subsidies available to those engaged in large–scale arable or livestock production, and when growers were not compensated for these losses, several decided that they could no longer afford to operate.

The SLHDA has also benefited from the marketing efforts of the Food Link Group (FLG) to link producers on Skye to local businesses including hotels and restaurants, and to facilitate dialogue between producers and consumers. The FLG is a voluntary association of producers and consumers, which emerged out of the Soil Association's Food Futures programme. It holds regular meetings which bring together different actors involved in local food production, including retailers, horticulturalists, chefs, and meat producers. A number of the SLHDA members are also members of the FLG and use the group's van to transport their produce around the island. This transport has proved crucial on Skye, since the island is very large, and many producers could not otherwise afford to move their goods. Customers who buy SLHDA produce distributed by the van tend to be in the commercial sector (hotels and restaurants), whereas members of the public buy SLHDA produce from a weekly market stall in the island's main town of Portree.

Positioning the SLHDA within theoretical frameworks

The SLHDA aims, according to the association website, to 'promote the use of local produce and products and to replace "imported" products and skills' (SLHDA 2008a). This is an approach to alternative food politics that goes beyond embedding food *products* in place, and seeks to reinvent the *processes* by which food is produced and consumed in a specific local area. In addition, the activities undertaken by the SLHDA linking horticultural producers with local schools underline the commitment to forging new connections between producers of food and consumers. The activities of the Food Link Group also explicitly aim to reconnect producers and consumers in a specific local area, both by providing the physical distribution network necessary (the Food Link Van) and by encouraging communication and discussion between the two groups through meetings. These combined activities on Skye represent, therefore, an emergent local food network that fits Fonte's (2008) 'reconnection perspective'. The projects on Skye were initiated by grassroots groups, and appeal to social and environmental concerns for greater self-sufficiency and community control of food systems, rather than simply using the qualities of 'local' or place-embedded production to increase the economic value of food products for export. Through this perspective, Fonte (2008: 204) explains, 'local food is seen as a way to both protect the environment and respatialise and resocialise food, establishing direct contact between farmers and consumers and generating a link of trust and reciprocal benefits between them'. Moving towards a local and *economically* sustainable food system is the primary concern for SLHDA members. By reinvigorating horticultural production on Skye and through expanding connections with consumers and local schools, the work of the SLHDA not only demonstrates that Skye can produce food, but reminds consumers of the benefits of a local food system, and re-embeds that food system in the local *place* and the community. Although organic production results in greater environmental sustainability on a local level, SLHDA members' engagement with the environmental aspect of sustainable food systems focuses on reducing imports of produce that could be grown on the island, rather than on encouraging organic production on Skye. At the local level, the SLHDA is concerned simply with increasing horticultural production for sale on the island, rather than encouraging a particular method of production.

The emergent local food system on Skye fits the model that some have suggested is more dominant in North America – an approach to alternative food politics that is 'radical' rather than 'reformist' in character and that addresses *processes* rather than simply changing food *products* (Goodman 2003, Maye et al. 2007). The evidence from Skye, therefore, suggests that this geographical dichotomy is oversimplified, and resonates with the argument made by Holloway et al. (2007: 5) that there should be 'ways of thinking about food networks which retain the sense of diversity and particularity of different food networks'. Holloway et al. (2007) seek instead to bring an understanding of power as a process central to all social relations to debates around alternative food politics. The ways in which power is exercised through the food system is a central concern for the SLHDA. A primary aim is to 'displace unnecessary, imported horticultural products and skills' (SLHDA 2001) and to develop a self-sustaining, learning network where all the knowledge required to develop and support horticulture on Skye is owned by its members. In this way, power is seen as being exercised through the transfer of knowledge and through knowledge-based practice. The source of knowledge also emerges as a key issue, both in terms of the *type* of knowledge, expert or tacit, and in terms of its geographical location, from within the community, or from outside a bounded local place.

Knowledge dynamics in the emergent local food network

When the SLHDA was founded, the skills and knowledge required to develop horticulture on Skye had largely been lost, and the pool of tacit knowledge around food production had declined, replaced by some knowledge about agro-industrial farming techniques and stocks of imported food from the mainland and beyond. The network began, therefore, with only a small group of growers and with considerable help from external sources.

When the founders of the SLHDA started to cultivate small plots to demonstrate the potential for horticulture on Skye, they drew upon knowledge from both external and the limited local sources. A horticultural specialist from the national Scottish Agricultural College (SAC) advised the growers on the technical aspects of growing, including treatments such as pesticides and fertilisers, and was able to advise on the varieties that would grow in the soils and climate of Skye. Locally, an established grower with horticultural experience provided basic practical knowledge on soil preparation and growing conditions. This knowledge was shared with new growers through a mentoring process. More experienced members helped new growers prepare their land, teaching them the basics of cultivation and providing support by visiting and talking to them regularly by telephone. This mentoring support is still provided to new members by the network and has proved essential to the success of the project.

The support of the SAC was crucial to the initiation of the project. There was little knowledge of local fruit and vegetable varieties since agricultural production on Skye had previously focused on livestock rearing rather than horticulture. The knowledge around plant varieties suitable for horticultural development on Skye came from the SAC, who also advised that the use of technologies such as polytunnels would be critical in allowing more 'fragile' produce, such as salads, to be grown on the island. Several growers reported that the generalised knowledge provided by from the SAC was inappropriate for the very specific and difficult growing conditions on Skye, and furthermore, SLHDA members were producing on a much smaller scale than the SAC was used to. The expert knowledge of the SAC was, therefore, combined with the small pool of tacit knowledge on Skye, producing new combinations of tacit and codified knowledge that were embedded in the specific local place. As a consequence, local growers came to be recognised as 'experts'. The SLDHA now sources considerably less expertise from national sources.

The process of combining the expert horticultural knowledge from the SAC with a growing pool of tacit knowledge of SLHDA growers represents the construction of a new local knowledge culture, based around horticultural practice embedded in place. The process of incorporating the external, expert knowledge of the SAC has led to some codification within this emerging knowledge culture, with the SLHDA producing newsletters and annual reports sharing information. Knowledge of the success of particular varieties is also shared informally, however, and the decision has been made not to codify this knowledge through the production of a database.

Indicative of the politics associated with the formal codification of such knowledge, it was found that suppliers take offence if their varieties are not included, and members were concerned that they might lose access to those suppliers. The network still uses advice from the SAC around new technologies to build capacity within the group. For example, one grower attended a course on soil analysis at the SAC and on their return to Skye, began to train other members in what they had learned. Trial and error also plays an important part in knowledge development around varieties, treatments and technologies; knowledge which is again shared through the network via mentoring, meetings and the newsletter. Beyond the SAC, there is little support available for horticulturalists at the national level, where meat, fish and cereal production are prioritised.

The role of knowledge in the construction of food quality

The concept of food quality has emerged as a particularly important driver of the production and consumption of local foods on Skye. As Ilbery and Kneafsey (2000: 217) have observed, 'the concept of "quality" is one which is contested, constructed and represented differently by diverse actors', and following food scares such as the BSE crisis, the perceived 'quality' of food products is of increasing importance to consumers. Constructions of food quality are often closely linked to constructions of nature and the degree to which food chains are embedded in local places (Murdoch et al. 2000: 107). The discourse surrounding local foods on Skye demonstrates the roles that different knowledge types play in different constructions of food quality, and highlights differences in constructions of quality between locally marketed food and food produced for sale to distant markets.

It is useful first to consider the role that knowledge plays in constructing food quality in the conventional food system. In agro-industrial food production quality is regulated through the application of rules and standards to agricultural production, governing a wide range of issues including inputs, growing conditions, processing, storage and labelling. Food labels provide the bridge between the regulation of agricultural production and the provision of information regarding that regulation to the consumer. The knowledge transferred to consumers by food labelling plays a significant role in the construction of food quality for foods purchased in supermarkets providing information about, for example, the presence or absence of genetically modified ingredients, the use of pesticides, whether organic production practices have been used, and the conditions in which livestock have been reared. While some food labels appeal to loose and poorly-defined constructions such as 'traditionally-reared' or 'sourced from a family farm', many form part of rigorous, codified systems of regulatory knowledge based on expert scientific knowledge. Examples include nutritional information such as percentages of recommended daily allowances (per cent RDA), gram or microgram content of fats, sugars, vitamins and minerals, or diet-related messages such as the '5-a-day' campaign to encourage fruit and vegetable consumption.

These messages help form constructions of quality based on the safety of food, and its nutritional content and value.

For consumers concerned about the sustainability of food systems, however, a much wider range of metrics contribute to constructions of food quality. In addition to knowledge of the food itself, these constructions of quality require that consumers consider the food chain through which the product was produced, distributed and retailed. Where in conventional food systems regulation and labelling ensure the quality of food products, local food systems also construct food quality through the social interactions and relationships that form between local producers and consumers. In these instances, the consumer might feel reassured that a food product is of high quality, not because of the assurances of a label reporting scientific knowledge, but because of a personal relationship with the producer and their production practices. Here, tacit knowledge contributes to a construction of quality that requires consumers to be more closely involved with the food system.

Construction of food quality within the SLHDA involves a combination of tacit and expert knowledge. Codified expert knowledge is represented by the rules and regulations surrounding growing, handling and processing, implemented by regulatory bodies such as Environmental Health, and the Soil Association, which certifies organic produce. Formally, any grower producing under the umbrella of SLHDA has to register with the Environmental Health office, so certain standards regarding production, harvesting and storage must be met. The network provides growers with training around health and hygiene issues. In addition, the organic growers in the network who are registered with the Soil Association must meet certain standards and criteria in order to be certified as organic producers. Tacit knowledge regarding constructions of quality by growers is based on the growers' knowledge of their products and how to grow it, and the knowledge that local award-winning hotels value and use their produce. Most importantly, the quality of food grown on Skye is frequently tied to constructions of Skye itself, which is viewed by tourists and local consumers as a wild, natural and unspoiled landscape (Árnason and Lee 2003).

Alongside the re-skilling of growers coordinated by the SLHDA, consumers have developed new knowledge of food produced on Skye. Buying locally produced food mobilises a different type of knowledge, since the codified scientific information that would be provided by food labels in a supermarket is not present on fruits and vegetables sold at a farmers' market. Instead, consumers negotiate questions of quality through knowledge about what fresh food looks and feels like, and through social relationships with producers. Skye's small size means that many consumers know, or are aware of, the people who are producing food, and this knowledge of where their food comes from and the ability to see it growing directly has helped established relationships of trust. Alongside trust, relationships based on reciprocity and exchange contribute to constructions of quality. The SLDHA coordinator argues: 'this is a small place and people soon get to know about the quality of people's produce. It encourages growers to produce at the top end of the market' (Personal Interview). Social relationships on the island also benefit the producers, since growers cannot afford to grow and market similar products without risking flooding the market. The size of the island and the networks of communication within the SLHDA raise awareness of the types of product being grown and, in addition to building a body of horticultural knowledge, encourage members to grow a variety of different produce. The different micro-climates on Skye mean that every product is slightly different and is harvested at different times. The fact that growers are producing unique products is evidenced by the fact that customers are able to recognise both the varieties and the produce of particular growers, and that some are more successful than others.

The SLHDA market stall in Portree forms the focus for a farmers' market during the summer months, and is staffed by the growers themselves. This has become a forum in which local people can discuss with growers different varieties, the quality of the produce, and the best methods to cook it:

> You do find people getting together and discussing varieties, techniques, growers and so on. It's an informal exchange of ideas. People who buy off the stalls discuss varieties and how things should be cooked. Buyers ask them how they cook things and growers need to know. So buying food becomes a social occasion, rather than a faceless interaction. (Personal Interview SLHDA Project Coordinator)

Through the process of getting to know growers personally and having the opportunity to speak to them directly about their produce, growers and consumers have established relationships of trust. By knowing which grower has produced certain foods, consumers are able to make their own decisions about the quality of certain varieties and the quality of certain growers. In a study of UK farmers' markets, Kirwan (2006: 310) concludes that 'it is the ongoing social interaction between producers and consumers which allows for more individualised and locally contingent quality evaluation'. In accord with Kirwan's findings, the market for locally produced food on Skye ensures quality primarily through social interaction rather than through regulation; a process through which producers and consumers redevelop the skills and tacit knowledge necessary for the operation of a local food system. Fonte (2008: 219) emphasises the value of a 'peer relation of learning among actors' in expanding local control of knowledge dynamics. The social interactions that are central to the emerging local food system on Skye demonstrate the processes of knowledge negotiation that Fonte describes:

> local actors are trying to rebuild local knowledge through networking as equals, shared experiences, discussions and observation. Scientific knowledge may be a starting point but it needs to be evaluated, adapted and integrated, according to local circumstances. (Fonte 2008: 219)

Although the SLHDA focuses on producing food for local consumption, the emerging food network on Skye is tied to the island's burgeoning tourist industry (Árnason and Lee 2003). With the enthusiasm of local businesses and SALE's Food and Drink Coordinator, food has become an integral part of Skye's tourism experience, and food marketing on the island has drawn on the images of Skye as wild, natural and pure used in tourism marketing to emphasise the close relationship between the food and the spectacular environment in which it is produced. As Árnason and Lee (2003: 31) note, 'consumers are being invited to buy a product or service because it is produced or located in a particular area with particular qualities', in Skye's case, food that is 'fresh' and 'natural'. The Food Link Group, whose website is headed 'Stunning scenery, spectacular food' (Skye and Lochalsh Food Link Group 2008), has been instrumental in enabling local hotels and restaurants to purchase local produce using the award-winning Food Link Van project. The van has approximately 60 business customers and has a Marketing Coordinator to encourage growth. Produce is picked up and delivered twice a week in summer, and once a week in winter, and now moves over £90,000 of produce annually representing a significant increase in the amount of food staying in the local economy (Skye and Lochalsh Food Link Group 2008), demonstrating the local food sector's contribution to the rural sustainable development agenda. The SLHDA has also benefited from SALE's promotion of Skye produce through the press, awards systems, and staging of festivals and events. In 2005, Skye hosted the annual congress of the Slow Food Movement, demonstrating its success in promoting itself as 'food destination'.

Conclusion

The emerging local food network on Skye seeks to challenge the dominant agro-industrial model of agriculture, and to build a local food system that is embedded in the landscapes and communities of Skye. By reconnecting producers and consumers with a more sustainable food system, this case study fits the 'radical' model described by the agro-food studies literature, and demonstrates that such politically driven food system change is taking place in Europe as well as North America. Local food production on Skye makes an important contribution to sustainable rural development on the island, no longer simply through food and drink exports but through the development of local markets for the island's produce. This shift has played a significant role in maintaining and enhancing traditional skills, encouraging self-sufficiency, retaining and circulating money within the local economy, and in supporting the island's burgeoning tourism industry. Skye's reputation as a place that produces 'quality' local food is growing, as evidenced by the fact that the island recently hosted the first UK congress on Slow Food.

To bring about the growth experienced in local food production and consumption on Skye, local actors have initiated the reconstruction of islanders' local food knowledge. In the context of the producer and consumer de-skilling caused by the conventional food system, training and knowledge transfer has been required to rebuild local tacit knowledge of Skye food over which islanders feel they have ownership. The SLHDA has taken advantage of expert training and knowledge provided by the SAC, but have used and reproduced such knowledge in a way that builds local bodies of tacit knowledge relating to food. Over time, reliance on such 'outside' sources of knowledge has decreased, as growers become 'local experts' through knowledge sharing and training within the community. Alongside these developments, consumers have regained a body of food knowledge and skills by building new social relationships with growers; processes facilitated by the Food Link Group and the farmers' market in Portree. The enthusiasm of local businesses, including award-winning hotels and restaurants, has supported the SLHDA's expansion, and demonstrates the key role that the local food network plays in the island's tourism sector.

The quality of local produce on Skye is constructed in different ways depending on the market. Food produced for spatially extended markets relies on nationally understood indicators of quality, including regulation, but also on an understanding that food is from a particular place. The branding of Skye as a pure, unspoiled natural environment has been key in this regard (Árnason and Lee 2003). Locally, quality is negotiated through social relationships, face-to-face contact between suppliers and consumers, short supply chains and by a burgeoning social intercourse around products and varieties. This verbal interaction between suppliers and consumers is part of a situated process of knowledge construction through which a local knowledge base about where food comes from and how it is produced has developed. The fact that Skye's top restaurants support and use local food has significantly increased pride in local produce, and is reversing what some actors felt was a long-held belief that food from outside the island was inherently better than locally grown produce. Markers of food quality such as 'organic' are now less important to local actors than the fact that food is local, and demand for locally grown produce now far outstrips supply.

The development of the local food network on Skye has been driven by local actors, and has taken place in the absence of significant leadership or support at the national level. Despite Skye's success in developing its food sector – its internationally renowned restaurants, its hosting of the Slow Food Congress and its backing of innovative and award winning projects such as the Food Link Van – local actors feel that the Scottish Government has failed to recognise those achievements. There are signs that national interest in local food networks is growing with the development of the Scottish Government's national food policy, but it seems unlikely that the dominance of the agro-industrial model will decline in the near future. In the context of this policy support vacuum at the national level, local and regional actors will continue to be the key players in the development of local food networks in Scotland, as has been the case on Skye.

References

Allen, P., FitzSimmons, M., Goodman, M. and Warner, K. 2003. Shifting plates in the agrifood landscape: the tectonics of alternative agrifood initiatives in California. *Journal of Rural Studies*, 19(1), 61–75.
Árnason, A. and Lee, J. 2003. *Crofting diversification: networks and rural development in Skye and Lochalsh*. Scottish National Report, RESTRIM Project. Arkleton Centre for Rural Development Research, Aberdeen.
DuPuis, E.M. and Goodman, D. 2005. Should we go 'home' to eat?: Toward a reflexive politics of localism. *Journal of Rural Studies*, 21(3), 359–371.
Feagan, R. 2007. The place of food: mapping out the 'local' in local food systems. *Progress in Human Geography*, 31(1), 23–42.
Fonte, M. 2008. Knowledge, food and place: a way of producing, a way of knowing. *Sociologia Ruralis*, 48(3), 200–222.
Goodman, D. 2003. The quality 'turn' in alternative food practices: reflections and agenda. *Journal of Rural Studies*, 19(1), 1–7.
Goodman, D. 2004. Rural Europe redux? Reflections on alternative agro–food networks and paradigm change. *Sociologia Ruralis*, 44(1), 3–16.
Guthman, J. 2004. Back to the land: the paradox of organic food standards. *Environment and Planning A*, 36(3), 511–528.
Highlands and Islands Enterprise. 2007a. *Sector profile: agriculture, May 2007*. [Online]. Available at: http://www.hie.co.uk/HIE-economic-reports-2007/ Sector%20profile%202007%20-%20agriculture.pdf [accessed: 20 January 2009].
Highlands and Islands Enterprise. 2007b. *Sector profile: food and drink, May 2007*. [Online]. Available at: http://www.hie.co.uk/HIE-economic-reports-2007/ Sector%20profile%202007%20-%20food%20and%20drink.pdf [accessed: 20 January 2009].
Holloway, L., Dowler, E., Tuomainen, H., Kneafsey, M., Venn, L. and Cox, R. 2007. Possible food economies: a methodological framework for exploring food production–consumption relationships. *Sociologia Ruralis*, 47(1), 1–19.
Holt, G. and Amilien, V. 2007. Introduction: from local food to localised food. *Anthropology of Food*, [Online]. Available at: http://aof.revues.org/index405.html [accessed: 5 November 2009].
Hunter, J. 1991. *The Claim of Crofting*. London: Mainstream Publishing.
Ilbery, B. and Kneafsey, M. 2000. Producer constructions of quality in regional speciality food production: a case study from south west England. *Journal of Rural Studies*, 16(2), 217–230.
Ilbery, B. and Maye, D. 2005. Alternative (shorter) food supply chains and specialist livestock products in the Scottish-English borders. *Environment and Planning A*, 37(5), 823–844.
Jaffe, J. and Gertler, M. 2006. Victual vicissitudes: consumer deskilling and the (gendered) transformation of food systems. *Agriculture and Human Values*, 23(2), 143–162.

Kimbrell, A. 2002. Introduction, in *Fatal Harvest: The Tragedy of Industrial Agriculture*, edited by Kimbrell, A. Sausalito, CA: Foundation for Deep Ecology and Island Press, 1–6.

Kirwan, J. 2004. Alternative strategies in the UK agro-food system: interrogating the alterity of farmers' markets. *Sociologia Ruralis*, 44(4), 395–415.

Kirwan, J. 2006. The interpersonal world of direct marketing: examining conventions of quality at UK farmers' markets. *Journal of Rural Studies*, 22(3), 301–312.

Lochhead, R. 2009. Shaping Scotland's farming future: the need for a new contract. Speech to the Oxford Farming Conference: 5–7 January 2009. [Online]. Available at: http://www.ofc.org.uk/images/stories/File/Lochhead%202009.pdf [accessed: 20 January 2009].

Macaulay Land Use Research Institute and University of Aberdeen Department of Agriculture and Forestry. 2001. Agriculture's contribution to Scottish society, economy and environment. [Online]. Available at: http://www.scotland.gov.uk/Resource/Doc/158216/0042826.pdf [accessed: 20 January 2009].

Maye, D., Kneafsey, M. and Holloway, L. 2007. Introducing alternative food geographies, in *Alternative Food Geographies: Representation and Practice*, edited by Maye, D., Holloway, L. and Kneafsey, M. Oxford: Elsevier, 1–20.

Morgan, K. and Murdoch, J. 2000. Organic vs. conventional agriculture: knowledge, power and innovation in the food chain. *Geoforum*, 31(2), 159–173.

Murdoch, J., Marsden, T. and Banks, J. 2000. Quality, nature, and embeddedness: some theoretical considerations in the context of the food sector. *Economic Geography*, 76(2), 107–125.

Polanyi, M. 1966. *The Tacit Dimension*. London: Routledge.

Qazi, J. and Selfa, T. 2005. The politics of building alternative agro-food networks in the belly of agro-industry. *Food, Culture and Society: An International Journal of Multidisciplinary Research*, 8(1), 45–72.

RSPB and Scottish Crofters' Union. 1992. *Crofting and the Environment: A New Approach*. Inverness: Highland Printers.

Scottish Agricultural College 2003. *Agricultural Forecasts for the Highlands and Islands Enterprise Area*. [Online]. Available at: http://www.hie.co.uk/HIE-HIE-economic-reports/HIE-sac-hie-agric-report-03.pdf [accessed: 20 January 2009].

Scottish Executive, Environment and Rural Affairs Department 2007. *Economic Report on Scottish Agriculture 2007 Edition*. [Online]. Available at: http://www.scotland.gov.uk/Resource/Doc/177540/0050493.pdf [accessed: 20 January 2009].

Scottish Executive. 2001. *A Forward Strategy for Scottish Agriculture*. Edinburgh: Scottish Executive.

Scottish Government 2008a. *Food and Drink Scotland: Key Facts*. [Online]. Available at: http://www.scotland.gov.uk/Resource/Doc/228568/0061860.pdf [accessed: 20 January 2009].

Scottish Government 2008b. *Choosing the Right Ingredients: The Future for Food In Scotland.* Edinburgh: Scottish Government.

Skye and Lochalsh Food Link Group 2008. *A Taste of Skye and Lochalsh.* [Online]. Available at: http://www.tastelocal.co.uk/ [accessed 20 January 2009].

SLHDA – Skye and Lochalsh Horticultural Development Association 2001. *Constitution.* [Online]. Available at: http://www.horticultureskye.co.uk/documents/ constitution.pdf [accessed: 20 January 2009].

SLHDA – Skye and Lochalsh Horticultural Development Association 2008a. *About SLHDA.* [Online]. Available at: http://www.horticultureskye.co.uk/about-slhda.php [accessed: 20 January 2009].

SLHDA – Skye and Lochalsh Horticultural Development Association 2008b. *Activities.* [Online]. Available at: http://www.horticultureskye.co.uk/activities.php [accessed: 20 January 2009].

Vos, T. 2000. Visions of the middle landscape: organic farming and the politics of nature. *Agriculture and Human Values*, 17(3), 245–256.

Watts, D.C.H., Ilbery, B. and Maye, D. 2005. Making reconnections in agro-food geography: alternative systems of food provision. *Progress in Human Geography*, 29(1), 22–40.

Chapter 4

Local Food Production in Sweden: The Eldrimner National Resource Centre for Small-Scale Food Production and Refining

Karl Bruckmeier

Introduction

In the northern Swedish region of Jämtland traditional forms of local food production are being revitalised through a project for small-scale food production and refining. This project for small-scale refining of food in the handicrafts tradition is today under implementation by a national resource centre known as 'Eldrimner'.[1]

The following analysis, based on the results of a case study in the CORASON European research project (Bruckmeier, Engwall, Höj-Larsen 2006), raises the question of how local food production and consumption can be maintained in a modernised agro-food system. Which natural and social conditions for the products should be taken into account by the producers and the consumers? Eldrimner started off by building territorial production-consumption networks where local farmers and food producers could come into contact with consumers in the region. It developed into an extended network of local production and consumption through cooperation with other local food projects and outside experts in handicraft-based food production. With its support for local food production, organic farming, and new producer-consumer relations, it promotes the ideas of fair trade movements: small local producers should receive a fair price for their products, a price that enables them to make a living when they compete on the local and regional food markets with products from conventional and large-scale agriculture.

The project combines support for food producers with other activities: the building of producer-consumer networks, marketing of local food, training of producers, support for environmentally sound and organic farming, organising a rural movement within a network of other movements. In all these respects Eldrimner is rooted in the socio-cultural history and the traditions of local rural movements in the region. These conditions are reflected in the producer-guided discourse on local refining of food as well as in a critical debate, applying health

---

1 *Eldrimner* is a term from old Nordic mythology: the cooking pot in which the *Särimner* pig was prepared as food. Every time it had been consumed, *Särimner* was resurrected, symbolising the idea that food is forever available.

and environmental criteria, that is part of a nation-wide discourse on food quality and animal welfare. Such discourses expose the problems that arise in the course of applying knowledge and synthesising different types of knowledge of local food production and processing. The knowledge utilised in these processes is not sufficiently understood as having been built up from traditional production practices that existed before the modernisation of agriculture and industrial food processing. It is important that product-specific knowledge and technologies be framed by expertise in marketing and knowledge about food markets, along with the relevant legislation. These are necessary components of managerial knowledge and of the knowledge practices that are required by rural movements in critical confrontation with the dominant agro-food industry. Finally, mention is made of the three discourses that form the social context for local food production: the discourse on knowledge and the criteria for local food production; the discourse and practices of local rural movements of which the local producers are a part; and the national discourse on food security and food quality. The framing discourses are outlined in the first part of the paper, followed by a more detailed description of Eldrimner, its practices and the history of its development.

Preconditions: local food production in a disadvantaged area

The Eldrimner project (Box 4.1) is located in Jämtland in the northern part of Sweden, a remote and sparsely populated region adjacent to the Norwegian border. The population (c. 120,000 inhabitants) is dispersed over a vast area with only one major city, Östersund. The region has a small ethnic community of Sami people, who continue their cultural tradition of reindeer herding in the mountain area on the borders between Norway and Sweden. Public and private services, agriculture, forestry, manufacturing and tourism are the main economic activities. When all small-scale refining and indirect employment is taken into account, agriculture provides work for 10 per cent of the population. Forestry is another important part of the economy – the region provides 10 per cent of the trees felled in Sweden. Three per cent of the population are employed full-time in forestry. The average size of farms is smaller than the national average (21ha in Jämtland, 34ha national average) and the farms are widely dispersed. Due to the harsh climate with long, cold winters and short summers, agriculture is different from that in southern Sweden. There is hardly any production of corn and vegetables. The main agricultural items are dairy products and meat (from cows, sheep and goats), along with small amounts of egg, poultry and potato production. As early as the beginning of this decade around 57 per cent of the agricultural land in the region was being used for organic farming. Hunting and fishing play an important part in the household economy of farmers and the products from hunting and fishing (elk, reindeer, wild salmon, etc.) are to be classified as examples of small-scale refining and local food.

Most farmers combine forestry with agriculture. Pluriactivity is more or less the norm there, because it is difficult to make a living from any one activity.

Box 4.1 The Eldrimner case study in the CORASON Project

The organisational centre for the Eldrimner project is located in the village of Rösta, municipality of Ås, where important project activities are carried out, e.g. courses and seminars for food producers and the annual 'Särimner' fair. With its ideal of local food production in the handicrafts tradition, it has adopted a multifunctional conception of the rural producer who is not only a specialised farmer in the modern sense but is also, as in traditional agriculture, and in this region, a combination of farmer, fisherman and hunter. With its focus on refining of food it sets clear priorities: the important element in local food culture is the multifunctional producer; his economy and livelihood are to be supplemented by small-scale food processing that is not limited to one product. It is not just one dominant product originating in the region and marketable under a regional label that is supported by Eldrimner, but a variety of food products in small quantities. It is not local food in the sense of one place, one community, one designated area or one dominant product. The products come from local producers on individual farms scattered over the whole region. Local production and processing knowledge is subject to Eldrimner's food quality criteria as described below. Limitations on the number of producers, products and product quantities stimulate a search for new products. Eldrimner encourages producers to experiment with new ideas. Products that have played a key role in the project are cheese (especially goat's cheese), dairy products, meat products (especially sausages), smoked fish and products from vegetables, fruits and berries. For each of the products, different distribution routes need to be found to reach the consumers. Most products are sold in the region, on the farms, in small co-operative and retail shops, at farmers' markets, through personal networks of producers. The products find their way to distant consumers through a variety of channels – through mail-order delivery, through restaurants and the tourist industry, through health food shops, through consumer fairs, and also, for some products, through big retail chain stores.

The Swedish researchers in the European CORASON research project have been regularly studying the development of the Eldrimner project since the beginning of this decade. Different qualitative methods have been used in the case study: qualitative interviews with project members, analysis of documents, visits to producers, participation in seminars held by Eldrimner and at the Särimner fair, informal talks on a number of occasions with farmers, food producers and food processing experts.

Sources: Bruckmeier, Engwall, Höj Larsen 2006.

Given the climate and resultant natural conditions in this area, agriculture here cannot compete with the intensive and large-scale production systems of southern Sweden in terms of the quantity of production and the market share it is able to secure. Developing an alternative type of quality production based on local processing and refining is a way for the local producers not only to survive economically, but also to initiate more socially and environmentally sustainable systems of food production and consumption.² The idea of small production runs for high-quality products is not the exclusive property of this project. Most small-scale farmers and fishermen resort to various forms of niche production or to building their own supply and consumer networks. Of more specific relevance is how Eldrimner developed the idea under unfavourable economic and natural conditions and without large groups of sufficiently affluent urban consumers willing and able to pay higher prices for the products (such groups have been present in many regions that have seen successful development of organic farming in past decades). The history of the project shows how it is indeed possible for a new culture of local food production and consumption to develop even under such circumstances. The lack of industrial and other urban-based employment in the region can even be seen as a factor potentially favouring local initiatives and projects. The low population density and long distances between the scattered settlements and farms spread over the area render unrealistic the conventional idea of developing the regional economy through commuting and creation of industrial work places in urban centres. The rural economy must be developed through utilisation of the natural resources actually present in the region, which has a relatively high percentage of the population working in agriculture, forestry and food production.³

The social context: framing conditions and discourses

Two framing conditions have been important for Eldrimner from the outset: (a) the core ideas of small-scale food production and processing rooted in the tradition of agricultural production, and (b) the existence of local popular movements in rural areas of Sweden. A third has gained in importance in parallel with the growth of the project, namely (c) the national discourse on food quality and animal welfare, in the context of which Eldrimner has defended its alternative ideas on food

---

2 The point is not only that natural conditions are unfavourable for environmentally friendly production as practised in small-scale and organic agriculture. Under the climatic conditions of the region, and for the specific products that can be developed there, it is not necessary to apply the technologies and production factors that have prompted criticism of modernised and intensive agriculture: the massive use of agrochemicals, synthetic fertilisers, pesticides and mono-cultural production systems that lead to erosion and contamination of soil and groundwater.

3 Norrland, the extensive and sparsely-settled region of northern Sweden, has played an important role in the history of Swedish industrialisation. With its rich material and energy resources – water, wood, minerals (Sörlin 1988) it has functioned as an 'internal colony' and a natural resource base for industry. This applies for a period long past. Today northern Sweden is dependent on governmental support and policy programmes to maintain its economy and retain the minimum of population necessary for the upkeep of infrastructure and services. The majority of the population, whether agricultural, industrial or in some other category, is to be found in southern Sweden.

quality in opposition to the mainstream ideas of scientific agencies, policy makers and governmental sources.

   a. As a facilitator of small-scale production and processing of rural products, Eldrimner is striving to revitalise handicraft traditions in food production and refining. Its basic objective is to bring together small rural producers who are interested in applying, learning and revitalising knowledge that is today marginalised or forgotten, and specifically in relation to technologies and traditions of small-scale food production and refining. Food is produced using small quantities of local resources, with the refining of products carried out by local producers themselves (farmers, food processors, laypersons, supported by experts and other persons involved with local production). The differences between the handicrafts approach and currently dominant industrial food production technologies cannot really be described in terms of generalities; they are product-specific, in relation to production technologies and preservation technologies alike. The small scale and local character of production are only some of the many elements common to all local products. Certain products equally share specific production and quality criteria. This is true, for example, in cheese making. The local cheese made by Eldrimner producers does not include certain preservatives and other ingredients used in the modern dairy industry. One key criterion adopted by Eldrimner is that food should be produced and processed locally right through to the stage where the final product is purchased by the consumer. This is a difficult rule to abide by today, when standardised food quality regulations, hygiene laws and other control systems tend to discourage the cultural diversity of local production. To defend local production and processing practices in the standardised food control conditions that prevail in the modern agrofood-industry, the project has developed a range of support technologies, which are described below.

     For Eldrimner the focus is on the rural producer, with the consumers being seen as a secondary consideration. This is not an altogether adequate formulation of the situation, as considerable efforts are being made to develop new producer-consumer relations based on direct contacts, along similar lines to practices in the organic farming and fair trade movements. A culture of local food production must be complemented by a culture of local food consumption. Eldrimner's ideas about consumers are derived from critical observation of modern social trends *vis à vis* food consumption. The food consumer is not viewed uncritically, as in mainstream economics, where the ideological and market-based conceit of 'consumer sovereignty' is the prevalent assumption. Typical consumers in modern societies are urban private households of one person or a number of persons who buy most of their food in shops and rarely produce anything themselves. Their active involvement begins and ends with preparation of their daily meals.

Even that happens less and less. In Sweden, as in most European countries, a large proportion of food consumption takes place not at home but in public places: in kindergartens, schools, workplaces, restaurants. The dominant food culture in modern societies need not be taken as unchangeable fact: it can be influenced by new ideas and values, such as those associated with local and organic food production. The Eldrimner approach to discussing consumer culture includes an imperative to motivate local food consumption, offering opportunities for purchase of locally processed products but also endeavouring to make each consumer, at least to some extent, a producer again in the private household. For consumers once again to be producing some of their own food does not mean reverting to the lifestyle of pre-industrial subsistence production, where the majority of the population lived in the countryside and from agriculture. The Eldrimner strategy is oriented to maintaining and/or strengthening a local production and consumption culture in a modernised society. What it does not accept is the idea of the passive consumer guided only by prices in his buying decisions.

To the present day a widespread everyday culture of 'maintaining ones' rural ties' has been preserved in Sweden, which is characterised as a post-agricultural and post-industrial society. The period of industrialisation, with its rapid urbanisation, was brief, and many families retain memories of their rural origins two or three generations ago, something which has helped to perpetuate a misleading romantic image of Swedish agriculture to which scientists and ethnographers have also have contributed. The strong rural ties of a large proportion of the population can be seen from such simple facts as that most Swedish families own summer cottages in the countryside and typically go there to live during the vacation period; many people also engage in hunting, fishing or collecting fruits as recreational activities in the countryside and/or in the forests. Recreational fishing is especially popular, with many people practising it. A much greater quantity of fish is caught by recreational fishermen than by the professional fishermen of the small-scale coastal and lake fisheries. A significant number of licensed professional fishermen in Sweden who fish in private waters, primarily for their own subsistence, are classified as 'household fishermen' (Bruckmeier, Ellegård, Piriz 2005). All this goes to show that rural life and culture, albeit in the guise of 'leisure activity', have a strong influence on everyday urban life. The past is not so long ago, as it sometimes seems to those living in late modernity.

Given the existing food and consumption culture with its strong orientation towards rural, local and individually produced food, projects such as Eldrimner seem perhaps less exceptional than they otherwise would. They are even able to build on aspects of the mainstream culture in the country. But still it is difficult to reach consumers. Those who cannot be reached through personal networks or through proximity to the producers or through ideas of producing a proportion of their food themselves can be contacted, directly or indirectly, through the marketing of local products, through on-farm sales, through producer cooperatives[4] and producer retail shops. The conventional retail system for food in Sweden is dominated by a few national retail chains; it is reluctant to market local products as it is organised to sell a standardised set of products nationwide, not small quantities of local products.[5] To gain new consumers the local food, products have to be brought into consumption in restaurants and hotels, schools, company canteens, hospitals and so on. These are all instances of what might be called 'public food culture' and, as with rural tourism, when one is seeking to approach it, intermediate actors are all-important. These would include food journalists, researchers, restaurant and hotel managers, education and training institutions for the restaurant and hotel sector, health sector operatives not to mention sections of the media such as lifestyle journals. The seminars organised by Eldrimner and the annual Särimner fair offer possibilities for such contacts.

b. Local food production and processing as part of a rural culture has been enhanced in Jämtland by a strong tradition of local movements. These local movements have had a 'seedbed' function for many local projects in rural areas. Their discourse has not been primarily about food. What they have been seeking is ideas on how to retain rural populations, preserve the infrastructure and service functions necessary for continued habitation of rural areas, help local groups to develop their own ideas and projects for earning a living. The local movements have not been able to create more favourable conditions for rural development without any aid and this necessity for outside assistance is something that has applied for Eldrimner also. Eldrimner has received funding through different national and EU programs, above all through the Swedish Rural Development Program (see below). Among the regional and local institutions supporting the project are the municipality, which hosts the centre; the chamber of agriculture, which is in search of new roles for itself beyond its traditional task of supporting the modernisation of agriculture; not to mention other rural development

---

4   Rural cooperatives have also played an important role in Swedish agriculture, and the tradition is still strong today in the food processing industry, particularly in the dairy and meat sectors, the idea being that these are the shared province of farmers. But since Sweden's late entry into the European Union in 1995 there have been accelerating changes, with concentration of capital and of firms and privatisation.

5   Marketing and distribution are difficult given the relatively small amounts of each product that are put on the market, the great distances (in Sweden generally and in Jämtland, with its low population density, in particular), and the domination of the retail food market in Sweden by a few big chains. The wholesale chains demand that certain given quantities be delivered regularly, also applying quality standards that may not be compatible with those of the Eldrimner criteria. The project has therefore set up its own mechanisms for marketing and sales support, distributing information about farm sales and other local sales points in the region.

resource and knowledge centres such as the JiLu (Jämtland Institute for Rural Development), a regional centre for integrated rural development.

The rural movements that have emerged from local and community based initiatives, projects and networks since the 1980s have promoted a discourse on 'the capacity and power of local people' (Ronnby 1995). Population numbers have been in decline in Jämtland for a long time and it is not easy today to make a living from a single economic activity, but strong regional identity and solidarity ties are still to be found there. Close connections between people in local communities were built in the course of their struggles to survive economically and maintain local cultures. The community life, the local groups, the people's efforts to generate new rural development ideas and new projects were an important factor long before the EU formulated its policy for rural development with the LEADER projects, or launched the more recent policy of integrated rural development that includes such inputs.[6]

The rural discourse in which Eldrimner participates brings together the different kinds of expert and local knowledge to be found in the region, from producers, small enterprises and rural movements. The discourse dates back to the 1980s, the time of emergence in the countryside of the above-mentioned local movements that are now organised in the Swedish 'Popular Movement Council'. The movements in question have years of experience in developing local projects of all kinds. The movements, and the initiatives they take, have been unexpectedly strengthened through the new experience of international influence on rural policy that has come with Sweden's belated entry into the EU in 1995. This brought accelerated specialisation and concentration to the agro-food industry, but it also brought new opportunities for rural networks to obtain resources for local projects. What characterises the movements as autochthonous local and rural movements is that they are initiated by rural population groups and develop out of traditional forms of rural economy and rural livelihoods. They are less experimental and less concerned with introducing pioneering forms of a new local economy than are, say, the eco-village movements or the local exchange and trade systems (LETS). They simply concentrate on

---

6   In recent years a certain style of discourse has also started to emerge from rural development policy and power centres that abounds with references to 'multifunctional agriculture'. Initiated by the OECD and taken up by the EU, it puts forward ideas similar to those of the advocates of local food production and processing, albeit without a specific focusing on food. Behind a rhetoric of holistic agriculture under conditions of late modernity, new forms of income are being canvassed for the rural sector of the economy. There is a return to ideas on agriculture that date from a period prior to technical modernisation, specialisation and large-scale production. A host of other resource use activities are also projected that might serve to preserve the cultural landscape or the type of local product processing that have been part of traditional peasant cultures.

trying to rebuild local economies and livelihood systems 'from within' the wider social and economic systems. In Eldrimner the core of traditional rural knowledge is presented as what is called small-scale food production and refining in the handicrafts tradition. The knowledge is activated through a discourse with heterogeneous participants, both professionals (farmers, fishermen, hunters, butchers, dairy companies; cooks, sales persons, firm managers) and people without formal professional or economic roles. All are part of the local rural population and are themselves consumers in their own specific households. Children are involved because it is necessary for them to become socialised in local food consumption. Women are involved, often in their traditional roles as housewives, who as in the past still have the main responsibility for preparing food for the household. But they are also involved in specific forms of local food processing, such as processing and preserving fruits and berries. Elderly persons are involved through drawing on their personal experience and their memories of traditional forms of food production. All these people can be seen as practitioners embodying the everyday local culture of food production and consumption. Further involvement of the kind of intermediate actors mentioned above, specifically, is also important for dissemination of products, as well as ideas on local food culture. Finally politicians and bureaucrats must be pressed to become more committed to their roles in formulating, implementing and managing policy programs for rural areas or in controlling food quality.

The Eldrimner discourse involves local and non-local, rural and non-rural actors in a variety of different roles as knowledge bearers, supporters and producers. This too is evidence that the discourse in question is one of rural movements. The guiding themes go beyond food processing. Apart from the familiar 'rebuilding of handicraft-based food production systems', 'strengthening of rural producers in the agro-food system', 'traditional knowledge and technologies for food production and conservation' other subjects come up for consideration. These are the themes that impel social movements and motivate their collective action: 'empowerment' and 'local power' in rural development, 'cooperation' and 'networking', influencing everyday life and culture through debates on cultural values, lifestyles and consumption behaviour. The linkage of themes identifies Eldrimner as being part of a broader culture of local movements aiming at mobilisation of local rural actors and their participation in rural development. Above and beyond the political and organisational questions of the new local movements there is also projection of a more general dialogue on civil society.[7] With the Eldrimner network as one participant in that dialogue

---

7 Swedish civil society and the relations between the public and the private sectors are of a quite specific kind, with a national tradition that is also visible in agriculture and food production. There are a large number of interest groups, associations and social movements at local, regional and national levels (such that Sweden has been characterised

the idea is put forward that the rural civil society of the future, where food producers will be a minority not just in urban but also in rural areas, should be built on the basis of a firm role and identity for rural producers. Rural areas have throughout human history been the places of food production for society as a whole. The priority given to production processes and local producers embraces the idea of the rural areas of the future continuing to be the natural resource base for all of society: the centre of food production, of capacities and knowledge for resource use. Among political actors this is a contested idea. In the policy process and through official support for modernisation and industrialisation of the agro-food system, local producers and their knowledge have been marginalised, in a ceaseless endeavour to separate agricultural production from natural conditions and ecosystems. This from the ecological perspective is an incoherent and irrational project. Its destructive implications for the environment, for human health and wellbeing, for animal welfare, for community and work life, for everyday culture, are what have given rise to the counter discourse of the new social movements that is still gaining influence in the public and societal debate on food.

c. Both of the above-mentioned discourses can be seen as part of an alternative to the mainstream discourse in Sweden on food quality, environmental care and animal welfare. This mainstream discourse, with its predominating input from scientific actors (e.g. from the Swedish University for Agriculture, veterinary and other sciences) and from public bureaucracies (e.g. agencies for regulation of food and animal welfare) has intensified in recent years and was one of the phenomena examined by the European 'Welfare Quality' project (Bruckmeier and Prutzer 2007). In Sweden, nationally produced food is subject to strict quality control and both government agencies and the majority of citizens demand higher standards of quality than those that are set for most other European countries. There is particular interest in quality standards when it comes to meat, which comprises a large part of the food bought and consumed[8] in Sweden. Imported meat, indeed imported food generally, is regarded with scepticism and seen as threatening the quality standards for national food production, particularly in the light of the rapid proliferation of imported products that has taken place in recent years with the advent of the new international discount supermarket chains.

---

as a 'movement society'), but at the same time close cooperation between governmental and non-governmental organisations. This has been represented as a corporatist or 'statist' tradition in Swedish society (Demokratiutredningen 1997).

8 The Swedish agricultural administration has published a study (JV 2005) comparing rules and regulations for animal welfare in the Scandinavian countries. The comparison highlights differences between countries in the regulation of animal welfare in the case of pork and other meat producers.

The debate on animal care and welfare and its relevance to food quality is also at the centre of a long-established tradition in Sweden. Though dating back much further, it has since the 1940s developed into a permanent problematic, on the basis of which there has been constructed an entire bureaucratic system of quality control. In the 1980s a public debate on animal protection was followed up by the passage of animal welfare regulations that included prohibition of the use of animal cadavers in fodder. The regulations must be counted a relative success. Among their accomplishments is the fact that Swedish agriculture and food production did not suffer from the BSE crisis that shook the European countries in the 1990s. As far as forms and conditions of production are concerned, the food quality debate has now entered a final stage of intensive policy discussion on farm animal welfare. The whole development can be seen as a critical reaction to the practices in modern animal husbandry and meat production. Where it differs from the local food production problematic becomes evident if one examines its guiding idea of making food production transparent from the farm to the table. There is clearly no intention of changing the industrial agro-food system. All that is proposed are some improvements in working conditions and animal care so as to regain the trust of consumers that has been lost through the successive food production scandals. Mainstream discourse becomes more heated when there is discussion of the criteria for food quality. Then it becomes obvious that there are no generally accepted criteria for food quality and no consensus on what food quality means. The veterinary and hygiene regulations for food products and the standards and legal criteria that apply in EU countries cover only certain specific aspects of food quality. Different ideas on food quality and different criteria for it are to be found with traditional agriculture as opposed to organic agriculture or to modernised agriculture.

The Swedish local food debate is part of the controversial food quality discourse at national and international levels. At present there is strong support for local food production and organic agriculture from policy makers and society as a whole. It is also visible in the media. A recently issued new lifestyle journal, *Camino*, for example, is one notable instance of the media orienting to environmentally critical and socially conscious consumers. The critical question 'How can quality control be introduced to cover the entire food production and marketing chain?' cannot be answered simply by proposing a bureaucratic control system that is applicable for all forms of food production. Such a system could not solve the problems that have accompanied the globalisation of the food industry and the importation of food from many countries with different, typically lower, standards of food quality control. Even just for the countries of the EU it seems hardly possible – by means of a bureaucratic control system for quality standards, hygiene and quality control that functions differently in different countries and regions – to impose uniform criteria and control processes. The passionate controversy over animal care in Sweden, and also the European research project on Welfare Quality, with their quest for better standards of animal care in farming, are evident indicators of the crisis of modern agro-food policy. The rebuilding of a culture of local food production and processing, as in Eldrimner, may contribute to the development of new forms of more decentralised regional or local quality control systems, but these are not easily implementable through the centralised Swedish system of governance. The problem still remains of how to reconstruct the present large-scale agro-food industry in conditions where it is currently evolving into still larger and more capital-intensive specialised and concentrated organisations under the control of multinational companies. Local food production and organic farming are more and more being confronted with unmeetable quality criteria or are finding that their further development within the confines of the system is blocked. Ways out seem possible only in the long run when these movements are participating in larger critical discourses seeking to build sustainable social-ecological systems from the local to the global levels.

The development of Eldrimner

During the first phase of Eldrimner's development a strategy for revitalising local food production and processing was established in the region. What followed was not a straightforward process of continuous success and improvement but rather a constant fight to gain acceptance for local food on the basis of criteria other than the bureaucratic quality criteria that were generally recognised. The recipe for success appeared to be not so much the capacity to secure the moral support of consumers and the public as the capacity to persevere in the initial effort.

In its second phase Eldrimner, now a permanent resource centre, was required to find new strategies and new ways of disseminating the local food culture and reaching new, distant, consumers. This involved sharper confrontation with the mainstream economy and the 'expand or die' imperative that was imposed on private companies. The construction of capital-intensive own-distribution and retail chains; collaboration with the dominant operatives and their strategies of long-distance transportation and global importing and exporting of food; all this is incompatible with the idea of local food production and consumption. Such an arrangement presupposes a locally or regionally organised economy: not as a closed system decoupled from national and international flows of goods and resources, but as a decentralised economic system of markets, production and exchange processes that is much more based on local resource flows and much more controlled by local resource users.[9]

---

9 Intensive comparative research has been carried out on the rebuilding of local economic systems in the context of modern economies; see e.g. the research on 'common

First phase – Eldrimner as project

Eldrimner evolved out of a previous project entitled *Matora* and was externally funded for three years, between 2001 and 2003. After that it had to find further funding on its own initiative, something that became possible through its connection with the agro-environmental measures in the Swedish Rural Development Program (RDP), local food production being among their priorities also.[10] Local food production is not a specifically identified goal of the RDP, but from its problem diagnosis and the objectives it has formulated it can be deduced that a high quality of food and environmentally sound food production count as important prerequisites for the support it extends to agriculture and rural development. Eldrimner originated from outside the program and did not change its aims so as to adapt them to the funding conditions.

In the mountain area where many of the farms owned by local producers in the Eldrimner network are situated, goat farming is an alternative to dairy farming with cows. The return to small-scale production on the farm, of dairy or other refined products, can help the farmer to continue farming and help keep the local economy and community in existence, but outside support is still required. The farmers engaged in small-scale food production are often in receipt of additional funding through the RDP. Although the guiding idea is local food production and processing, all producers involved with Eldrimner are using environmentally sound methods of production and most of them have KRAV certification, KRAV being the Swedish certificate for organically grown products. That gives some indication of how organic and small scale production can go together, but it does not show that Eldrimner is, in terms of its own explicit aims, a project for support of organic farming. In the mid 1990s Sweden was not among the EU countries with a large organic production sector in its agriculture. However, as can be observed in Sweden, Finland, and to some degree in Denmark, the percentage of land being used for organic farming is rapidly growing in response to European agri-environmental policy, with rapid expansion having occurred particularly in the late 1990s (NVVSC 2000: 57).

Objectives of the project Eldrimners' overall goal from the outset has been to support the dissemination of practical knowledge in small-scale production and refining of food, making use of knowledge from traditions of local production

---

pool resource management' (Gibson, Williams and Ostrom 2005).

10 The RDP for the years 2000–2006 and then for 2007–2013 encompasses all national and regional measures derived from two EU regulations (Council Regulation EC 1257/99 and Commission Regulation EC 1750/99). Its predominant focus is on environmental measures for agriculture. The RDP is an ambitious programme, aiming to bring about conversion to organic farming. Within the programme period 2000–2006 the percentage of land under organic cultivation was supposed to double from 10 per cent to 20 per cent, thus achieving early accomplishment of an overall EU agro-environmental policy goal.

and handicrafts. There has been no precise definition of the target groups for the project. The project rather acts as a facilitator in the region for all persons interested in small-scale food refining. This facilitator role becomes evident with enumeration of the project's specific objectives:

- to provide a forum and meeting place for small-scale food-producers;
- to help with the launching of new businesses and with training beginners in food production;
- to encourage the exchange of practical knowledge between small-scale producers within the region and knowledge derived through adaptation of the practices of other regions (nationally and internationally);
- to provide a permanent support base for producers and help develop small enterprises;
- to strengthen local communities while at the same time monitoring the local ecological conditions;
- to support environmentally sound food production without resort to substitutes.

The objectives are pursued through courses in small-scale food refining and through the application of specific production techniques in cheese making, pork butchering, and the refining of berries and vegetables. The courses also include subjects such as marketing, economics and food-production legislation. The most popular and well-attended courses have been those in food safety and regulations; basic and advanced cheese production; and refining of fruits and berries. The knowledge disseminated through the network and through the courses often comprises a blending of local traditional knowledge from one product and from this specific region with ideas and technologies developed as local knowledge in other rural areas and other countries. This highlights an important feature of the project: 'local' is not a static category, either in time or in place: it implies continual adaptation and improvement of knowledge; local production technologies for specific products can come from other countries. Joint excursions and other forms of co-operation with small-scale food-producers in France and Germany are an established part of Eldrimners' practice. Food producers from those countries have taught on Eldrimner courses. As the course topics indicate, local knowledge needs to be linked with other kinds of knowledge required in the market situation: knowledge of food marketing or food policy; country-specific knowledge of legislation, food production, food regulation. Scientific knowledge in the agricultural and health sciences may also be relevant. To manage the different forms of knowledge in accordance with clear criteria and clear priorities, Eldrimner has developed a detailed set of criteria for defining food quality.

Definition of 'quality' At Eldrimner, following a differentiated set of criteria (see Box 4.2), the quality of locally produced food is highlighted as the key desideratum

Box 4.2   Quality aspects of local food

The debate in Eldrimner covers the following aspects of food quality:

1. A primary product of high quality in terms of a clean production environment: the small producers involved in Eldrimner think that the sparsely-populated areas of northern Sweden possess a 'natural advantage' for production of quality food because there is no problem of dangerous pollutants in this region. Pollution is associated mainly with large-scale modern agriculture, i.e. with something that is a very long way from here.

2. Small-scale production methods, technology, quantities: the methods and technologies applied in small-scale production are not closed technological systems. Human and personal skills are important in food production. As a project representative put it: 'The hand should always be part of production'. Manual labour in the food production process is perceived as a mark of quality, a sign of the exclusiveness that comes with producing only small quantities.

3. Good craftsmanship: small-scale food production should be based on a combination of traditional practices and methods (as formerly found in handicraft production) and modern expert knowledge of health and food-safety standards for health. Time is a key component of such craftsmanship – most small-scale production and refining processes are time-consuming. High quality is also contingent on the time allowed for the different stages of the production and refining process (ripening of cheese etc.).

4. The 'product story': the unique story of how each product has come into existence provides an index of its value in easily understandable form. The story should provide an insight into the expertise involved in the product's manufacture, the way that tradition, craftsmanship, local materials and identified producers have come together to create the end result. Consumers should know where and how the products have been made, and the products should connect to local tradition and history.

5. The 'moral dimension': the less tangible element that could be called 'moral quality' is also to be detected in the reflections of the participants. The essence of moral quality is perhaps encapsulated in the ethic of individual production conveyed by the message that 'small is better'. This does not mean only that small-scale production is better ecologically. It also means that it is better economically and socially, i.e. better for the local communities and for the rural inhabitants. The idea of sustainability is also entailed in the moral dimension, though members of the project would not necessarily use the term. The agro-industrial food production systems are regarded as harmful to ecosystems and society, draining rural areas of their economic resources and reducing their social capabilities.

Sources: Bruckmeier, Engwall, Höj Larsen 2006.

in small-scale production. The criteria are not static but are the subject of a continuing discussion.

The quality aspects mentioned in Box 4.2 – apart from the moral ones – are explicit in the Eldrimner project and its products. They are elaborated and discussed in seminars, courses, fairs and everyday communication. The moral aspects discussed more indirectly, in personal communication and in reflections on the 'project philosophy'. The quality aspects of small-scale production are the key theme in Eldrimner, in all the courses and all the applications. But the practice of local food production and processing requires further active implementation to become permanently effective.

Networking activities of the project The project includes elements that are at present seen as prerequisites for sustainable rural development – co-operation between resource users, networking, training, awareness building, mobilisation of local people, utilisation of local knowledge and local resources, creation of new producer-consumer networks, exertion of influence on the discourse concerning food quality. All these activities are conducted in a climate of awareness of gender relations: women are active participants in the project, both as managers and as producers.

To meet the food-refining needs of local small-scale producers, farmers and entrepreneurs, a host of different activities are carried out in support of the production, processing and marketing of dairy and meat products, fish, vegetables, berries (see Box 4.3), and other commodities. Difficulties met with in food refining are a by-product of the prevailing 'de-localised' food production and quality control regulations, which have been developed around the assumption that food production involves a limited number of large industrial production units – specialised farms, slaughterhouses, dairy firms, bakeries, transport firms, that can easily be controlled by bureaucratic authorities and subjected to standardised rules and norms. Many techniques that have been part of traditional small-scale production and processing of food in the countryside such as slaughtering animals on the farm and the on-farm processing of meat products cannot easily be carried out today if one is to abide by the hygiene regulations and other legal requirements underpinning the dominance of the agro-food industry. 'Production at home' is difficult, for example, where this means slaughtering animals and producing sausages on the farm. Moreover dairy products are not always produced on the farm but in dairies close by. Refining the food from small-scale producers and their companies requires a support system with its own knowledge and its own technology. The Eldrimner project is able to provide such support for the producers in the region.

The projects' knowledge management Eldrimner's guiding idea is encapsulated in the formula 'small is better' and most of the knowledge-use and 'good practice' criteria that are utilised by the local food production and processing network can be derived from this assumption. Eldrimner's strength resides not only in its anchoring in local society and its ability to mobilise local people and human resources, but

Box 4.3 Eldrimner's practice in facilitating local food production and processing

Besides organizing training courses in production and preservation technologies, Eldrimner provides help with the procedures involved in starting and managing small enterprises. It is involved with preparing an inventory for small-scale refining of rural products. The project helps provide rural people interested in small-scale refining with opportunities to engage in it, helping them to find ways of putting their products on the market. There are a number of different installations that are at the disposal of small producers. The Rösta dairy (managed by Rösta agricultural college) holds courses in cheese making. The dairy has a small shop selling local products. A production facility for pork butchering and other types of food processing is also in operation, having been established more recently. For small-scale cheese production, Eldrimner entered into a wide-ranging exchange of knowledge on small-scale cheese production with cheese makers in Provence, southern France, after the participants discovered that there were many similarities between traditional methods used in both regions, and that the problems encountered were likewise similar. Every year study-groups from Eldrimner visit Provence and other small-scale producer regions to exchange information on the techniques in cheese making and on marketing strategies and market development.

In response to the problem of the wide dispersal of producers in individual farms over the whole region, Eldrimner has developed a mobile production unit with approved equipment for different kinds of refining (dairy production, pork butchering, refining of berries and vegetables). The production unit is designed to suit the needs of small-scale refining. It can be rented inexpensively by producers on a weekly basis. Only persons living in the region are entitled to rent the production-unit, which can be transported to the producer's farm but can also be used at the Eldrimners installation in Rösta. A set of self-policing regulations have been formulated, with a view to safeguarding food safety.

The annual Särimner fair is organized to support the marketing of products and give producers, retailers and consumers an opportunity to meet face-to-face. The fair includes exhibitions, food sales, product quality assessments, seminars and discussions on the future of local food producers and small-scale food businesses. Every year a product contest is held to choose the best examples of small-scale products in different food categories. In their marketing the winners publicize this badge of approval from a group of experts as a way of demonstrating the quality of their products.

Through self-organized activities, courses and knowledge mediation, and the development of a certification of small-scale products, Eldrimner is endeavouring to become less dependent on national and EU funding. The problem of how consumers can become aware of locally processed products when these are not either marketed locally or made available through producer-consumer-networks is something that is increasingly preoccupying Eldrimner. Certification is being widely canvassed as a solution and many of the small-scale producers are pushing for product branding. There are many practical problems with such certification: how should 'small-scale' be defined? What methods can be included? Should all primary products be regionally produced? Should the certification be extended to cover the organic category? To deal with these questions, a process for developing regional certification is under

> way, with a working-group that is also a debating forum. The certification that is gradually being developed will probably include references to methods of production as well as to the size of the enterprise. As a certification system (KRAV) already exists for organic products, it could be argued that this aspect not be included in regional certification. Certification implies information about the knowledge and technologies used in production.

Sources: Bruckmeier, Engwall, Höj Larsen 2006.

also in the fact that the project and its activists are able to construct an international 'grassroots network' for rural development Europe-wide as a mechanism for exchanging knowledge, learning, and broadening experience. The network makes it possible for the project to accumulate knowledge, test it, and disseminate it through the producer network in accordance with the above-mentioned dynamics of local knowledge use.

The predominating knowledge in Eldrimner is local knowledge of a specific kind – knowledge of food processing as it was carried out prior to the progressive modernisation of agriculture throughout the twentieth century, the result of which was an 'intellectual expropriation' of local producers and farmers and their tacit knowledge of agriculture and food production. Today this core knowledge must be re-discovered and re-vitalised. If it is no longer available in the region or present as local knowledge there, it must be sought out in other areas and from other sources. This knowledge moreover needs to be further elaborated and to be brought into a synthesis with other knowledge that is necessary today for the production and marketing of food. As far as the bearers of such knowledge are concerned, different kinds of expert knowledge and local knowledge, and less scientific knowledge, are what is called for, as indicated in the introduction above.

How the distinct knowledge requirements are to be identified and brought into synthesis is something that began to acquire added importance in the second phase of the project, during which it attracted more interest and attention from scientific institutions also. Local producers' practice of experimenting with production techniques and with new ideas and products with a view to improving them leads to a situation where the knowledge base becomes more diversified and knowledge has to be assessed more systematically (for a more detailed description of this second phase, see below). When experts, scientists, universities, etc. become involved, the focus shifts to one of seeking out contacts primarily in the region, so as to maintain the project as one dominated by rural actors and their interests.

The success of Eldrimner in its first phase   The project succeeded in compiling, documenting, applying and disseminating local knowledge on food processing. Some of the lessons of the experience are more reconfirmed than learned, the new lessons being those reported from earlier examples of rural movements in the region (Ronnby 1995: 244 ff). They are: that local people and rural inhabitants should take the initiative for projects is an important prerequisite for their success; that local communities and municipal administrations should create an enabling environment by supporting the projects with administrative, legal, etc. advice and support is also important. External support and stimuli from national governmental and/or EU institutions and policies can be helpful for rural projects, but external support can neither replace local actors nor assume the dominant and directing role in this process.

Second phase – Eldrimner as a National Resource Centre

The project has developed from a regional resource centre for small-scale food production and refining into a national centre, co-existing with two other regional centres. There was some competition and lobbying also for the other centres, but finally it was Eldrimner that succeeded in acquiring the higher status. It was its basic ideas and the mode of performing the work and disseminating knowledge that characterised the project's prior development that gained it its success in becoming the national resource centre. As national centre Eldrimner continues with its seminars, through its product network, with the annual Särimner fair, which is a forum for local producers and for the processing industry, to work for the restaurant and tourist economy, and for consumers.

Creation of support structures and new networks   MATKULT (a Swedish acronym meaning 'food culture') is an association of local entrepreneurs producing in, or otherwise working with, the Eldrimner network. The association was formed to support the project at a time when it was not sure how it could survive and develop further. Above and beyond this regional support activity it was also important that Eldrimner should succeed in obtaining the support, even if only temporary, of the Swedish Ministry of Agriculture (especially in the years 2002–2003 when new funding had to be found) and of some political parties represented in the Swedish parliament. Such support structures do not remain the same but change over time, with a changing membership. They are nevertheless necessary for the further development of the centre.

With spatial extension of the producer networks, they began to seem less concentrated and less dependent on locality, place and region, becoming by contrast more dependent on the interests and demands of producers, retailers and consumers throughout the country. This need not be understood as a loss of the local tradition and knowledge of rural development that Eldrimner is supporting. The core ideas of the project remain those of small-scale artisan knowledge, the local focus, and tradition-based production and processing of food. The experts who are so important for the continuity and survival of the project remain, as before, local producers and artisans. For the producers a key consideration is that the focus should be kept on local food-production and the knowledge and experience required for it. So far the extension has not given them cause for anxiety. But it is important that even as a national centre Eldrimner should continue to serve at every stage of the process as a facilitating organisation for small-scale production. The involvement of more producers in other regions can be seen as a sign of growing interest in the ideas of Eldrimner. Support structures and activities analogous to those described for the original project in Jämtland will now have to be created elsewhere. The network's expansion operates in effect as a mechanism for international linkage of the experience of small-scale production. It establishes contacts between local food producers, wherever they are to be found, generating co-operation. Within the Eldrimner network there is strong solidarity with other small-scale producers in rural Europe. Co-operation and sharing knowledge improve their chances of overcoming obstacles to local food production.

Consumers There has already been discussion of the basic ideas on consumers and consumption behaviour that Eldrimner is putting forward. The consumers to have been targeted include local people in the Jämtland region, tourists visiting the region, and consumers in urban areas. Eldrimner's transformation into a national centre necessitates co-operation with new actors: food retailers, restaurants, gastronomic services reaching non-local consumers.[11] As the project has grown stronger, as more small-scale producers have joined it and started to become more successful, the role of urban consumers has grown in importance. The fact that the products represent a rural history and tell a story, the exclusiveness suggested by the small volumes produced, the superior quality: these are all features with the potential to capture the interest of up-market restaurants and well-off consumers in Stockholm and other metropolitan areas. Marketing strategies now need to address the problem that large traders and their food-distribution chains are reluctant to include local products in their inventory. For the food-chains to accept the local products quantities have to be larger, necessitating more transporting of food. Both these requirements are incompatible with the principle of local production and consumption of food. The inclusion of distant urban consumers need not, perhaps, affect the organisation or objectives of Eldrimner, as long as the process remains subordinated to the original ideas of the project and the trends towards bureaucratisation that are inherent in large projects and networks are kept limited. To ensure that the original ideas of Eldrimner will continue to apply in future, clear principles of knowledge management will be required. What is the core knowledge? Who can deliver it? What is supporting knowledge? Who can provide it?

Types of knowledge utilised in Eldrimner As can be seen from the above project-descriptions, several types of knowledge are drawn upon, corresponding to different stages in a ranking of priorities.

---

11   To some extent local retailers in the region, including in its capital Östersund (for example retailers specialising in delicatessen products) are already selling local products from Eldrimner farmers. But new forms of marketing need to be developed if locally grown and processed products are to be brought to the consumer via other channels, e.g. the gastronomic and tourism sector.

a. *Traditional local knowledge on food production.* Local knowledge and skills in food production have largely vanished, even among rural populations.[12] To re-vitalise such local knowledge Eldrimner was obliged not only to go to the region in question and seek out the knowledge, e.g. from elderly people in the local communities. Local knowledge also had to be obtained from many other sources, from contacts with local producers in other countries, from literature, from archives, from information search via the network, from personal contacts which often resulted in new members with special knowledge joining the project. These processes provide a good illustration of the dynamic quality of local knowledge: like other forms of knowledge it continuously evolves, it is tested, modified, adapted, it takes in new ideas, it rejects others as outdated or no longer practically usable, it incorporates elements of scientific or bureaucratic knowledge (Eldrimner producers are obliged, for example, continually to monitor the legal and bureaucratic requirements for food quality and observe them when making their products). To use an ecological metaphor: local knowledge possesses more of the qualities of a flow than of a stock resource.

b. *Tacit and lay knowledge.* Another dimension of local knowledge, including its modern forms, is its character as tacit knowledge, and in this connection lay knowledge as well as expert (handicraft) knowledge of food production techniques is of central relevance for the project. When local knowledge includes knowledge not acquired through formal and certified education and training, it has to be developed, elaborated and disseminated as it is through the Eldrimner network. Combining knowledge from other rural areas and sharing experiences with local producers from other European countries have been the primary mechanisms by means of which a more systematic approach has been developed and producer knowledge maintained. Because of its characterisation as 'traditional', such knowledge is often seen as lay knowledge, despite the fact that it includes a very specific kind of expertise, as has always been the case with peasants' and farmers' knowledge. The affinities with lay knowledge may be seen as symptomatic of its being part of everyday culture and practices for which food production and consumption are paradigmatic examples.

---

12 Local knowledge is not a homogeneous and exact concept, as evidenced by the anthropological debates about local and indigenous knowledge (Ellen et al. 2002). Local knowledge is often distinguished from scientific or universal knowledge through its specificity. It is place-, time- and culture-bound knowledge. But this too is inexact, though it does capture one important quality of local knowledge, which also applies for food production. Both are rooted in cultural practices: they comprise action-bound knowledge. One distinguishing feature of the local knowledge for food production is its dynamic quality of change and adaptation, its modification through practice and experience, its interaction with other forms of knowledge.

c. *Expert knowledge with a practical focus.* The boundaries between lay and expert knowledge become hazy when it comes to local food production. The term 'expert' has a specific meaning in the Eldrimner network: experts are not scientists or bureaucratic-managerial experts from governmental institutions who exercise power over rural development projects and the deployment of knowledge. The requisite managerial and marketing knowledge is imparted in courses for the local producers themselves. For Eldrimner an expert is a person with such knowledge: with expertise and proficiency in specific traditional and artisanal modes of food production. Such expertise cannot be acquired in purely formal ways through education and vocational training. It presupposes access to traditional local knowledge in rural areas. The expert knowledge utilised in the project has a practical purpose: it is to be applied directly in the production processes. Experts in different areas of food-production are invited to lecture on the different quality aspects mentioned above as well as on methods and regulations. Seminars, workshops and fairs are arranged for the purpose of training small-scale producers. Printed educational material on the various aspects of small-scale refining is produced by Eldrimner.

d. *Scientific knowledge.* The participants in Eldrimner are somewhat sceptical about scientific knowledge in local food production. The focus on application and product-specific knowledge makes more generalised or universal knowledge superfluous. What we are talking about here is not an across-the-board scepticism about science. It is a scepticism about the bureaucratised form of science, whose form of knowledge replaces the practical experience of producers and the local knowledge utilised in small-scale, artisan production. Scientific knowledge about food production is not so much needed for the production and refinement processes adopted by Eldrimner participants. The problem is rather that the local knowledge and the traditional technologies frequently have to be protected from the scientific knowledge that has gone into the dominant food production policies and regulations. This scientific-technical knowledge is first and foremost the knowledge required to direct and control production process *in absentia* and from a distance, in accordance with the idea of automated production systems and laboratories. A new instance of interaction with science emerged when Eldrimner itself became the object of scientific interest to be studied by researchers from a variety of institutions ranging from agricultural universities to social-scientific departments, with a focus on the cultural or process aspects of the project. The recent step upwards from a regional to a national centre entails some expectations that Eldrimner will be willing to collaborate more closely with the Swedish Agricultural University and the Swedish Farmers' Association. This idea is however contested by the small-scale producers in Eldrimner who want to continue with food-production from a practical and local perspective.

The main factors in Eldrimners' success   A number of internal and external factors already mentioned help account for Eldrimner's' success. The contributory factors for success generated by the project itself might be summarised as follows:

a. Eldrimner built a technical and knowledge infrastructure for local food processing that can be used throughout the region by all local producers.
b. Eldrimner works with rural and local producers and did not start from the conventional ideas of 'what the market takes' or 'what consumers want'. It is taken for granted that consumers have to rediscover the importance of quality and relearn the value of locally produced and locally processed food.[13]
c. From the outset Eldrimner has projected an entirely specific conception of the value of local food and the advantages of producing it. The pioneers of the project, who still occupy the key decision-making positions, describe this type of production as 'handicraft production'.

Above and beyond such internal factors there is also discussion of certain external aspects. Eldrimner has focused considerable attention on the limiting and supporting conditions in this connection. Although not directly linked to food production, social preconditions such as relevant local traditions, prior experience, active and interested rural populations, together with natural conditions such as the climate and topographic and soil forms, all helped to put the project on the road to success. Eldrimner counts as a project embedded in the local culture and in nature, this being one precondition for local food production and processing to become permanent, albeit not yet dominant.

Conclusions

When it comes to formulating conclusions it seems more important to start from the contextual factors outlined above than it is to articulate a precise definition of local food and decide on its quality criteria. The project has succeeded in linking product- and place-specific knowledge successfully with a long-term strategy of monitoring the natural and social context, in this case a remote area in northern Sweden.

1. Projects for local food refining can make it possible for farmers and local producers to remain in the rural region, strengthening the local economy, reducing the necessity for commuting or for emigration from rural areas. When it comes to promoting local food production, the area's traditionally strong regional identity and its specific culture of rural development are evidently of more importance than the level of producers' and farmers' formal

---

13   This re-learning process has many affinities with the task of re-learning the virtues of environmentally friendly agricultural production, implying for consumers development of a more comprehensive awareness, 'disalienation' and appreciation of the environmental quality of food.

education. Such traditions serve to reinforce the efforts at self-help, assisting with the discovery of new ways of continuing food production at a time when many farmers are giving up. They are also conducive, and this is important, to growth of a culture of local food consumption to parallel that of production.

2. The knowledge that is important for the Eldrimner project and for local food production in general as part of a sustainable rural economy includes the type of local and experience-based knowledge that is marginalised and suppressed today in agriculture and food production by the scientific and bureaucratic variety of knowledge. Local food production knowledge, in contradistinction to local ecological knowledge, is not place-specific in the straightforward sense of being unique: it may embrace knowledge from different areas and countries where similar products are produced. A combination of different types of knowledge may come into play when local knowledge is deployed. But specific kinds of local and expert knowledge are here more important than scientific knowledge, which in the case of food is relevant to more general problems of food quality and food regulation.

3. How much of a region's food production can be maintained as local production and processing, or re-established as local production and processing? Eldrimner's transformation into a national resource centre does not in itself provide the basis for a detailed answer. It is, however, congruent with what must be the answer: that it is everywhere a limited part, albeit in the Jämtland region probably less limited. But modernisation, rationalisation, specialisation, promotion of large-scale agriculture, the saturation of markets, all these factors place limits on its expansion.

4. To extrapolate from Eldrimner the ingredients for a more general governance strategy in the case of local food production and distribution leads one in the direction of sustainable rural development, though neither the Eldrimner project nor the movement as a whole discuss their activity in these terms. Through its practice Eldrimner supplies the ideas that have become widespread in rural development, in Sweden and indeed all over Europe: using experience built up from local projects to enhance the role of rural inhabitants, local producers and local resource users, strengthening their participation in locally-based resource management strategies as one core element of 'sustainable governance'. These ideas are very suitable for application in food production. Whether they work as well for other sectors of production and resource use is a subject on which views differ.

Eldrimner's implementation of the core ideas of local production and food refining seems to have reached the limits of what it can do within the present market system. The remaining barriers to development can probably only be overcome on the assumption of further changes in the wider society involving issues that go beyond food production. The holistic nature of the project, as indicated at the beginning of this paper, permits it to conceptualise a future where it will be able to develop as part of a broader network of rural movements. The discourse on local food and its significance in a globalising food culture is gradually metamorphosing into an ecological discussion on food production and consumption. The keynote of that discussion is the interaction of social and ecological systems, a consideration that must guide any study of the natural resource base at local, regional, national or global levels for the purpose of achieving its sustainable management. The revitalisation of local knowledge that is taking place at Eldrimner can be viewed as part of a process that has been described by ecologists as 'enhancing social-ecological memory'. It embraces a variety of different roles for knowledge producers and users as they assimilate and retain knowledge, interpreting it and making sense of it, networking and facilitating, becoming stewards and leaders, visionaries and inspirers, innovators and experimenters, entrepreneurs and implementers, followers and multipliers (Berkes, Colding and Folke 2003: 368f).

One question about such knowledge processes that remains unanswered has to do with the future bearers of local ecological knowledge, given the gradual eclipse of such traditional sources as small peasants in the course of the modernisation process. The dramatic changes in European agriculture that occurred in the second half of the twentieth century, with the disappearance of most farmers and small producers, have resulted in a largely unremarked loss of ecological knowledge. Rather than confining oneself to regenerating a minority culture and a niche market for local food produced by fewer and fewer farmers, it behoves the activist to ask: who can be the future producers of local food in an agro-food system that has reduced farmers to the status of dependent deliverers and users of knowledge and technologies under the control of agro-business corporations? This raises local food production questions that are at present unanswerable. Local food has to be produced today in a situation of continually deteriorating natural resource quality and on a diminishing genetic resource base in terms of biodiversity. Many old, locally adapted species of plants and animals can no longer be used in agriculture. Agro-biodiversity is rapidly diminishing, or is only conserved in the archival forms of seed banks or agricultural museums and is not available for food production. The trend is being accelerated through patenting and control of seeds, species and varieties by private corporations. How can local food of high quality be produced, for example, with chickens that are only selected for rapid growth and meat production? How can the risks of genetic engineering of food under the only criterion of producing more effective high yield varieties be controlled and allow for local diversity of food products?

Eldrimner is active during a time where there is already public debate over whether national food production should be maintained in Sweden (Royal Swedish Academy of Agriculture and Forestry 2004). A future without national production would not be seen as desirable for rural development by most participants in rural policy debate. But the discussion is highlighting an important fact: the ideas that are being evoked for support of Swedish food producers and food production in globalising markets are similar to those that have been put forward by Eldrimner for strengthening of local food production: concentrate on quality production (irrespective of the criteria being laid down for quality), concentrate on environmentally-friendly food production, on animal welfare, on higher farm income, on improved processing and food refining, on co-operation between producers, and on co-operation with retailers (Royal Swedish Academy of Agriculture and Forestry 2004: 34ff).

References

Berkes F., Colding, J. and Folke, C. (eds.) 2003. *Navigating Social-Ecological Systems*. Cambridge: Cambridge University Press.
Bruckmeier, K. and Prutzer, M. 2007. Swedish pig producers and their perspectives on animal welfare: a case study. *British Food Journal*, 109(11), 906–918.
Bruckmeier, K., Ellegård, A. and Píriz, L. 2005. Fishermen's interests and co-operation: preconditions for joint management of swedish fishery at local and regional levels. *Ambio*, 34(2), 101–110.
Bruckmeier, K., Engwall, Y. and Höj-Larsen, C. 2006. *Local Food Production and Knowledge Dynamics in Rural Sustainable Development*. Manuscript: CORASON-project, report work package 6; Gothenburg University, Human Ecology Section.
Demokratiutredningen, 1997. Study on democracy in Sweden mandated by the government. Available at: www.regeringen.se [accessed: 13 November 2009].
Ellen, R.F. 2002. Déjà vu, all over again and again: reinvention and progress in applying local knowledge to development, in *Participating in Development: Approaches to Indigenous Knowledge*, edited by Sillitoe, P., Bicker, A. and Pottier, J. London and New York: Routledge, 235–258.
Gibson, C., Williams, J. and Ostrom, E. 2005. Local enforcement and better forests. *World Development*, 33(2), 273–284.
JV (Jordbruksverket), 2005. Merkostnader och mervärden i svenskt jordbruk: Report 2005: 3, Stockholm.
NVVSC (Naturvårdsverket and Statistiska centralbyran) 2000. *Naturmiljön i siffror 2000*. Stockholm.
Ronnby, A. 1995. *Den lokala kraften*. Människor i utvecklingsarbete. Liber Utbildning, Stockholm.
Royal Swedish Academy of Agriculture and Forestry, 2004. Skall vi ha någon livsmedelproduktion i Sverige i framtiden? *Kungl. Skogs- och Lantbruksakademiens Tidskrift*, 143(19).
Sörlin, S. 1988. *Framtidslandet*. Carlsson: Stockholm.

PART II
Valorising Local Food and Local Knowledge

Chapter 5

From the Local to the Global: Knowledge Dynamics and Economic Restructuring of Local Food

Isabel Rodrigo and José Ferragolo da Veiga

Introduction

The successive CAP reforms along with EU and national rural development policies have been relevant tools in enabling the agricultural sector and rural areas to fulfil other functions besides the productive one and in creating opportunities for breathing life into rural economies. In consequence, both the agricultural sector and many EU rural areas have experienced complex changes. The term 'post-productivist dynamics' (Marsden 2003), a notion much vulgarised through the 1990s, continues to be used in the literature in connection with such changes.

The increasing attention paid to quality in food production along with extensification, promotion of sustainable farming and dispersion of production patterns, (i.e. the opposite of intensification and territorial specialisation), are identified as key characteristics of the new post-productivist forms of production and organisation that are emerging in many rural areas. Placing emphasis on 'the local environment and environmental protection for its own sake' the post-productivist model embodies a challenge to the agro-industrial model (Marsden and Sonnino 2005: 50). But far from having an identical impact on all of the EU's rural areas such changes are overwhelmingly concentrated in certain areas, such as the UK and the advanced economies generally (Wilson and Rigg 2003: 681).[1]

In addition to the new functions and values attributed to nature and the casting of the countryside as *loci* of consumption, instances of re-qualification of local resources leading to differentiation within the food economy rather than to a standardisation, have begun to emerge. They are identified in the literature as new forms of food governance or alternative food supply chains and are seen

---

1 In the current academic discussion this fact is seen as placing in question not only the validity of the conceptualisation of recent shifts in agrarian priorities as a post-productivist transition (Evans, Morris and Winter 2002, Hoggart and Paniagua 2001, Burton and Wilson 2006, Mather, Hill and Nijnik 2006), but also its capacity to adequately address all contemporary circumstances and particularly those of developing regions (Wilson and Rigg 2003).

not only as evidence of the post-productivist model but also as contributing to consolidation of the so-called sustainable rural development paradigm (Marsden, Banks and Bristow 2000, Marsden 2003a, 2003b, Marsden and Sonnino 2005; Fonte 2006). In the alternative agro-food systems the emphasis is on food diversity and authenticity, in contrast to homogeneity and mass produced foodstuffs; on quality rather than quantity and on the 'local' with all the relevant implications in terms of ecological particularity and contextual dependence.

The literature identifies three alternative modes of food governance, corresponding to 'particular types of interrelationships between the state, the market and ordinary people' (Kjærnes, Jacobsen and Dulsrud 2008): organic produce, quality production and short supply chains. The market for organic production, both certified and non-certified, has been growing, particularly in the northern countries of the EU. The quality production system has to do with foods that conform to defined standards of quality regulation (ISO, HACCP, GlobalGap and so on), by virtue of which they legitimise their presence in quality markets. It includes also foodstuffs that are accredited under EU legislation, namely under the Protected Designation of Origin (PDO), Protected Geographical Indication (PGI), and Traditional Speciality Guaranteed (TSG) systems, as well as under national legislation schemes. The short supply chains model pertains to markets structured around new forms of social relationships characterised by proximity and trust between farmers/producers and consumers. While organic produce and the 're-invention' of the short supply chains mode of food governance are particularly well represented in the northern member-states of the EU, the quality production system is more widely diffused in the Mediterranean countries (Marsden 2001, Fonte 2006, 2008).

The different geographical centres of the new models of food governance are paralleled in the two different perspectives, identifiable in the ongoing debate on 'relocalisation of food production and consumption' (Fonte 2008: 200). Focusing on the types of political, economic and social interests in the midst of which the local food networks emerge, develop and consolidate, the debate identifies two perspectives within the European alternative agro-food context. While the 'reconnection perspective' is identified as often developing 'in a context of longstanding export-oriented agriculture and the loss of food culture', most characteristically in northern Europe (and North America as well), the 'origin of food perspective' develops more frequently in the territories which were 'the latecomers to industrial development and never fully completed their "great transition"' (Fonte 2008: 202–3), namely Mediterranean rural areas (Portugal, Spain, Greece and southern Italy).

The two perspectives, on the one hand, illustrate the strategic differences whose origins can be traced to the differences between the rural development approach implemented in the agro-food systems of the northern member-states and those of the south, and on the other effectively refute the notion of a single EU rural development model. Far from being effaced, the historically conditioned distinctions between the EU's northern and southern societies (Hoggart, Buller and Black 1995) continue to be reflected in differentiations of current dynamics. The greater level of disputation over rural territories by competing stakeholders and interests, the greater social pressures to force farmers to adopt environmentally friendly agricultural practices, the higher social valuation, consumption and demand for rural amenities by urban people, the higher level of concern about food quality and safety, the greater purchasing power of consumers[2] in the EU's northern societies than in the societies of the Mediterranean, all these are relevant factors that help to explain the differences in rural dynamics and in the approaches to rural development between the EU's northern countries and the southern/Mediterranean ones (Rodrigo and Veiga 2008).

The two perspectives on rural development use different tools in redefining the relationships between food production and nature and in stressing the local embeddedness of food chains. Efforts in the reconnection perspective rely on 'grass roots initiatives for relocalising the food system' and on rebuilding the links between farmers and consumers, whereas in the origin of food perspective reliance is placed on quality certification of local food production 'in relation to values associated with territory, tradition and pre-industrial production practices' (Fonte 2008: 202). The focus of this chapter will be on the latter strategy.

The effect of pursuing the above-mentioned two strategies has been, schematically, in the reconnection perspective the development and consolidation of short supply chains and in the origin of food perspective a range of regional speciality products protected by the EU designation of origin certification (Fonte 2008: 202). It is nevertheless important to note that in the rural areas of the EU's Mediterranean member countries the certified foodstuffs 'are *only one part of a much bigger and diffused universe* that comprises various forms of certified and non-certified local food, rooted in a pre-industrial tradition that was marginalised but never became completely extinct' (Fonte 2008: 202; italics ours). In fact, although the principal theme of this chapter is the characteristic approach to rural development that is based on the relationships between food production and nature, i.e. the approach adopted by the southern/Mediterranean countries, also identified below as *local food for distant markets*, there is also analysis of the above-mentioned diffused universe, here called *local food for local markets*. Through the analysis of a case study, the chapter aims to question how and to what extent the strategy dominant in the south contributes to sustainable rural development.

The following comments in this section focus on the other conceptual framework employed in the case study analysis.

At the analytical level, the local is assumed to be 'a form of social contingency' (Marsden and Smith 2005). In the context of the case study it denotes a space of encounter between different actors, in possession of specific knowledge systems, pursuing distinct goals and strategies and taking advantage of particular opportunities.

---

2   In 2005 the GDP per inhabitant in purchasing power parity (PPP) (EU15 = 100) was, 78 in Greece, 66 in Portugal, and 90 in Spain (Eurostat. Available at: http://epp.eurostat.ec.europa.eu; accessed: 15 June 2007).

Knowledge is understood to be 'the ways in which people categorise, code, process and impute meaning to their experiences' (Long 2001: 189). The notion thus embraces both the scientific, or what 'until recently, was held to be (…) the best and the only consistent way of producing reliable knowledge of the world' (Kloppenburg 1991: 529), and non-scientific, or local, forms of knowledge as well (Long 2001; Bruckmeier and Tovey 2008).

Those forms of knowledge, in the possession of different types of actors, support the two distinct local food networks which are to be analysed. The scientific/technical/managerial forms of knowledge support the *local food for distant markets*, i.e. the certified food network. By contrast the *local food for local markets* network, or the short supply chain, is rooted in local forms of knowledge. Knowledge 'is *local* in the sense that it is derived from the direct experience of a labour process which is itself shaped and delimited by the distinctive characteristics of a particular place with a unique social and physical environment' (Kloppenburg 1991: 528). The distinction between tacit and lay knowledge in the concept of local knowledge will also be considered (Tovey and Mooney 2007: 102–3, Fonte, Introduction to this volume). To sum up, it is assumed that there are different forms of knowledge, each with its own particularities and applications.

This chapter focuses on an agro-food network and process centred on quality production. Its main goals are to explore the contribution of the food certification strategy to sustainable rural development and to illustrate the complex relations between the local and the non-local in the certification process of a traditional local food product. This will involve initiating a case study of the conversion process of a local product traditionally manufactured for the local market, Barrancos Cured Ham (*Presunto de Barrancos*), into a local product produced for the global market, Barrancos Cured Ham PDO. In examining this transition, attention will focus on characteristics of the agro-food network, including the types of actors involved, their objectives, the strategies pursued and the forms of knowledge mobilised by the actors concerned, as well as the way in which such forms have changed and interacted over time. Empirically the analysis is based on qualitative interviews. The geographical and socio-economic setting is the municipality of Barrancos, which is located in the sub-region of Baixo Alentejo, a NUT III of the Alentejo region (NUT II).

The chapter is divided into five sections, including this introduction. Section two provides a brief overview of the main characteristics of the certified agro-food system in Portugal and in Baixo Alentejo. The third section describes the case study on the basis of the main empirical findings from fieldwork. Section four analyses how the local traditional food processors have reacted to the certification process, outlining their current situation. The final section summarises the main conclusions.

The certified agro-food system: a broad description

The agro-food system (farming and the agro-food industries) is not only characterised by a significant degree of variation at the local, sub-regional, and regional level but is also closely linked to the diversity of the landscape and the cultural pattern. Its importance in the economy as a whole has gradually declined in terms of the employment it provides and especially in terms of the farming component. Its economic and social significance is nevertheless still great in Portugal.[3]

In addition to the agro-food industries there is a network of agro-food producers covering the whole country. These are mainly micro and small businesses producing different types of foodstuffs (undifferentiated and geographically differentiated) for consumers in local and distant markets. The distinctions occur within each individual industry, and good examples of it are the meat-based products industry (cured sausage and ham) and the dairy products industry (cheese). The dispersed pattern of location of these agro-food networks is explicable in terms of the necessity for them to be located near the source of raw materials and in terms of the close ties with food traditions that prescribe that they should be located near local consumers.[4]

Following the introduction of the 1992 European Community regulation (Reg. EEC 2081/92), Portugal, as other EU member states particularly those located in the Mediterranean basin, began the process of identification, typification and protection of the names of geographically differentiated agro-food products (Fonte 2008, Fragata 2003, Marsden 2001). As it is well-known, France and the so-called French *produits du terroir* played an important role in securing approval for that regulation despite the opposition of the northern countries (Fragata 2003: 451).

Up until 1992, in Portugal only a few wines had benefited from the specific protection deriving from the delimitation of regions of demarcated production.[5] Between 1992 and 2006 the number of certified agro-food products has steadily risen. In 2006 there were 110 new protected product names,[6] not including wines; 93 of these enjoyed EU protection (54 PDO, and 39 PGI) and 17 enjoyed protection at the national level: three Designation of Origin (DO), 11 Geographical Indication (GI) and three Guaranteed Traditional Speciality (TSG).[7]

---

3   In 2001 the agro-food system accounted for 7.2 per cent of gross value added at market prices and 11.8 per cent of total employment (GPPAA 2003).

4   For more details about the agro-food system in Portugal see Veiga and Rodrigo (2006).

5   The Port wine 'demarcated region', dating from 1756, is the oldest.

6   In 2006, 152 other new names were awaiting approval: 20 from the EU and 132 from national authorities (IDRHa 2006).

7   In 2006, the PDO products were: olive oil (5); fruit (13); meat (14); ham (1); cheese (11); cottage cheese (1); and honey (9), the PGI products were: fruit (7); meat (11); cured ham (1); cured sausage (19); cheese (1). The products enjoying DO national protection

The broad range of products with protected names, to be found in all parts of the country, are geographically differentiated thanks to the existence of local knowledges and local skills, reflecting the historical differentiation of farming systems. They derive basically from different local varieties of olives (olive oils), fruit (fruit-based products), flora (honey), and native livestock breeds combined with local knowledges about preservation and/or processing of the products. In addition to these products, others are produced on a cottage-industry basis, either just for local consumers or for customers in distant markets, with whom links have been established by virtue of personal association and tradition.[8]

The growing number of certified food products are benefiting from the policies for protection of agro-food product names with regional characteristics, implemented by virtue of national and EU incentive schemes, including the LEADER programme, and as a result of growing demand. These factors have also encouraged new investors, including large commercial organisations, to become involved in the sector. Yet, despite the significant number and diversity of certified food products, the overall contribution to national product is low, something attributable in part to failures of organisation in some of the certified agro-food networks that have been set up or revived. As a result, many products have not yet penetrated non-local markets, at the national or international level.

Certified food production has nevertheless made a positive contribution to rural development, even if only because of other roles they fulfil, whether environmental (conservation of local genetic resources; adoption of traditional farming systems and agricultural sound practices) or cultural.

At the regional and local level food certification has made a positive contribution to rural income. Many regional and local olive oil and meats, under powerful threat from competition with more productive breeds and varieties, were previously sold as undifferentiated products. The consequence of this was that many local breeds of animals and plants began to be socially devalued, leading many of them to the verge of extinction. With certification and the process of endowing products with guarantees of quality, the above trends began to be reversed; product differentiation transformed a weakness into a strength, empowering producers and their organisations.

The recent increased production of processed products with traditional characteristics has had as its natural accompaniment a return in the use of native breeds and varieties (which has also been supported by EU policies, namely the agro-environmental programme). At the same time, traditional environmentally sustainable farming systems and agricultural practices, which are closely

---

were: olive oil (1); fruit (1); ham (1); GI: potatoes (1); meat (1); ham (2); sausages (7); products with national TSG status were: fruit (1); meat (1); sausages (1) (IDRHa 2006).

8 In such cases no need is felt for certification. A good example is bread (in the north, different kinds of bread: cornmeal bread and rye bread; in the south and west, wheat bread) which is highly differentiated at the local level, depending on local food traditions.

interconnected with traditional food production, have also been recovered. These changes have, in turn, strengthened rural and territorial identities.

Despite the positive contributions just enumerated, the certification of local food products has also had negative effects. Among others – and in particular – a number of cured sausage products and dairy products (sheep's cheese) that have until recently been produced by small producers on a cottage industry basis have disappeared. Very often associated with local commercial activities (such as butchers' shops or grocery stores) these products have been characterised by a high degree of differentiation from place to place. The introduction of increasingly rigorous food hygiene and safety norms, coinciding with the introduction of the food certification schemes, which are under strict legal regulation, led on the one hand to the closing down of the vast majority of those producers and on the other to changes in the previous food networks. Faced with the new norms and control, some producers have set up larger organisations that are able to comply with the legal provisions, others have chosen to ignore the law and carry on producing on an informal and illegal basis, while others simply give up and abandon production.

The relative heterogeneity of local foodstuffs, producers and markets, and the fact that not all the food products traditionally consumed are currently certified (mainly due to certification costs),[9] help to explain the differing levels of interest in food certification among producers. In fact the economic advantage that may result from certification is not guaranteed to all producers and products, either because in the markets in which they are sold – local or even national markets – there is no requirement for such certification, or because markets for high-quality products are very small, leading to exclusiveness and rapid saturation, and so impeding the development of the industry.

Just to provide some contextualisation for this case study, let us pause for a brief comment on the certified agro-food sector in Baixo Alentejo.

The continuing decline of farming in Baixo Alentejo[10] is in contrast to the relative growth that the agro-food industries have been experiencing in the sub-region.[11]

---

9 In the case of the 'Queijo da Serra' cheese, with a well established national market, different strategies toward certification coexist and compete (Dinis 1999).

10 In the 1950s and in 2001 the agricultural sector employed, respectively, 73 per cent and 14.4 per cent of the active resident population of Baixo Alentejo (INE 1950 2001). A similar trend has been experienced at the national level. In the 1950s Portugal was essentially a rural and agricultural country. Agriculture and forestry accounted for 27.0 per cent of GDP at market prices and 46.3 per cent of total employment (1955). By 1980 these percentages had dropped to 9.4 per cent and 18.4 per cent, respectively. By 1986, when Portugal joined the EEC, there had been a further decline to 8.5 per cent and 15.2 per cent. The figures for 1990 were 7.8 per cent and 12.4 per cent and in 2002 they were 3.5 per cent and 9.5 per cent (Banco de Portugal, *Séries longas para a economia portuguesa*, www.bportugal.pt; GPPAA 2003; INE 1950 and 2001).

11 In 2001 the Baixo Alentejo agro-food system accounted for 20 per cent of employment in the Alentejo region (of which agriculture accounted for 15.7 per cent and the agro-food industries 4.3 per cent (CCDR 2004).

In line with the country-wide scenario, the number of agro-food certified products has also been rising in the sub-region. In 2006 there were 14 protected names, of which 9 enjoyed EU (8 PDO and 1 PGI) and 5 national protection (2 DO, 2 GI and 1 TSG).[12] These are basically products deriving from local varieties: olives (olive oil), and native breeds of cattle (Alentejano and Mertolenga breeds), sheep (regional merino) and pigs (Alentejano breed), where the level of production is high or supply is concentrated, leading to success in distant markets.

The case study: local food for distant markets

A local agro-food production system, whose main product is the *Presunto de Barrancos* (Barrancos Cured Ham), was selected for a case study.[13] Focusing on the 'origin of food' perspective, it traces the transformation process from a local product traditionally produced for the local market (Barrancos Cured Ham) – *local food for local markets* – to a local product for the global market (Barrancos Cured Ham PDO) – *local food for distant markets*. Cured ham is one of a range of products deriving from the Alentejano pig and forming part of the local cultural and gastronomic tradition.

Before proceeding it is important to note that the required local/lay knowledge involved in traditional Barrancos Cured Ham processing and marketing was not lost, nor did pre-industrial/traditional characteristics of non-certified local food disappear. Continuing on a *local food for local markets* basis, the existing Barrancos Cured Ham local network underwent an adjustment process to the new circumstances, with local actors losing significant bargaining power and local representativeness through the certification process.

The study area: a broad portrait

The municipality of Barrancos is located in the northeast of the Baixo Alentejo sub-region, bordering the Spanish regions of Andalusia and Estremadura.

In Barrancos there is a striking rocky outcrop (350 metres in height) in the vicinity of the Serra Morena hills in Spain, contrasting with the wide plains that characterise the landscape of the sub-region. It is an important element in

---

12 The PDO products were the following: 'Azeite de Moura' (Moura Olive Oil); 'Carnalentejana' (Alentejo-breed beef); 'Carne Mertolenga' (Mertolengo-breed beef); 'Carne da Charneca' (beef); 'Carne de Porco Alentejano' (Alentejo-breed pork); 'Presunto de Barrancos' (Barrancos Cured Ham); 'Queijo Serpa' (Serpa Cheese); 'Mel do Alentejo' (Alentejo Honey). The PGI product was 'Borrego do Baixo Alentejo' (Baixo Alentejo Lamb) (IDRHa 2006).

13 The analysis of the case study is based on official documents and on primary data gathered through interviews with actors involved at Évora University, UNIAPRA and AGRICERT.

the unique micro-climate of Barrancos and plays an important role in the ham-curing process (MADRP 2006). The predominant farming practices and the agro-silvo-pastoral system, known as *montado*,[14] a typical example of farmland of high natural value (EEA 2004), are in general terms adapted to the shallow schist soils and the climatic conditions of the municipality and the sub-region.

In line with the trends in the sub-region, the agricultural labour force has also been declining since the 1950s in Barrancos. In the 1950s the sector employed 67 per cent of the active population of the municipality but by 2001, this had dropped to a mere 9.3 per cent, at the same time that the agro-food industries have seen significant growth. In 2001 the agro-food system accounted for 14.1 per cent of total employment in Barrancos (farming employed 9.3 per cent of the working population and the agro-food industry 4.8 per cent). However, given the recent growth experienced by the sector, the importance of the agro-food industry in the municipality will currently be greater. In 2006 there were 6 protected product names in Barrancos, all EU-based (5 PDO and 1 PGI).[15]

Barrancos is closer to the Spanish settlement of Encinasola in Andalusia than to other Portuguese villages and its inhabitants have developed a specific culture and identity. The existence of a local dialect, *Barranquenho*, and the specificity of its individual cultural and gastronomic features, such as the cured ham and sausage produced from Alentejano-breed pork, are illustrative of its mixed cultural and gastronomic influences, from the Alentejo and from the Spanish regions of Andalusia and Estremadura.

The traditional farming system

The Alentejano-breed pig (*Sus ibericus*) has constituted the basis of the local diet over the centuries due to the range of products it supplies and its ease of preservation, using simple techniques that make possible year-round consumption. Being the most profitable and the best suited for raising by means of the extensive montado[16] farming system, the Alentejano-breed pig was very important in economic terms in Baixo Alentejo until the end of the 1950s, when its number

---

14 The equivalent of the Spanish *dehesa*, a *montado* is a human-made agro-forestry-pastoral ecosystem adjusted to the local climate that consists of scattered tree cover dominated by cork-oak (*Quercus suber*) and/or holm-oak (*Quercus Ilex spp. rotundifoliae*), with pastures and agricultural fields usually in a large-scale rotation scheme that includes several years of fallow. Shrubs sprout frequently and are either cleared out or artificially kept at low densities (Pereira and Fonseca 2003: 3).

15 PDO products were: *Carnalentejana* and *Carne de Mertolengo* (meat); *Carne de porco Alentejano* (pork); *Presunto de Barrancos* (Barrancos Cured Ham: a unique product, ownership of which is in the hands of the municipality); *Mel do Alentejo* (Alentejo Honey). The PGI product was *Borrego do Baixo Alentejo* (Baixo Alentejo Lamb) (IDRHa 2006).

16 The Alentejano pigs greatly benefit from grazing the holm-oak acorn in the *montado* farming system, which gives a unique taste and flavour to the meat and is particularly praised in the cured ham.

started to decline due to the African Swine Plague (Peste Suína Africana) leading it almost to extinction. Along with changes in consumer preferences with meat, the disease played an important role in the decline of the extensive pig breeding system. Extensive livestock production shifted mainly to sheep and cattle, with extensive production in the case of pigs being replaced by intensive breeding in confinement using highly productive breeds and cross-breeds, whose meat, being less fatty, was also more in line with the new consumer preferences.

More recently, the legal imposition of food hygiene and safety rules, the closing of the municipal abattoir in the early 1990s, leading to the centralisation of slaughtering at a single abattoir,[17] and the legal banning of family pig-slaughtering have also contributed to a reduction in the number of Alentejo-breed pigs produced and a fall in the number processed. This scenario is obviously not specific to Barrancos, but reflects changes occurring all over the south of Portugal, where holm-oak and cork-oak *montados* are located. Despite the marked decline of the Alentejano-breed pigs, some large farmers have continued to breed them for self-consumption, for the Portuguese market, which has been shrinking, and also for the more important Spanish market, which slowly began to attribute new economic and cultural/gastronomic value to the *porco ibérico*.

The slaughter of the pig and the processing and preservation of meat and fat were traditionally carried out each year in December or January, by the members of the financially more comfortable families, on the basis of local/lay knowledge. The slaughter constituted a social ritual involving the sharing of effort and the fostering of solidarity within the community. Cottage-industry artisans processed the meat for the local market, attracting consumers from neighbouring municipalities from whom they took orders.

The cottage-industry artisans of Barrancos, who resisted the above-mentioned changes, began to produce cured ham and sausage products for the local market; loyal customers in neighbouring municipalities began to consume cross-breed pork. Notwithstanding the efforts to hold the line against standardisation, by the late 1980s the number of cottage-industry artisans had dropped and only a small number of cured hams was being produced.

Cured ham and sausage from Barrancos owe their specific qualities to the natural characteristics of the area (namely, to the relief of the land combined with the climate, the genetic characteristics of the Alentejano pig and the acorns on which it was fed), as well as to the local/lay knowledges involved in traditional cured ham processing. These knowledges convey an accumulation of insights and skills acquired from learning-by-doing, transmitted through generations, taking advantage of and rooted in these local natural characteristics. Both factors have allowed for the perfecting of a technique of a distinct type of production. Unlike the cured ham of north and northeast Portugal, the Barrancos ham and sausage are cured in the open air rather than smoked.

---

17 Located in Beja (North Alentejo NUT III).

Since the 1990s the Alentejano-breed pig has been reviving. On the one hand with the complete eradication in 1989 of the *African Swine Plague* it became possible to resume the traditional extensive pig breeding system and to begin to export the production, whether in the form of pork meat or transformed products. On the other hand, changes in consumer preferences have led to an increase in demand for Alentejano pork and processed products (cured ham and sausage). Pig-raising has thus been boosted and the number of animals has risen sharply.[18] In 2006 a total of 24 certified products were produced from the Alentejano breed, 19 of which enjoyed EU protection (two PDO and 17 PGI), while five enjoyed national protection (one DO and four GI) and were awaiting the granting of EU protection (IDRHa 2006).

Local food for distant consumers: contingencies, actors, knowledge and networks

An initiative to set up the Barrancos Cured Ham network was launched with the involvement of the University of Évora in the late 1980s when a lecturer at the Department of Zootechny chose Barrancos Cured Ham as subject matter for a PhD thesis (Nunes 1993).[19] The research work received vigorous support from the outset from the Barrancos municipal council, being undertaken at a time when the food quality debate was getting under way in Portugal and the EU generally.

Research work began with the selection of a local cured ham manufacturer to take part in the study from among three producers nominated by the council of Barrancos. The selection took into account the kind of seasoning used in the manufacture of the ham, a key element in imparting to it the traditional high quality recognised in the local area. In the first year (1989–1990) Alentejo-breed pigs and cross-breed pigs were purchased by the University of Évora and fattened on the *montado* of the University. The animals were later slaughtered in Barrancos and cured hams produced by the local manufacturer at his plant. During this first year the university expert's intervention consisted in recording the various stages of manufacture and listing the unforeseen occurrences, without involvement in the technological matrix.

In the second year (1990–1991) a prototype manufacturing unit was set up in Évora to evaluate whether or not the climatic conditions of Barrancos, and the breed of the pig, were having an influence on the characteristics of the ham. By the end of the second year the study had concluded not only that the pig-breed was important, i.e. the Alentejano breed made it possible for ham of better quality to be produced, but also that climatic conditions (the local microclimate in the vicinity

---

18   The number of piglets registered rose from 1,350 in 1995 to 41,737 in 2004 (http://www.uniapra.pt, accessed January 2006).

19   This university, located in the Alentejo NUT II, has also been involved in the promotion of quality regional food products and has influenced their promotion in other ways also.

of the Serra Morena hills) influenced the characteristics of the ham. There were, however, still problems to be solved, namely those of hygiene and pest control.

In the third year (1991–1992) scientific answers to these questions were being sought. As regards the former issue, hams were no longer being produced at the local manufacturer's factory, but rather at special premises made available by the Barrancos municipal council. But pest control remained an unsolved problem. The way that local manufacturers solve it traditionally is by increasing salting time, producing ham with a saltier flavour. Though local consumers were (and are) accustomed to this salty taste, it was not pleasing to non-locals. Bearing in mind the relevance of consumer taste *vis à vis* this saltiness when it comes to determining the products' potential for penetrating new (non-local) markets, the university expert changed some of the quality parameters of the product's traditional processing.

By the end of the third year the study had obtained scientific results which enabled an application to be made to official bodies for certification of the ham.[20] All the necessary particulars had already been registered in the specification book, having been practically completed on the basis of the records made by the lecturer. In the meantime solutions were found to the problems related to the drying of hams and the management of mould for non-local markets, especially the Spanish, which is much more demanding than the Portuguese one.

Table 5.1 shows the stages in production of the ham, also indicating the technological changes introduced by scientific knowledge. The information covers only the duration of the stages and the regulation of temperature, humidity and ventilation.

The Barrancos Cured Ham PDO network was gradually set up with the involvement of the lecturer both on the organisation of production and processing.

In the early 1990s two Alentejo-breed pig-raisers' associations were established in Alentejo: the *Associação de Criadores de Porco Alentejano* (ACPA) based in Ourique (Baixo Alentejo) set up in 1990, and the *Associação Nacional dos Criadores do Porco Alentejano* (ANCPA) set up in Elvas (north Alentejo) in 1991. In 1992 they amalgamated to form the *União das Associações de Criadores do Porco Alentejano* (UNIAPRA). According to the lecturer UNIAPRA was created to avert the proliferation of small stockbreeders' associations and to provide an

---

20 The name 'Barrancos Cured Ham' was first officially protected by national law in 1992 (Portaria no. 431/92) as a 'collective brand with indication of origin' (Decreto-Lei no. 262/87). Later, in 1995, and now within the legal framework established by Reg. (EEC) 2081/92 (which was integrated into Portuguese legislation by Normative Resolution 293/93), it was granted the Protected Designation of Origin status. Current official definition of 'Barrancos Cured Ham' can be found in IDRHa (2003). It should be noted that in this definition the traditional characteristics of aroma and taste, which were referred to in first definition for protection purposes (a pleasant, sharp smell, *sui generis*, and a slightly salty, spicy flavour), were toned down, and the ham was designated as being heavier (hindquarter with a fresh weight of over 6 kg).

Table 5.1 Production stages in the production of Barrancos Cured Ham

Production stage	Duration	Technological change
Curing		
I – Salting	1 week	Temperature and humidity control: < 6° C and > 80%
II – Washing	2 to 5 days	Temperature: 3 to 5° C
III – Storage	25 to 45 days	Temperature and humidity control: 3 to 6° C and > 80%
IV – Drying – Maturing	6 months (minimum)	Ventilation and temperature control: between 18–22° C and 30–35° C
V – Ageing in cellars	6 to 18 months (minimum)	

Source: Collected fieldwork data.

administrative body to take responsibility for the Alentejo Breed Section of the Portuguese Book of Pig Genealogy. The production of Alentejano-breed pigs was thus organised on the basis of two main nuclei of large-scale breeders: Ourique/Castro Verde and Elvas/Arronches.[21]

Given that the natural conditions (the existence of a microclimate) and the available traditional/local/lay knowledges tended to confine Barrancos cured ham processing to this municipality, a proposal was put forward by the lecturer to set up a co-operative society aimed at bringing together local manufacturers under the umbrella of a single larger-scale organisation. But this initiative never got off the ground due to lack of interest on the part of the local manufacturers.

Barrancarnes-Transformação Artesanal Lda. was set up in 1992 as a limited liability company. It was based in Barrancos and included 11 partners, among them the lecturer and a butcher from another Baixo Alentejo settlement. Barrancarnes benefited from public support (investment aid) for companies whose products enjoyed Designation of Origin status. It began operations in December

---

21 In 1995, when Barrancos Cured Ham was granted designation of origin status, the Producers' Group was called UNIAPRA and UNIAPRA in turn named ANCPA, one of the founding organisations of UNIAPRA, as the private organisation responsible for control and certification. Later, in 2000, in view of the concentration of power in the hands of the breeders' associations and the fact that this situation was not in conformity with the general criteria for agro-food product certification bodies stipulated in Portuguese Norm EN 45011, the Producers' Group renamed itself ACPA and ACPA in turn appointed the *Certificação de Produtos Alimentares Lda.* (AGRICERT) as the body responsible for the Certification of Barrancos Cured Ham PDO. But the new situation did not significantly alter power positions in the network, as will be analysed below.

1994 and marketed its first cured hams in 1996, making use of research work already carried out as well as drawing on local/lay knowledge provided by the two partners residing in Barrancos and the commercial experience and managerial knowledge of the butcher. Training courses were held for the staff contracted, and the company entered into a partnership agreement with the University of Évora to continue the research work on cured ham.

In the first few years cured hams were sold to restaurants only in Évora and Lisbon. It was only later that the products came to have labels with product details on them and hams began gradually to be distributed to markets located in other parts of the country.

Pilot exportation initiatives were also carried out. In 1998 the company increased its capacity in response to growing consumer demand and achieved the capability of producing around 3,000 Barrancos Cured Hams PDO per year. In 2001, again facing the need to increase capacity, the company was bought by the Grupo Amorim (a large Portuguese group originating in other *montado* production, namely of cork, but expanding into other ventures such as finance and real estate). Between 2001 and 2005 the new owner of Barrancarnes carried out a series of investments, including construction of an abattoir for pigs in the Reguengos de Monsaraz municipality and construction of a new unit for the processing and drying of hams in Barrancos, with capacity for processing 20,000 to 25,000 pigs and producing 40,000 to 50,000 cured hams per year.

In order to guarantee the supply of the raw material (Alentejano breed pigs), and bearing in mind the need to get all actors on board in pursuit of its objective of making the PDO synonymous with its own brand, the company sought to establish partnership agreements with individual pig-breeders as well as with Alentejo pig-breeders' associations. In 2004, the Associação Interprofissional para a Defesa do Porco de Raça Alentejana (APRA) was set up, bringing together both processing units, in particular Barrancarnes, and individual Alentejano pig-breeders, the *confrarias* (gastronomic interest groups), and the National Consumer Protection Association (DECO). The Alentejano pig-breeders' associations did not join APRA, choosing instead free negotiation of market prices, not only in Portugal but also, and particularly, in Spain, where the demand price was (and still is) primarily subject to negotiation.

By exploiting the favourable position on the market they enjoy by virtue of the certification process and making full use of their negotiating potential, especially with Barrancarnes, which has the largest processing capacity for Alentejo pork, the Alentejano pig-breeders associations have secured rising prices for the pigs they sell. The prices have risen parallel with the growth in demand for diversified products, the consumption in Portugal of PDO fresh meat, the development of other processing industries in Portugal (cured ham and sausage), the emergence of the fledgling biological pork market, and the exportation of live animals to Spain, which is the most important market for Barrancos Cured Ham. The Spanish market serves the useful function of facilitating absorption of production, given the low number of pigs slaughtered in Portugal, and in a way it is also complementary to the Portuguese market, since the Spanish market demands pigs weighing 160kg[22] or more while the Portuguese market accepts pigs of under 160kg. Barrancos Cured Ham PDO has in the meantime achieved a high level of value in the market, indicating that there is a considerable potential profit margin available and that high prices are obtainable through negotiations including suppliers and consumers of the raw material.

At the commercial level, Barrancarnes is seeking to penetrate the international market and has signed contracts with hypermarkets. Barrancos Cured Ham PDO won the *coq d'or* in 2003, at a blind tasting event in Paris. It was selected for the 'Guide des Gourmands 2004' and, in the following year, won the *coq d'or* again. Since then it has been sold at the Petrossian delicatessen in Paris. Barrancarnes' main objective has been to penetrate the Spanish market, something considered indispensable in the light of the new higher level of output. Given the difficulty of this task owing to the specific characteristics of the market, in late 2005 the Grupo Amorim sold its majority shareholding in Barrancarnes to a Spanish group.

In the meantime, three other cured sausage-producing companies have been set up in Barrancos, of which only the *Boleta Barranquenha*, founded in the late 1990s by an ex-Barrancarnes partner residing in Barrancos, now produces cured ham as well, albeit in much smaller quantities than the cured sausage. The other two cured sausage-producing units were built from scratch by local people who were not ex-cottage-industry artisans. They ended up abandoning cured ham production mainly due to the difficulty of penetrating the relevant market in the face of competition from Barrancarnes. The latter is now practically the only company that produces and markets Barrancos Cured Ham PDO, though at prices that are more than double those of local cottage-industry ham.

In 2006, the principal actors of the Barrancos Cured Ham PDO network were: Barrancarnes; the Alentejano pig-breeders' associations (ACPA, ANCPA, and UNIAPRA); consumer groups (the *confrarias* and DECO); and public institutions (University of Évora; Comissão de Coordenação e Desenvolvimento Regional do Alentejo; Direcção Regional de Agricultura do Alentejo; and Instituto de Desenvolvimento Rural e Hidráulica). The focus of their activity was the PDO and the Alentejano-breed pig. The network is now extensive, but is poorly organised. The national territory covered is a vast space for the production of the raw material, i.e. the whole *montado* area of the south of Portugal.

Within the network, power is shared between the raw material producers and their associations, which exercise control over the Alentejano-breed pig (responsibility for the genealogic book), and the ownership of the brand name (PDO) – it is nominally owned by the state, but the context is one of a competitive market in which Barrancarnes is just one among several clients – and the processors,

---

22 There are also imports of pigs from Spain. However, as the predominant breeds in Spain are either cross-breeds or thoroughbreds which do not figure in the Genealogic Book of the Alentejano Breed, they cannot be used in the manufacture of products with designation of origin and geographical indication.

over which Barrancarnes exerts disproportionate market power, meaning that it is now difficult for new processors to enter the market. Nevertheless, the move to reach out to global markets has brought new actors into the network, not only new consumers, with different tastes and preferences, but also new producers of Barrancos Cured Ham.

Although processing takes place only within the municipality of Barrancos, there are interests allied to Barrancos Cured Ham PDO that are based outside the local area. It is naturally essential to attract outside investors if there is to be wealth creation and generation of employment in the local area. The impact of the certification process in terms of direct and indirect creation of jobs has evidently been great. In 2006 Barrancarnes employed 30 people, with the other three businesses employing a total staff of around 25. The setting up of the four companies has also had a direct impact on agro-silvo-pastoral operators supplying the raw material. Farming incomes are now higher due to the increased value of the Alentejano breed of pig and these factors have had an indirect impact on the economy of the municipality, making a positive contribution to regional and local rural development. The fact that the PDO is associated only with the municipality provides it with some guarantee of economic stability. But the current level of attractiveness will be maintained only as long as opportunity exists in the market. That this is so is clear from the successive changes that have occurred at Barrancarnes over a short period of time. The whole process is undoubtedly being propelled by consumers with great purchasing power.

The role of EU and national policies has been crucial, not in terms of production of the raw material because the pig-breeding sector is not supported by the CAP, but in terms of support for investment in farms and processing plants that has come with the system of protection of the names of agro-food products introduced by Reg. (EEC) no. 2081/92. By monitoring and facilitating the work of the university expert, the local authority created the conditions for the setting up of new factories. But the rapid growth of the industry meant that its capacity for response to market demands was outstripped.

The traditional actors and their network: local food for local markets

The cottage-industry artisans who rejected the proposal made by the lecturer that they should set up a co-operative society still produce cured ham in Barrancos. How many of them there are is unknown: they produce illegally, operating on an informal basis, for the local market and by order for customers in neighbouring municipalities. Their cured ham is obviously not certified and is sold at a price less that half that of Barrancarnes' Barrancos Cured Ham PDO. The municipal council has tried to persuade them to legalise their operations and set up shops selling directly to the local market (within a radius of 30 km).

Social, cultural and economic factors help to explain the refusal of the cottage-industry artisans to form a co-operative society and their preference for the current *status quo*.[23] Firstly, they are advanced in years, with low educational qualifications or even with no schooling at all, and lacking in managerial capital. These are all factors that limit individual initiative. Secondly, each of these participants in artisan activity is convinced that it is he who makes the best product,[24] and he therefore does not want to share the 'the technological secret' with the others. In other words they are lacking in trust, a necessary precondition for work on a co-operative basis. Thirdly, because their business has always been informal in character, they do not wish to be too well-known. Finally, they may not possess the economic capital required for investment in industry. Or they may simply think that the co-operative would not be the best solution for them, either because in Portugal there is no tradition of cottage-industry co-operatives, so that they have no other examples to follow and learn from. Or, simply because it does not fit with their world view or social identity.

In any case they see their own activity as being economically viable. Otherwise the municipal council would already have convinced them to invest in better equipment and buildings (if the small scale is maintained, not much money is required for the necessary investments), to legalise operations and set up shops to sell directly to the local market (local residents and visitors).[25] To some extent the local small cottage-industry artisans are identifiable with the old/updated rural element or with rural protagonists currently providing, to cite Ploeg, 'an intriguing "travelling" of the peasant principle' (van der Ploeg 2005: 1).

Bearing in mind the sociology of food and recent debate on the relationship between 'resistance, agency and autonomy' (Ploeg 2005 and 2007, Friedland 2008, Long 2008) the individual response of the local ham processors can be assumed to be exemplifying what Friedland calls a 'spontaneous *resistant agency*'. As he notes, 'individuals and collectivities, whether consciously or not, may express agency in ways that pose no threat to existing ideas and practices, or in ways that *reject existing hegemony*' (Friedland 2008: 62; italics ours). This may be the other face of the 'peasant agency' or 'peasant principle'.

Either because (objectively) they cannot afford to become part of a co-operative society or because (subjectively) they reject the hegemonic methods being imposed by outsiders, the local cottage-industry manufacturers have been

---

23 During the fieldwork it was not possible to interview any of the local small cottage-industry artisans because of the illegal character of their operations.

24 This presumption is widespread among petty producers (see also Fonte, in this volume).

25 There is an example of this kind of arrangement in the north of Portugal (the municipality of Vinhais) where a much of the local small cottage-industry has been legalised and currently sells sausages at the annual fair, *Feira do Fumeiro*. But these sausages are also sold on the home market. Tibério, Cristóvão and Fragata (2001) note the importance of the *Feira do Fumeiro* in strengthening personal trust and proximity between producers and consumers and in promoting consumer appreciation and social valorisation of local tradition and the gastronomic patrimony.

marginalised from the certification process. In consequence their relations with raw-material producers have become more strained, since they have suffered a loss of negotiating power, the ham they produce is now worth less, and they are therefore forced to buy at lower prices cross-breed or Alentejano-breed pigs which do not meet the requirements, in terms of weight or acorn feeding, for supply to the processing industry.

Conclusion

Taking a specific case study as its basis, the chapter analysed the process for certification of a traditional local food and the typical strategy adopted by the southern/Mediterranean MS in implementing the rural development approach rooted in alternative agro-food systems.

Both the traditional Barrancos and the PDO cured ham are produced in the same rural space: the municipality of Barrancos. The former takes advantage of EU and national policies for the protection of the names of agro-food products with regional characteristics. It is a modern-day local food, processed and promoted for the most part by a limited liability company and aimed at reaching distant markets. The non-certified cured ham is a traditional local food which continues to be produced, for the local market, by local cottage-industry artisans.

The case study analysed focuses attention on the following issues, which will be further commented on below: the contribution of the food certification process to sustainable rural development, and the complexity and consequences of the process.

On the question of rural development, the certification process has been contributing to local and regional development in economic, social and environmental terms. Economically, the setting up of the processing companies involved in production of the certified food has been of central importance for the creation of jobs (directly and indirectly) in the local area. Moreover, by virtue of that process the value of the Alentejano breed of pig has increased, with direct positive impact on the farm income of the regional suppliers of raw materials.

Socially, through the jobs it has created and through the positive impact it has had on the economic stability of raw material suppliers, the certification process has been helping to counter the exodus from rural areas and from farming. The latter aspect is of crucial importance, as the pig-breeding sector is not supported by the CAP. Farmers and other actors involved in the process of certification have of course benefited from other EU and national policies to the extent that they have secured support for investment in farms and processing plants. That is at the core of the distinction between the strategies adopted by the EU's northern and its southern/Mediterranean member states in implementation of a rural development approach rooted in alternative agro-food systems. While in the former case the strategy is anchored primarily in civil society and grassroots movements, in the latter it is based on public policy, without which the certification of the Barrancos

Cured Ham would probably never happened, as illustrated by the above-mentioned scenario of the national certified agro-food system.

Finally, by contributing to the revival of the Alentejano breed on the basis of the extensive farming system, the food certification process brought benefits in terms of conservation of the *montado* ecosystem, protection of the landscape and environmental conservation.

As indicated, the certification process is the successful outcome of the encounter, in a local space with specific natural characteristics, between different types of actors (local and non-local; individual/private and institutional), in possession of specific knowledge systems (traditional/local/lay and scientific/technical/managerial), in pursuit of distinct goals and strategies, availing themselves of specific opportunities. Such opportunities arise by virtue of natural characteristics of the local space and the traditional/local/lay knowledges that emerged to take advantage of the unique micro-climate of the Barrancos municipality, and became consolidated. The policies referred to above represent would-be means of exploiting them, notably in traditional cured ham processing as carried out by the local cottage industry. To enumerate these elements is to hint at the complexity underlying the certification process, and the importance of the local area as a space of contingency.

Not neglecting the importance of the other factors referred to above, the appropriation process of traditional/local/lay knowledges, transmitted over generations through the method of 'learning by doing', by non-local actors who, in turn, have introduced changes to it, was of crucial importance to the birth of the certification process. The changes introduced take the form of technological innovations for the production of PDO Cured Ham, its promotion, and consolidation of a new food network embracing distant markets. In consequence of the certification process the local food for local markets arrangement was required to undergo adjustment to the new circumstances. The local cottage-industry artisans lost significant bargaining power and local representation. Notwithstanding its contribution to sustainable rural development, the certification process has thus produced 'winners' and 'losers', in this way contributing to the emergence of new asymmetries of power grounded in the interactions between different systems of knowledge. Far from being neutral, it can result in the establishment of new power relations and new hierarchies, as has indeed occurred at the local level.

References

Bruckmeier, K. and Tovey, H. 2008. Knowledge in sustainable rural development: from forms of knowledge to knowledge processes. *Sociologia Ruralis*, 48(3), 313–329.

Burton, R. and Wilson, G. 2006. Injecting social psychology theory into conceptualisations of agricultural agency: towards a post-productivist farmer self-identity? *Journal of Rural Studies*, 22, 95–115.

Comissão de Coordenação e Desenvolvimento Regional do Alentejo (CCDR) 2004. *Plano Regional de Inovação do Alentejo, Relatório Final 2.ª Fase*, Dezembro, Évora.

Dinis, I. 1999. *Denominações de Origem e Desenvolvimento Rural: O Caso do Queijo da Serra*, Dissertação de Mestrado, Instituto Superior de Agronomia, Universidade Técnica de Lisboa.

European Environment Agency (EEA) 2004. High Nature Value Farmland. Characteristics, Trends and Policy Challenges, EEA Report, No. 1, Luxembourg.

Evans, N., Morris, C. and Winter, M. 2002. Conceptualizing agriculture: a critique of post-productivism as the new orthodoxy. *Progress in Human Geography*, 26(3), 313–332.

Fonte, M. 2006. Slow food's presidia: what do small producers do with big retailers?, in *Between the Local and the Global: Confronting Complexity in the Contemporary Agri-food Sector*, edited by Marsden, T. and Murdoch, J. London: Elsevier, 203–240.

Fonte, M. 2008. Knowledge, food and place: a way of producing, a way of knowing. *Sociologia Ruralis*, 48(3), 200–222.

Fragata, A. 1999. *Elaboração Social da Qualidade em Produtos Agrícolas Tradicionais. Carne Mirandesa, Queijo de Niza e Melão Tendral*, Programa de Investigação para Prestação de Provas Públicas de Acesso à Categoria de Investigador Coordenador, Oeiras, INIA-EAN.

Fragata, A. 2003. Da qualidade dos produtos agrícolas tradicionais. Elementos para a sua elaboração social e técnica, in *Portugal Chão*, edited by Portela, J. and Castro Caldas, J. Oeiras: Celta, 449–462.

Friedland, W. 2008. Agency in the agrifood system, in *The Fight Over Food. Producers, Consumers and Activists Challenge the Global Food System*, edited by Wynne, W. and Middendorf, G. University Park, PA: The Pennsylvania State University, 45–67.

Gabinete de Planeamento e Política Agro-Alimentar (GPPAA) 2003. *Agricultura Portuguesa, Principais Indicadores 2002*. Lisboa: Ministério da Agricultura, do Desenvolvimento Rural e das Pescas.

Hoggart, K. and Paniagua, A. 2001. The restructuring of rural Spain? *Journal of Rural Studies*, 17, 63–80.

Hoggart, K., Buller, H. and Black, R. 1995. *Rural Europe: Identity and Change*. London: Arnold.

Instituto de Desenvolvimento Rural e Hidráulica (IDRHa) 2003. Available at: http://www.idrha.min-agricultura.pt – 2003 figures [accessed: January 2006].

Instituto de Desenvolvimento Rural e Hidráulica (IDRHa) 2006. Available at: http://www.idrha.min-agricultura.pt, [accessed: January 2006].

Instituto Nacional de Estatística (INE) 1950. *Recenseamento da População*.

Instituto Nacional de Estatística (INE) 2001. *Recenseamento da População*.

Kjærnes, U., Jacobsen, E. and Dulsrud, A. 2008. Consumer choice as a mode of governance: the case of farm animal welfare in Europe, paper presented at the

Second Sustainable Consumption Conference: Sustainable Consumption and Alternative Agri-food Systems, Arlon, Belgium, May 27–30.

Kloppenburg, J. Jr. 1991. Social theory and the de/reconstruction of agricultural science: local knowledge for an alternative agriculture. *Rural Sociology*, 56(4), 519–548.

Long, N. 2001. *Development Society: Actor Perspectives*. London: Routledge.

Long, N. 2008. Resistance, agency, and counterwork: a theoretical positioning, in *The Fight Over Food: Producers, Consumers and Activists Challenge the Global Food System*, edited by Wynne, W. and Middendorf, G. University Park, PA: Penn State Press, 69–89.

Marsden, T. 2001. *New Communities of Interest in Rural Development and Agrofood Studies: An Exploration of Some Key Concepts*, Paper presented at the Workshop: Rethinking food production-consumption: integrative perspectives on agrarian restructuring, agro-food networks and food politics, University of California, Santa Cruz, Nov/Dec.

Marsden, T. 2003a. The conditions of rural sustainability: issues in the governance of rural space in Europe, in *The Reform of the CAP and Rural Development in Southern Europe*, edited by Kasimis, Ch. and Stathakis, G. Aldershot: Ashgate, 19–37.

Marsden, T. 2003b. *The Condition of Rural Sustainability*. Assen: Royal Van Gorkum.

Marsden, T. and Smith, E. 2005. Ecological entrepreneurship: sustainable development in local communities through quality food production and local branding. *Geoforum*, 36, 440–451.

Marsden, T. and Sonnino, R. 2005. Rural development and agri-food governance in Europe: tracing the development of alternatives, in *Agricultural Governance. Globalization and the New Politics of Regulation*, edited by Vaughan, H. and Lawrence, G. London: Routledge, 50–68.

Marsden, T., Banks, J. and Bristow, G. 2000. Food supply chain approaches: exploring their role in rural development. *Sociologia Ruralis*, 40(4), 424–438.

Mather, A.S., Hill G. and Nijnik M. 2006. Post-productivism and rural land use: *cul de sac* or challenge for theorization? *Journal of Rural Studies*, 22(4), 441–455.

Ministério da Agricultura, do Desenvolvimento Rural e das Pescas (MADRP) 2006. Guia de 2006 – *Guia dos Produtos de Qualidade DOP/IGP/ETG/DOC/IPR/AB/PI*, Lisboa, Instituto de Desenvolvimento Rural e Hidráulica.

Nunes, J.L.T. 1993. *Contributo para a Reintegração do Porco Alentejano no Montado*, Dissertação de Doutoramento, Évora, Universidade de Évora, 276 p.

Pereira, P.M. and Pires da Fonseca M. 2003. Nature vs. Nurture: the making of the Montado ecosystem. *Ecology and Society*, 7(3), 21 pp. Available at: http://www.consecol.org/vol7/iss3/art7 [accessed: 5 January 2006].

Ploeg, J.D. van der 2005. Empire and the Peasant Principle, Paper presented at the plenary session of the *XXI Congress of the European Society for Rural Sociology*, Keszthely, Hungary, 22–26 August.

Rodrigo, I. and Veiga J.F. 2006. *Portugal WP8 Country Report on* 'Innovatory Economic Development' *for the CORASON Project*. Available at www.cosason.hu [accessed 15 October 2009].

Rodrigo, I. and Veiga J.F. 2008. Portugal: natural resources, sustainability and rural development, in *Rural Sustainable Development in the Knowledge Society*, edited by Bruckmeier, K. and Tovey, H. Aldershot: Ashgate, 203–221.

Tibério, M., Cristóvão A. and Fragata A. 2001. Produtos tradicionais e construção da qualidade: o caso das designações protegidas Salpicão de Vinhais e Linguiça de Vinhais IGP, *Actas del IV Colóquio Hispano-Portugués de Estúdios Rurales, Associação Española de Economia Agrária* (AEEA) *e Sociedade Portuguesa de Estudos Rurais* (SPER), Santiago de Compostela, 7–8 Junho 2001.

Tovey, H. 2008. Introduction: rural sustainable development in the knowledge society era. *Sociologia Ruralis*, 48(3), 185–199.

Tovey, H. and Mooney R. 2007 CORASON Final Report. Available at www.cosason.hu [accessed 5 October 2009].

van der Ploeg, J.D. 2007. Resistance of the third kind and the construction of sustainability. Paper presented at the *Plenary session on sustainability* of the ESRS Conference, Wageningen, Holland, 27 August.

Veiga, J.F. and Rodrigo, I. 2006. *Portugal WP6 Country Report for the CORASON Project*. Available at www.cosason.hu [accessed 5 October 2009].

Wilson, G. and Rigg J. 2003. 'Post-productivist' agricultural regimes and the South: discordant concepts? *Progress in Human Geography*, 27(6), 681–707.

Chapter 6
The Construction of Origin Certification: Knowledge and Local Food

Maria Fonte

Introduction

The 'local food paradigm' has developed from a critique of the 'conventionalisation' of organic farming, exemplified above all in its growing specialisation, intensification and scale of production (Padel et al. 2007). Certification is seen as useless and even harmful, given that it raises costs for producers and prices for local consumers and aims at de-localising consumption, a trend which, on the contrary, the local food initiatives want to reverse (Tovey, in this volume). Certification and labelling schemes are perceived as being 'market-oriented' and therefore as impediments to radical transformation of the food system (Watts et al. 2005, Goodman 2003, Winter 2003).

Re-connecting producers and consumers at the local level is evidently an alternative to a globalised food system, especially under 'food desert' conditions, where a market for local food no longer exists and food has become a placeless good (Fonte 2008). But from a more general perspective the question is unavoidable as to whether such re-connection may be the *only* alternative to globalised food production. How much of the food economy in a given region can be locally produced and locally consumed? Can local food systems be expected to be the only or even the primary source of food provisioning (Anderson and Cook 2000: 244, Feagan 2007: 35)? How, finally, should the rest of the food system be organised if it is to conform to principles of holistic sustainability and food sovereignty?

In this chapter the perspective adopted is that of an 'underdeveloped' area of southern Italy, the territory of the Aspromonte National Park in Calabria, where a traditional food culture and a local market for food still survive, but are threatened by marginalisation of the local economy and depopulation of the local community. Can origin certification be of any help under these conditions? What are the prerequisites for re-vitalisation of local food provisioning and local community?

From a market perspective, in a quality food economy where quality is not directly observable but derives from the specificities of the place or the methods of production, information asymmetry becomes a problem. Oosterveer (2007) has analysed the dramatic changes in the production and consumption of food in the last 30 years and the failure of nation states to govern the globalised corporate food system. He presents certification and labelling as governance tool in the 'space of place' and 'the space of flows' (Castells 2006).

According to this author ISO and HACCP norms[1] and WTO-based regulations[2] reflect a placeless model of governance, regulating the circulation of food in the 'space of flows'. But these norms, dealing as they do with the characteristics of the product, are intrinsically incapable of responding to citizen-consumers' place-based concerns regarding food, such as effects of the methods of production on the environment, on health, on local communities. The de-globalisation of food production promoted by 'local food' initiatives on the basis of such concerns, on the other side, is considered limited in scope. By consequence a variety of NGO-initiated labelling schemes (organic, free-range eggs, fair trade and so on) are being promoted and may help satisfy the evident growing need for reliable information about food production practices (Oosterveer 2007: 55–56).

In this approach, process and place-based certification and labelling are a response to the information gap generated by the separation of the place of production from the place of consumption, separation which may be seen both from a spatial and from a social and cultural perspective. At the same time, certification and labelling need to be seen also from another perspective. They are not only vehicles for information or instruments of communication between producers and consumers, but also the result of a process which may involve third parties (experts, certification agencies and so on) and set in motion a new dynamic and an interaction between different actors and ways of knowing. Our objective in this chapter is to unpack the black box of certification and analyse the role local knowledge plays in its construction.

In the CORASON project we posited a practical distinction between scientific, managerial and local forms of knowledge, a distinction summarised in Box I.1, in the Introduction to this volume. Monitoring the process of the 'Aspromonte National Park Product' label construction, we shall attempt to shed light on the way a new knowledge network is constructed, with a focus on the interplay between local and expert knowledge.

The data and information are derived from in-depth interviews with protagonists in development of the certification (the President of the Park, the Mayors of the Park municipalities, the IGEA (Control Agency for the Ecological Guarantee of Agrofood Process) experts, representatives of farmers associations, local farmers) during 2005 and 2006. A revision was also carried out of the large amount of documentation on local food produced at local level during the process of construction of the certification system.

---

1   ISO (International Organization for Standardization) is the organisation in charge of developing standards based on voluntary involvement of all interests in the market-place. HACCP (Hazard Analysis and Critical Control Point) is a systematic preventive approach to food and pharmaceutical safety, based on a science-based inspection methods.

2   The reference is here especially at the Sanitary and Phytosanitary Agreement of the World Trade Organization.

Unpacking certification: local knowledge and alternative agriculture

Standards and certification are a new instrument of governance in the global food value chain. As food systems and especially retailing industry become more oligopolistic, quality rather than price becomes the key factor in competition. Competition over quality demands new institutions, in the first instance certification and labels.

Many authors argue that certification (especially third party certification) is used strategically by supermarkets for the purpose of gaining access to niche markets, coordinating their own operations, discriminating between suppliers, saving on costs for control and even devolving responsibilities for safety risks while providing quality and safety assurance to their consumers (Farina and Reardon 2000, Giovannucci and Reardon 2000, Reardon et al. 1999, Konefal et al. 2005, Hatanaka et al. 2005). It is also recognised that certification systems can be a useful tool in building trust between actors in the commodity chain, by providing independent parties' assurance that the product meets the appropriate processing and production standards. The actions of social movements, finally, may guarantee that social and environmental standards are taken into account by producers and retailers, irrespective of where the commodity has been produced (Hatanaka et al. 2005, Hatanaka and Busch 2008, Oosterveer 2007). This vision of certification is clearly functional to the 'space of flows'.

Our interest here is focused into the 'space of place' and specifically into local food. That is why our attention will be directed to one particular category of certification, that is Geographical Indication (GI), attesting the provenance of the product. There are different approaches to the GIs of food (Barjolle and Thévenod-Mottet 2003, Josling 2006). Some consider them analogous to individual or corporations trademarks. Under these circumstances GI becomes something like an instrument of product differentiation, a private label, especially if does not imply any obligations in terms of methods of production and place of provenance.

The EU interpretation of the certification of origin is more restrictive. It designates *territorial identity* as a quality sign, imposing strict norms on methods of production as consolidated in the history and tradition of a particular geographical area. As laid down in Regulation 92/2081/CE (now 510/2006) article 2, origin designation identifies the region of provenance of the product, but also 'the quality or characteristics ... which are essentially or exclusively due to a particular geographical environment with its inherent natural and human factors'. *Territory* in this case embraces both 'natural and human factors', that is the physical place (a set of physical-chemical and pedological features), the specific methods of production, the characteristics of the product and the circumstances under which it is consumed. The certification of origin thus implies obligations in relation to the provenance, the methods of production and the organoleptic properties of the product, which are all important elements in the cultural identity of the product and the territory and the basis of its special reputation. In this respect, labels of origin belong to the region where the product is obtained and are considered collective rather than private goods (Barham 2003).

Certification of food origin may be considered a compromise between domestic, industrial and market conventions (Boltanski and Thévenot 1991). It responds above all to the needs of producers and consumers, but it also protects a specific method of production and quality products in an imperfect market, where lack of information may lead, through market mechanisms, to disappearance of the superior good[3] (Akerlof 1970). Producers gain from the protection extended to the product's reputation and the incorporated local knowledge and are to some extent shielded from falsification and counterfeiting (agro-biopiracy). Consumers are protected against misleading information on the territorial identity of food. Insofar as public interests are being protected, certification can also be seen as embodying a civic convention. Typicality in fact derives from local varieties and breeds and from traditional techniques of production which are often more environmentally friendly and have been conserved in small farms in marginal areas. To valorise typical products combines the positive effects of protecting income and employment in marginal areas, revitalising local communities, conserving traditional systems of production and biodiversity and protecting the landscape, food and cultural diversity.

To sum up, certification works as a *sui generis* intellectual property regime for a collective good: territorial quality. The benefits accrue to generations of farmers in communities which, in specific localities and over time, have developed innovative local knowledge, new methods of production and new local varieties, thus contributing to enhancement of the reputation of the protected products.

As a form of intellectual property protection, certification offers the possibility of valorising the differential rent deriving from the territorial and cultural reputation of the product. But unlike patents, which establish a regime of private property, origin and provenance certifications are collective rights, upholding the collective nature of the product's territorial reputation. The collective nature of the right precludes it from accruing to only one enterprise or farm, or to an exclusive group which can buy or sell it. It does not, of course, preclude the eventuality that economic strategies (and power relationships) might be used in the territory, even by extra-local people, to appropriate the good and exclude local producers (the legitimate owners of the rights) from benefiting from it (Rodrigo and Veiga, in this volume; Adamsky and Gorlach, in this volume).

How is origin certification constructed? From the perspective of knowledge, certification may be seen as a process of social negotiation involving multiple actors with different knowledge, but also different objectives, status and power

---

3   The mechanism described is the following: in case of quality differentiation, when qualities are not immediately perceivable, producers, adopting *moral hazard* behaviours, may sell inferior goods at higher price. Deceived consumers will not consider anymore price as a reliable vehicle of information about the product quality and will only buy the inferior good at lesser price. By consequence, the market for the superior good will disappear.

resources. Our question is: are these different forms of knowledge *per se* carriers of different power relationships so that the encounter between local and scientific/managerial knowledge will necessarily bring to the expropriation of the former? Can certification serve the purpose of valorising local knowledge, as opposed to facilitating its expropriation?

As with regard to local knowledge, the CORASON project draws a distinction between *tacit* and *lay knowledge* (Tovey and Mooney 2007, Bruckmeier and Tovey 2009, Fonte Introduction to this volume). In the case of local food, lay knowledge denotes the technical knowledge utilised by farmers, producers and consumers for growing or preparing food in the specific agro-ecological and social context in which they operate. It includes knowledge of production and preparation techniques, local varieties or breeds, local natural environmental processes and product properties. With the development of industrial agriculture and its technocratic structures, in both private and public sectors (van der Ploeg 1986, Benvenuti et al. 1988), lay knowledge was generally devalued and discredited, so much so that ultimately it came to be characterised as *traditional*, meaning with that outdated, static and no longer useful.

Relocalisation of food in a way embodies a contestation of the dominance of scientific knowledge and new emphasis on local forms of knowledge, radically inverting the historical trend that has led to current domination of the agro-industrial food system in Europe (Marsden 2003). How does the interaction among different actors embodying different forms of knowledge affect the new knowledge process? Under which conditions does local knowledge manage to become an integral part of the new knowledge process and be re-vitalised, overcoming its historical condition of inferior status and powerlessness?

We will pursue these questions in the analysis of the construction of the 'Aspromonte National Park' (ANP) certification system, commencing with a brief description of the economic and social context of Aspromonte, a mountainous area located at the southernmost extremity of Calabria (see Figure I.1, in the Introduction to this volume).

Aspromonte, Calabria, Italy: marginalised traditional agriculture and the persistence of a local food culture

Calabria (population of about 2 million) is one of the 20 administrative regions of Italy and the worst off in terms of many economic indicators: poverty index, unemployment rate, per capita gross domestic product (GDP).

From the 1950s to the 1980s, Calabria, along with all of southern Italy, was an area of special government intervention (through a development program called the *Cassa per il Mezzogiorno*), which was in this region entirely unsuccessful. It remains to this day a region with no big industries; agriculture and industry account for less than a quarter of the regional GDP, employing the same proportion of the working population. It is the public sector that sustains employment in the region. The

Calabrian economy specialises in traditional products (especially agro-food) and is closed to international exchange (exports and imports represented respectively 1.2 and 2.2 per cent of the region's GDP in 2005, according to the National Institute of Statistics (Istat) official data). The firm structure is extremely fragmented: 70 per cent of the non-agricultural enterprises are 'individual' enterprises, which employ only one person. Economic infrastructures are lacking.

Local institutions are as weak as the economic structure. Development efforts are hindered by the widespread presence of criminal organisations, an inefficient bureaucratic apparatus and fragile local government institutions. In the 1980s and early 1990s Aspromonte was notorious as operational base for one of the most ruthless kidnapping rings in Italy.

In Calabria generally and Aspromonte in particular, agriculture is all-important in terms of employment opportunities and contribution to the family income. As in many southern European regions, agriculture here is characterised by the hegemony of the agro-industrial model in the more productive plains and the persistence of traditional agriculture in the internal, mountainous and hilly areas that comprise the largest part of the territory (about 90 per cent). Agricultural activities in such areas are traditionally based on the presence of small household farms, whose products are partly sold in local markets, partly self-consumed or exchanged in gift or barter relations. A local food culture has been preserved because of the late entry of Italian women into the workforce and low rates of female employment outside the home.

About one fourth of the families in Calabria earn part of their income from a farm. Production for self-consumption is very important in Aspromonte, as it is in other similar mountain areas of the Mediterranean. According to a survey conducted for the Aspromonte National Park (PNA 2004b), in 2001 85 per cent of people owning or using land in the Park territory reserve part of their production for self-consumption, with an average of three types of product per family (oil, vegetables, fruit and cheese being the most prevalent self-consumed products). The effects of this phenomenon are relevant not only from an economic perspective,[4] as a way to sustain family income in an area where not much money is coming in, but also in social and community life. Some of these products are in fact sent as gifts to relatives and close friends in other areas (PNA 2004b).

There is a great deal of regional variation in Italian cuisine. Though many culinary specificities have vanished since the 1950s, others remain quite strong. In any case, despite undeniable tendencies towards homogenisation in food consumption, resistance and countertrends are also strong. Changing EU agricultural policies and re-orientation towards valorisation of the quality dimensions of the agricultural process and product tend generally to favour this resistance. The new European model of an agriculture based on multifunctionality

---

4   Self-consumption is associated above all with the 'quest for authenticity' (42 per cent); followed by 'economic necessity' (18.6 per cent) and 'maintaining food traditions' (17 per cent).

and quality was readily adopted in Italy by both national and regional authorities and by the agricultural professional associations – most notably by Coldiretti, one of the most representative farmers' associations.

Also important is the diffusion of a grass roots movement for the protection of traditional agriculture and local food (Cavazzani 2008). This movement is both rural and urban in composition and comprises very different civil-society actors (Slow Food, the different associations for organic agriculture, groups of self-organised urban consumers, Civiltà Contadina, groups of seed savers and so on). Two distinct, partially overlapping tendencies are to be identified in this movement, oriented, on the one side, towards the rehabilitation of traditional agriculture and traditional family food in the name of protection of local culture, local development and biodiversity (Civiltà Contadina, seed savers, Slow Food Presidia, the organic movement); on the other towards traditional local food as the high-quality ingredient in the new filière of fashion food, linking small traditional farmers with famous restaurants and gourmet boutiques (Gambero Rosso, in part the same Slow Food).

While well-versed in the tradition of local food and rich in products with a great potential for valorisation, the south of Italy lacks the organisational resources, private or public, for effective institutional management of quality policies by local authorities. Such organisation is indispensable for the attainment of access to the added value to be derived from the GI protection system.[5]

Relatively few products from Calabria and Aspromonte are protected by designation of origin certification in accordance with European regulations. The high number of the region's traditional products is evident, though, from the many entries that southern Italian regions have in the *National Register of Traditional Products*, an instrument that was initially devised to help with the preservation of traditional techniques of food transformation, granting exemptions from the requirements of the 1993 EU Hygiene Directive (Council Directive 93/43/EEC). *Traditional products* are defined as those agro-food products 'whose methods of transformation, conservation and ageing are consolidated in time, homogenous in the concerned territory, according to traditional rules, for a period not less than 25 years'. The National Register of Traditional Products was established by law in the year 2000 and is periodically updated. In 2009 it contained more than 4,000 products, 272 of which come from Calabria.

---

5   With its 173 entries in September 2008, Italy, together with France, Spain and Portugal, is amongst the European countries with more Protected Designation of Origin (PDO) and Protected Geographical Indication (PGI) products. It is worth noting that in 2007 about 64 per cent of the total production value of protected products derives from five historical products among which only one is produced in a southern Italian region (Parmigiano Reggiano, Grana Padano, Prosciutto di Parma, Prosciutto San Daniele, Mozzarella di Bufala Campana,). In addition, 62 per cent of the production value of the PDO/PGI products comes from only two regions, Emilia Romagna and Lombardia (ISMEA 2007).

In the Calabrian context, the Aspromonte National Park's initiative of creating a label for local products was challenging in many respects. Instituted in 1994, the ANP played a very innovative institutional role in the 1999–2004 period, under the presidency of Tonino Perna, a university teacher and very active member of civil society (Grando 2007). The initiative to create a Park label was part of a strategic management of the Park as a laboratory of sustainable rural development (Grando 2007: 58–59), aimed to valorise local resources that incorporate both material and immaterial assets: landscape, artisan products, forests, typical food, music, the minority language 'Greek of Calabria' (Fonte et al. 2007).

While the construction of the certification system was completed, it was never implemented for various reasons. First because at the expiring of the mandate of Tonino Perna, at the end of 2004, changed political conditions at local and national level didn't lead to his re-confirmation as Park President (Grando 2007). Actually a long period of instability followed in the political governance of the ANP. Secondly, because of conflicts between the Ministry of Environment and the Ministry of Agriculture with respect to the legitimacy of the Parks' certification system. In fact, the Parks' authority to promote the use of their name as sign of quality, recognised by the national law 394/91, is contested by the Ministry of Agriculture, because of the differences in criteria used by each Park, the lack of a nationally recognised control structure and the confusion it may ingenerate with the European GI labels (Silvestri 1999).

Even so, the case study is useful for the way it sheds light on the main steps in the construction of a certification system at local level, clarifying and identifying some of the conditions for interaction among different knowledge traditions in this process.

The construction of the 'Aspromonte National Park' label as civic engagement

Under Italian law 394/91 (Framework Law for Protected Areas) which constituted new national parks, a park 'may license … the use of its name and its logo to local services and products complying with required quality characteristics and satisfying park objectives'.

The aim of this norm is to valorise agro-food and artisan products and rural services (excursions, hospitality), produced in accordance with the institutional objectives of the national parks. What is involved is thus a specific form of process certification, aimed at informing consumers not only of the provenance of the product but also of the methods employed in its production. The desideratum is that they should guarantee respect for the environment, nature and landscape of the park area. The Park certification was intended as a locally regulated certification that could, in many cases, supplement origin, organic or other forms of certification.

In its 2004 Economic and Social Plan,[6] ANP recommends an integrated programme of interventions, articulated in a variety of actions,[7] for valorisation of typical products (Project 15), and a project for 'creation and promotion of the Park quality label' (Project 17), aimed at 'disseminating among consumers the concept of "uniqueness" of the Park products and associating the product with the area of origin'. Another objective is 'to activate commercial mechanisms inside and outside the Park area' (ANP 2004a: 178).

The label was intended to apply not only to the typical agro-food products, but to an entire range of products and services that comply with the Park requisites of environmental sustainability, typicality and distinctive quality. The project was charged with high symbolic value. As told before, Aspromonte had acquired a bad reputation as the territorial base for many criminal organisations and illegal activities. The project of valorising the local Park products was part of a series of initiatives aimed at restoring the image of the Park and implementing a new sustainable model of development[8] (Grando 2007).

The certification project thus represented a response first and foremost to civic and symbolic values. It aimed at:

1. Imparting to the product and thence the consumer the image of the restored Park territory:

> Following the activity of promotion carried out by the Park over many years, the denomination 'Parco Nazionale dell'Aspromonte' now exercises a strong attraction for thousands of consumers, evoking pleasant images and memories ... The symbolic meaning of this denomination and its relevance for the territory can be employed as a means for valorising the many food production activities that constitute the basis of a well-conserved rural landscape and various specific ecosystems in the Park (PNA 2004c:1)

---

6   The Economic and Social Plan is the instrument through which a National Park may programme its initiatives for the social and economic promotion of the Park territory (L. 394/91, art.14 comma 2).

7   The actions are articulated as: Implementation of an integrated system for the valorisation of typical products; construction of a network of actors for their valorisation; definition of production protocols; creation of a central platform for the packaging, trading and distribution of products; devising of a marketing plan for their promotion, for a total cost of 1,400,000 Euros (ANP 2004a: 167).

8   The reconstruction of the cultural identity of the Park communities was the central objective of the Park Authority, under the presidency of Tonino Perna (1999–2004). To this end the Charter of the Aspromonte Civilisation was compiled and the North-South Caravan established a delegation of mayors from Aspromonte who went visiting other national parks in Italy in order to disseminate the principles and objectives of the Charter.

2. Improving the image of the Park through the quality of its products. Agricultural activities play a fundamental role in preservation of the peculiarities of the Park landscape and may make a contribution

> to the execution of the tasks and duties assigned by law to the Park. That this is so may not be unrelated to the different forms of organic and sustainable agriculture still being carried out on the Park lands. A certification programme promoted by the Park may facilitate communication with those consumers who are paying attention to the natural beauties of Aspromonte and are interested in their preservation (PNA 2004c: 1).

An attempt was made to activate a virtuous circle: imparting the restored image of the Park to the products, while the distinctive qualities of the products contributed to improving the symbolic value of the territory. Construction of Park products certification was intended as a civic action.

The *Guidelines* for the definition of the product specifications stress *three dimensions* of quality: traditionality, environmental sustainability and social responsibility. The entire production cycle (starting from the production of raw materials) must take place within the Park territory, in accordance with traditional practices and local culture; local services, products and their production methods must correspond to the Park objectives of preserving and valorising the environment and satisfying defined social and ethical concerns and considerations of legality.

It was, however, deemed difficult to codify social responsibility in the short term (work conditions and labour organisation are, for example, largely informal on family farms and as such very difficult to codify).[9] The criteria used for characterising production protocol were therefore focused in the first instance on the product's origins and on environmental sustainability. To obtain the licence for use of the Park logo, products and inputs of the production process were required to be of local origin and to be the result of traditional and organic[10] methods of production, as codified in the production protocols.

These general criteria were established by the Park as an institution – the Park Authorities comprise the President and the Council of Mayors (representatives of all municipalities whose territory is included in the Park). The certification guidelines and the production protocol for each process and product were then negotiated with local social actors.

The IGEA (*Ispezioni e controlli per la Garanzia Ecologica dei processi Agroalimentari*), a local agency for ethical and organic certification in the field

---

9 Their presence in the Guidelines was, in any case, aimed at having social and ethical concerns included in the Park programme.

10 Organic methods are mandatory in the areas located in the highly protected A and B zones; low environmental impact methods in the other areas (C and D).

of agro-food production,[11] was appointed by the Park to take responsibility for drawing up the Guidelines and constructing the production protocols for most known typical local products. IGEA operators have all a formal education in agricultural studies, at high school or university level. Many of them completed their university studies outside Aspromonte and even outside Calabria, but they are all of local origin, that is they were born and grew up in Aspromonte or in a nearby area. They led the consultation process with local actors, especially the professional agricultural associations.

There were three particularly debated topics in the negotiation process: a) the scale of the territory to be included in the Park certification programme, b) the norms for environmental and safety characterisation of the production processes and c) the selection of the *model farms* to function as points of reference for traditionality, that is for compliance with typicality norms.

As for the boundaries of the territory, the associations of agriculture professionals required definition of a pre-park area that would not compromise the possibility of certification: it was established that some of the production/transformation activities could be carried out not only within the administrative borders of the Park, but also in the pre-park area, which extended in a 5 kilometre-wide belt around the park borders. Firms applying for the certification were required to be located in the park area or in the pre-park belt and to demonstrate that all the production cycles were carried out and raw materials produced on this territory. A few exceptions were allowed for traditional products with a long-standing history, such as the Cittanova stockfish, a typified preparation of the dry fish imported from Norway.

In relation to environmental characterisation of the production process, it was decided that organic agriculture should be taken as a benchmark. The use of integrated pest management was restricted in areas C and D (the areas of the Park area with less strict environmental constraints) and only for processes difficult to convert to organic methods.

As far as hygiene requisites are concerned, it is well-known that many traditional production processes do not comply with European norms. In order to save the traditional character of these production systems, it was agreed that they should be accepted for certification,[12] employing ad hoc criteria to establish systematic controls and principles of sound practice.

Finally, as regards formal codification of a production protocol for each food chain, a crucial point in negotiation was the selection of the model farm or the model firm. They were selected in a negotiation process, on the basis of being successful farms or firms with deep roots in the region and representative of what

---

11   Promoted in 1999 by the Calabria section of AIAB (*Associazione Italiana per l'Agricoltura Biologica*, an association for the promotion of organic agriculture), the IGEA also functions as a control agency in the field of agro-food production certification.

12   As mentioned before, in Italy, traditional products, classified as such in a national list, may be exempted from application of the EU hygiene directive.

is collectively considered the 'local tradition'. Once selected, their production practices and product characteristics were codified as a point of reference for the production protocol and the quality of the typical product. Observation of farmers and artisans in their practice as well as farmers consultation is usually the basis on which the various steps of the production process are formalised (see also Rodrigo and Veiga, in this volume).

It is often implied that certification leads to the homogenisation of local production practices. This is true, but only to a certain extent. Actually the existence of the same typical product in an area is a sign that local practices have consolidated around some common characteristics; otherwise there could be no typical product. A good production protocol codifies the essential commonalities: the characteristic of the raw materials (types of cultivars, of yeast, of milk and so on), but leaves scope for variation in the individual recipes (for example, quantities of flavouring or spices are usually comprised in a range or not defined at all; see Kvam, in this volume, Buciega et al., in this volume).

In order to identify the main traditional products of the Park area, IGEA experts collected local documents and other nationally-circulating documents and publications (for example tourism and enogastronomy guides), communicated with organisations of agricultural professionals and agricultural unions, and consulted and interviewed farmers and other local people, especially the elderly.

Finally 45 typical products were identified and assigned to one of 14 categories.[13] For each product the most important characteristics were described: production areas and calendar, production and transformation techniques, consumption traditions, territorial peculiarities, historical facts. This information is summarised in Table 6.1 reproduced for a few products by way of example.

Typicality is codified through two characteristics, 'traditionality' and 'territoriality'. The first involves the temporal, that is how long the product has been associated with the territory; the second the spatial dimension, that is the exclusiveness of the product's association with the territory. Information is also provided on the diffusion of the product (from 'widespread' to 'on the verge of extinction').

Furthermore, following initial description of the quality dimensions, production protocols were redacted according to the specificities of each individual chain. In accordance with the ANP specifications 'the licence permitting use of the logo can be demanded by firms which carry out the final steps of the production process (and/or labelling) and qualify their suppliers', that is check that product and process characteristics of the previous production stages comply with the requirements of certification. This implies promoting bottom-up local coordination of the filière, involving planning, self-regulation and mechanisms for improvement in organisation of the entire chain. The logo licensee is also assigned

---

13 Cheeses, olives and derivatives, fresh pasta, bakery products, biscuits, desserts, fruit, vegetables and derivatives, honey, fresh meat and its preparation, salamis, wines, fisheries, liqueurs.

Table 6.1 Examples from the ICEA table identifying typical products in Calabria

Product	Present situation	Traditionality	Territoriality	Source of information
Capocollo (a type of salami)	Quite widespread	Archaic (> 5 centuries)	Also prevalent in other areas	PDO products list
Goat ricotta cheese	Small quantities	Archaic	Also prevalent in other areas	Local, informal
Canolo rye bread	On the verge of extinction	Traditional (> 50 years)	Exclusive to the area	MIPAF decree 14 June 2002
Palizzi Igt wine	Small quantities	Archaic	Also prevalent in other areas	Traditional products list
Aspromonte potato	Widespread	Recent (> 25 years)	Interesting local variety of cultivars prevalent in other areas	MIPAF Decree 14 June 2002

Source: Personal elaboration from the ICEA table (PNA 2004c).

the task of documenting the use of local traditional recipes. The local producer is then mobilised to collect information, documenting the traditional knowledge (written and oral) in such a way as to leave leeway for a flexible definition of *typicalness*.

It should be noted that construction of certification and production protocols becomes a crucial tool for transformation of the traditional local systems and the activation of local economic actors. Technical specifications assign an important role to the actors (firms, farms, consortiums, cooperatives) applying for the certification. Certification in fact presupposes an action plan for the hygiene and safety of the products, an action plan for environmental protection, a program of controls on raw materials, intermediate and finished products, a self-regulation of the production process and the activation of mechanisms for improving the organisation and the attention to consumers (PNA 2004c).

Given that localisation in the Park territory is an essential requisite for certification, firms are motivated to seek local suppliers. Such is the case with Romeo, a firm producing pasta located at Gambarie in the Aspromonte Park territory, which replaced national flour suppliers with local ones with a view to obtaining the Park logo licence. Certification becomes an incentive to strengthen territorial networking relations at production level.

Construction of the Park certification system mobilises many actors within a new knowledge network: institutional actors (the Park Authority), IGEA experts, a non-profit association (AIAB), professional associations, agricultural unions, local producers and local consumers. It also activates local coordination of the filière and strengthens territorial relations.

Discussion: certification as a hybrid forum

About two decades after Kloppenburg's (1991) seminal work 'Local Knowledge for an Alternative Agriculture', I would like to reintroduce the question asked in that article: can we achieve a truly alternative agriculture without an alternative science? Kloppenburg's hypothesis was that an alternative agriculture requires an alternative science and that this can only be built by taking into account local knowledge.

Reflection on local knowledge should take as its starting point certain long-standing debates in anthropology and sociology of science. One important acquisition in these fields is the insight that local knowledge is not only contextual, that is linked to a specific place, but also socially situated, that is necessarily linked to a social context. But exactly the same is also true of expert knowledge. Latour (1987) has brilliantly demonstrated how scientific knowledge is typically situated in the social context of the laboratory and the web of relations researchers are able to create 'at a distance' (Jasanoff and Martello 2004, Sillitoe et al. 2002).

Anthropology and sociology of science therefore warn against a reification of 'forms of knowledge' and their antinomies according to which local knowledge would be subordinate, oral, practical, experience-based, intuitive, holistic, subjective, while scientific knowledge, by contrast, would be dominant, literate, didactic, analytical, reductionist, objective, positivist (Sillitoe 2002: 110). This kind of oppositional vision ends up creating categories of 'good' and 'bad', rather than fostering understanding of the potentialities and limits of both.

The black-boxes of 'local' and 'scientific' knowledge both need to be deconstructed. Local knowledge 'is not monolithic, but individually variable' (Sillitoe 2002: 120), linked to differences in gender and social roles. On the other side, scientific knowledge is not 'more objective' or more rational simply because is more formal and codified. Furthermore, in a globalised world 'local' and 'scientific' knowledge are not opposite poles: 'when we talk about scientific and indigenous knowledge we are referring to the overlapping yet variable understanding of individuals taken as bearers of these traditions, in some circumstances linking them together' (Sillitoe 2002: 120).

Rather than looking at local and scientific knowledge as dichotomies it should be possible to see them as the extremes of a continuum, very much like the urban-rural continuum. Rural sociologists are well aware that a linear vision does not exclude a dualistic representation of reality, with the two poles of the continuum (local and scientific knowledge) again becoming opposites. This is why Sillitoe (2002) proposes that local and scientific knowledge should be represented as a cluster of relations in a tri-dimensional 'globe of knowledge', in which interaction among different domains and traditions of knowledge is the norm.

If we accept that all knowledge is situated, no form of knowledge can claim universal validity. 'If the globalising process is resulting in a mish-mash of knowledge, to what extent can any society lay claim to owning any knowledge exclusively or using it to direct change?' (Sillitoe 2002: 132). Each form of knowledge has its elements of truth and may be important or inadequate in particular contexts and circumstances. This vision would exclude any hierarchical ordering of forms of knowledge and would lead us to support a dialogue among different knowledge traditions and their bearers.

In the history of agriculture there is no good tradition of a dialogue between traditions of knowledge, that is local and scientific knowledge. In the second half of the twentieth century, the typical 'green revolution' advocate saw his mission as one of persuading the farmer to abandon his techniques, his practices, his varieties, his knowledge, in order to adopt new varieties and with them a new technological package. Not farmers, but professional agronomists working in laboratories and on experimental field dominated the production of knowledge. Farmers were supposed just to adopt, or at best to adapt, the knowledge generated by the scientists. Only such adoption/adaptation could offer hope of success; the choice of non-adoption condemned one to marginalisation. A chasm opened up between traditional knowledge and scientific/expert knowledge. The result in terms of knowledge was that in the most economically successful areas local traditions of knowledge were lost, while in peripheral areas they were synonymous with a social and economic process of marginalisation and in many cases stopped evolving. The ascendancy of professional knowledge was the result not of its intrinsic superiority, but of the capacity to build a world-wide network of agronomists, extentionists, experimental stations and research centres able to impose its application everywhere, in this way creating the same conditions for its viability (Latour 1987).

The shortcomings of industrial techniques, their detrimental effects in terms of environmental damage, social exclusion and loss of traditional knowledge and biodiversity are today evident (IAASTD 2009, Kloppenburg 1991). They constitute the base on which the 'de-constructive project' (Kloppenburg 1991) is being carried out and a dialogue among different knowledge traditions is today invoked. Drawing on a variety of intellectual tendencies, including feminist critiques of science and knowledge (Harding 1986, Haraway 1988, Harper 1987), Kloppenburg projects three possibilities for re-construction of a new science of agriculture with the potential to reinstate the farmer in the position to which he/she is entitled: a reformed science, a successor science or a decentred science.

A *reformed science* assumes that local and scientific knowledge are complementary and that the former may be more or less translatable into existing scientific frameworks. Farmers' knowledge must be taken seriously, but it is still the scientists who will drive the process of reforming science.

According to the second perspective reform is impossible, since local knowledge constitutes a reality separate from scientific knowledge and there can be no complementarities between local and scientific knowledge. No translation is possible: what is needed is a radical epistemological reconstruction. Deconstruction of existing science will necessitate the emergence of a *successor science* on a new epistemological basis (see, for example, *Mode 2* science in Nowotny et al. 2001), implying a dismantling of the institutional and intellectual boundaries separating farmers and agricultural scientists.

Finally, the *post-modern perspective* assumes that a universal epistemological stance is impossible and would actually lead to a new hegemonism. Local and scientific traditions correspond to different ways of knowing: they are different and both are necessarily partial. Differences must be recognised and valued; productive interactions between them can be established through partial connections (Harding 1986). Ways of knowing need to be articulated so as to combine different forms of knowledge and make possible mutually beneficial dialogue. The problem is one of creating conditions in which these separate realities can be articulated, enter into dialogue with each other and inform each other. But no integration is possible.

According to this vision, we can look at the multifarious forms of cooperation among scientists and farmers (participatory development of new varieties, for example) both as creating the conditions for partial connections and interaction among different ways of knowing or as paving the way for the epistemological revolution that could lead to a 'successor science'.

Origin certification, especially the participated local forms of certification, may be seen as one more instance of this type of interaction. Whether the result of the process is appropriation of local knowledge by powerful actors or the re-activation of local knowledge as an important resource in food and local resource management is not written in the process, but it depends on the objectives pursued and specific conditions which shape the relationships among different actors.

The turn to local knowledge may make of certification process an epistemic network, totally different from the 'epistemic community' as intended by Haas (1992). While the latter is intended as a coalition of recognised professionals in a particular domain of knowledge, sharing normative and principled believes, common casual explanation and internally defined criteria for validating knowledge, the former may be intended as an hybrid forum (Callon et al. 2009) where experts and lay people may foster and explore new forms of interaction, starting from recognising equal dignity among multiple ways of knowing.

Aspromonte is a marginal area. Local knowledge has persisted here as traditional knowledge, its evolution curtailed by isolation and emigration. The participative process of construction of the Aspromonte National Park logo may be seen as the hybrid forum where interaction and dialogue among experts and local traditional farmers were experimented. There are specific conditions that may be conducive to the success of such a dialogue. Let us examine three such conditions in our specific case: objectives, actors, power relations.

Construction of the Aspromonte National Park label served not only the economic objective of rural development (increasing income from agro-food activities through valorisation of local products); it also served such social, ecological and civic objectives as restoring the pride of local people in their own identity, which had suffered damage owing to the criminal activities of the local mafia, restoring and preserving environmentally friendly agricultural techniques, mobilising local actors around issues of food production, valorising local food traditions.

Figure 6.1 illustrates the multitude of actors and 'actants' (that is, non-humans) involved in construction of certification. The certification process should be guided by principles and criteria capable of securing the foundations on which local knowledge is created and re-created: for example that local varieties continue to be used. Norms and regulations, should allow space for flexibility and variation in local techniques, so that local knowledge can evolve and be renewed.

Local farmers, producers, local varieties, local techniques, local consumers, may be considered bearers of the local tradition of knowledge, Park authorities and the IGEA experts the bearers of the 'expert system'. But it is in practice difficult to assign clear cut systemic roles to each actor. Representatives from farmers' associations, for example, comply with different roles, as professional bureaucrats, bearers of managerial knowledge, or intermediaries among managerial and local knowledge.

Let us also take a closer look at the other actors: the IGEA experts. They belong to the local community and have relatives among the farmers. Some of them farm a small plot of land themselves, and they have in any case learnt in their family to taste local food. On the other side, local farmers have mostly attended formal courses of study: one olive oil producer in the process of valorisation of his product was able to stimulate new university research into the potentiality of a local olive cultivar to yield high quality oil; one of the most locally recognised bergamotto[14] grower is a former lawyer who in his 30s opted for going back to manage his grandfather's farm. Furthermore, in this case, 'experts' and 'lay' people share the same tacit knowledge.

Following the specific threads of knowledge and distinguishing the actors and actants of the construction process for Park certification, we find evidence of different overlapping traditions of knowledge and perceive difficulties in classifying the various traditions of knowledge as polar antitheses. It would be more accurate to acknowledge that actors embodying hybrid forms of knowledge are the rule rather than the exception.

This could be considered a favourable condition for a successful dialogue among traditions of knowledge in the process of construction of a certification. It is not, of course, a guarantee that the dialogue will develop on an equal footing and with mutual respect.

In cases in which a stronger division among traditions of knowledge persists, the dialogue needs the creation of a special context. The case of the Palizzi wine, in Aspromonte, offers an example of such a situation. Palizzi, like many other towns in the south of Italy, has its own tradition of vine growing and winemaking, but in the 1960s winemaking passed into the hands of professional oenologists,

---

14 A citron variety cultivated for cosmetic use and specific to the Aspromonte area for the characteristic it acquires in this natural environment.

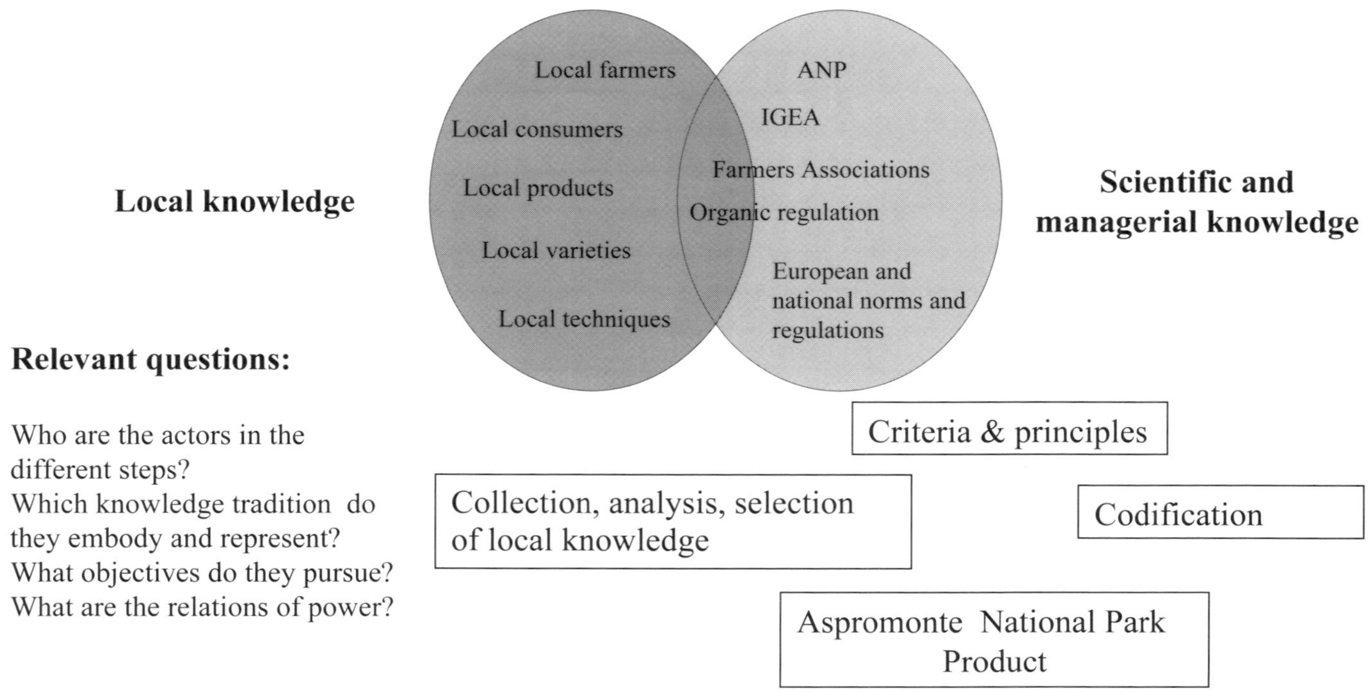

Local knowledge

Local knowledge (oval containing):
- Local farmers
- Local consumers
- Local products
- Local varieties
- Local techniques

Scientific and managerial knowledge

(oval containing):
- ANP
- IGEA
- Farmers Associations
- Organic regulation
- European and national norms and regulations

Collection, analysis, selection of local knowledge

Criteria & principles

Codification

Aspromonte National Park Product

Relevant questions:

Who are the actors in the different steps?
Which knowledge tradition do they embody and represent?
What objectives do they pursue?
What are the relations of power?

Figure 6.1 The construction of Aspromonte National Park certification as a hybrid forum

a process common to many European countries (Papadopoulos, in this volume, Buciega et al., in this volume). When the president of a local cooperative (an agronomist, also working for a farmers' association) wanted to start a dialogue about winemaking techniques between oenologists and local farmers he felt that there was a problem of communication. Local techniques are tightly intertwined with entire life-experiences of traditional farmers and critiques of technique may have been perceived as a personal offense by local winemakers: 'If I had told them [that I did not consider their wine of a good quality] I would have appeared as one who lacked respect for them ... I would never attempt to desecrate their work. A third person was necessary'. (Personal interview)

To get the dialogue under way the cooperative adopted a different approach. A famous oenologist from Tuscany was invited to stay few days in the village. Only after having established some personal contacts with farmers, was the foreign oenologist (thanks to his being at the same time a foreigner and a national authority in the field) in a position to evaluate and criticise the local winemaking practices without this being taken as a lack of respect. Finally, the dialogue had some positive effects, especially among young people.

It should be borne in mind that interaction and dialogue are not in themselves a sufficient condition for guaranteeing that asymmetrical power relations in the social, political and economic arenas will not prevail and lead to expropriation of local knowledge and the benefits to the locality that may accrue from it. Other conditions make this context favourable for dialogue and interaction in Aspromonte: there is no great polarisation in the economic structure of agriculture and agro-food; the economy and the territorial identity do not derive from a single dominant product but rather from a basket of products, each produced in relatively small quantities.

As we mentioned, implementation of Aspromonte National Park certification was interrupted by exogenous factors, and we can not follow up the social and economic consequences of its construction. Nonetheless following the certification process in its making sheds light on the interaction process of the many local actors and the different traditions of knowledge that is at the base of the construction of certification.

Conclusion

It may be fruitful to look at provenance certification not only as a vehicle of information to citizen-consumers, but also as an epistemic network or still better a hybrid forum, intended as a public space where to establish a dialogue and an interaction among different actors and different ways of knowing, functional to the project of an alternative agriculture.

In the construction of Aspromonte National Park logo for local food, the certification project was not finalised exclusively to local development, but rather intended as a civic action, meant to reassert the value of local identity, based on the specificity of place and culture. The Park as promoter of the initiative was a guarantee that social and environmental sustainability was a priority in the project. The presence of hybrid actors and overlapping traditions of knowledge was favourable to interaction among actors and knowledge forms. The negotiation, discussion and elaboration of the certification schemes led to a process of collective learning about local capacities and local resources and, especially, to recognising equal dignity to different forms of knowledge. There is no pronounced asymmetry of power resources among economic actors in the area, which probably prevented the formation of divisive conflicts on the topic of negotiation (area's boundaries, selection of exemplary farms, criteria for environmental sounds practices or the definition of technical protocols).

Rather, conflicts exploded at higher level (between national and local government) and were able to stop the experimentation of an important tool of local governance for rural development. Which reminds us that the 'local' is never totally self-contained, but always strictly intertwined to higher layers of society and economy organisation and governance and that the link between local and extra-local need to be taken into account in the construction of local food economies.

Notwithstanding the end of the Aspromonte National Park certification experimentation, this experience may teach that it is appropriate to analyse the provenance certification's construction as a hybrid forum, where local knowledge may gain equal dignity in the design of the future of rural communities and rural development. Further research may lead to discussions about the formal procedures needed to guarantee that certification processes function as space of dialogue among different ways of knowing, rather than as standardisation practices destined to expropriate local knowledge from local actors.

References

Akerlof, G.A. 1970. The market for 'lemons': quality uncertainty and the market mechanism. *Quarterly Journal of Economics*, 84(3), 488–500.

Anderson, M.D. and Cook, J. 2000. Does food security require local food systems? in *Rethinking Sustainability: Power, Knowledge and Institutions*, edited by J.M. Harris. Ann Arbor, MI: Michigan University Press, 228–248.

Barham, E. 2003. Translating terroir: the global challenge of French AOC labeling. *Journal of Rural Studies*, 19(1), 127–138.

Barjolle, D. and Thévenod-Mottet, E. 2003. Policies evaluation: general synthesis. *WP6 Report, DOLPHINS Project* (EU Concerted Action). Available at: http://www.origin-food.org/cadre/careport.htm [accessed: 11 February 2009].

Benvenuti, B., Antonello, S., De Roest, C., Sauda, E. and van der Ploeg, J.D. 1988. *Produttore agricolo e potere*. Roma: Consiglio Nazionale delle Ricerche – Istituto per la Ricerca nell'Agroalimentare.

Boltanski, L. and Thévenot, L. 1991. *De la Justification: Les économies de la grandeur*. Paris: Gallimard.

Bruckmeier, K. and Tovey, H. 2009. Conclusion: beyond the policy process: conditions for rural sustainable development in European countries, in *Rural Sustainable Development in the Knowledge Society*, edited by Bruckmeier, K. and Tovey, H. Aldershot: Ashgate, 267–287.

Callon, M., Lascoumes, P. and Barthe, Y. 2009. *Acting in an Uncertain World: An Essay on Technical Democracy*. Cambridge, MA and London: The MIT Press.

Castells, M. 2006. *The Information Age: Economy, Society and Culture*. Oxford: Blackwell Publishing.

Cavazzani, A. 2008. Innovazione sociale e strategie di connessione delle reti alimentari alternative. *Sociologia Urbana e Rurale*, 87, 115–134.

Farina, E.M.M.Q. and Reardon, T. 2000. Agrifood grades and standards in the extended Mercosur: their role in the changing agrifood system. *American Journal of Agricultural Economics*, 82(5), 1170–1176.

Feagan, R. 2007. The place of food: mapping out the 'local' in local food systems. *Progress in Human Geography*, 31(1), 23–42.

Fonte, M. 2008. Knowledge, food and place. A way of producing, a way of knowing. *Sociologia Ruralis*, 48(3), 200–222.

Fonte, M., Grando, S. and Sacco, V. 2007. *Aspromonte. Natura e cultura nell'Italia estrema*. Roma: Donzelli.

Giovannucci, D. and Reardon, T. 2000. *Understanding Grades and Standards and How To Apply Them*. Washington DC: The World Bank. Available at: http://www.worldbank.org [accessed: 24 November 2009].

Goodman, D. 2003. The quality 'turn' and alternative food practices: reflections and agenda. *Journal of Rural Studies*, 19(1), 1–7.

Grando, S. 2007. Tutela dell'ambiente e sviluppo locale: l'esperienza del Parco nazionale dell'Aspromonte. in *Aspromonte. Natura e cultura nell'Italia estrema*, edited by Fonte, M., Grando, S. and Sacco, V. Roma: Donzelli, 27–68.

Haas, P.M. 1992. Epistemic communities and international policy coordination. *International Organization*, 46(1), 1–35.

Haraway, D. 1988. Situated knowledges: the science question in feminism and the privilege of partial perspective. *Feminist Studies*, 14(3), 575–599.

Harding, S. 1986. *The Science Question in Feminism*. Ithaca, NY: Cornell University Press.

Harper, D. 1987. *Working Knowledge: Skill and Community in a Small Shop*. Chicago, IL: The University of Chicago Press.

Hatanaka, M. and Busch, L. 2008. Third-party certification in the global agrifood system: an objective or socially mediated governance mechanism? *Sociologia Ruralis*, 48(1), 73–91.

Hatanaka, M., Bain, C. and Busch, L. 2005. Third-party certification in the global agrifood system. *Food Policy*, 30(3), 354–369.

IAASTD, 2009. *Agriculture at a Crossroads, Global Report*. Washington DC: Island Press. Available at: http://www.agassessment.org/docs/ [accessed: 16 October 2009].

ISMEA, 2007. *Il mercato delle DOP e IGP in Italia nel 2007*, Roma: Ismea. Available at: www.ismea.it [accessed: 16 October 2009].

Jasanoff, S. and Martello, M.L. 2004. *Earthly Politics: Local and Global in Environmental Governance.* Cambridge, MA: MIT Press.

Josling, T. 2006. The war on terroir: geographical indications as a transatlantic trade conflict. *Journal of Agricultural Economics*, 57(3), 337–363.

Kloppenburg J. Jr. 1991. Social theory and the de/reconstruction of agricultural science: local knowledge for an alternative agriculture. *Rural Sociology*, 56(4), 519–548.

Konefal, J., Mascarenhas, M. and Hatanaka, M. 2005. Governance in the global agro-food system: Backlighting the role of transnational supermarket chains. *Agriculture and Human Values*, 22(3), 291–302.

Latour, B. 1987. *Science in Action: How to Follow Scientists and Engineers through Society*. Cambridge, MA: Harvard University Press.

Marsden, T. 2003. *The Condition of Rural Sustainability.* Assen: Royal van Gorcum.

Nowotny, H., Scott, P. and Gibbons, M. 2001. *Re-Thinking Science. Knowledge and the Public in an Age of Uncertainty*. Cambridge: Polity Press.

Oosterveer, P. 2007. *Global Governance of Food Production and Consumption: Issues and Challenges*. Cheltenham: Edward Elgar Publishing Ltd.

Padel, S., Röcklinsberg, H., Verhoog, H., Fjelsted Alrøe, H., de Wit, J., Kjeldsen, C. and Schmid, O. 2007. Balancing and integrating basic values in the development of organic regulations and standards: proposal for a procedure using case studies of conflicting areas. Final project report EEC 2092/91 Revision (Project number SSPE-CT-2004-502397). Available at: http://orgprints.org/10940/ [accessed: 24 November 2004].

PNA 2004a. *Piano Pluriennale Economico e Sociale*. Reggio Calabria: Ente Parco Nazionale dell'Aspromonte.

PNA 2004b. Tra Vitalità e Abbandono. Indagine sulle realtà socio-economiche del Parco, in PNA 2004a, 255–365.

PNA 2004c. Principi e criteri per la predisposizione e l'approvazione dei disciplinari di produzione dei prodotti agricoli freschi e trasformati del Parco Nazionale dell'Aspromonte, unpublished manuscript.

Reardon, T., Codron, J.-M., Busch, L., Bingen, J. and Harris, C. 1999. Global change in agrifood grades and standards: agribusiness strategic responses in developing countries. *International Food and Agribusiness Management Review*, 2(3), 421–435. Available at: http://www.ifama.org/tamu/iama/nonmember/OpenIFAMR/Archive/v2i3-4.htm [accessed: 24 November 2009].

Sillitoe, P. 2002. Globalizing indigenous knowledge, in *Participating in Development: Approaches to Indigenous Knowledge*, edited by Sillitoe, P., Bicker, A. and Pottier, J. London and New York, NY: Routledge, 108–138.

Sillitoe, P., Bicker A. and Pottier J. (eds.) 2002. *Participating in Development. Approaches to Indigenous Knowledge*. London and New York, NY: Routledge.

Silvestri, F. 1999. L' emblema del parco come strumento di promozione: dal marchio di prodotto al marchio d' area, *Parchi*, 27, 27–30. Available at: http://www.parks.it/federparchi/rivista/P27/index.html [accessed: 18 July 2009].

Tovey, H. and Mooney, R. 2007. *CORASON final report*. Available at: http://www.corason.hu/download/final_report.pdf [accessed: 16 June 2009].

van der Ploeg, J.D. 1986. *La ristrutturazione del lavoro agricolo*. Roma: Edizioni per l'Agricoltura.

Watts D.C.H., Ilbery B. and Maye D. 2005. Making reconnections in agro-food geography: alternative systems of food provision. *Progress in Human Geography*, 29(1), 38–46.

Winter, M. 2003. Embeddedness, the new food economy and defensive localism. *Journal of Rural Studies*, 19(1), 23–32.

Chapter 7
One Tradition, Many Recipes: Social Networks and Local Food Production – The Oscypek Cheese Case

Tomasz Adamski and Krzysztof Gorlach

Introductory remarks: setting up the problem

The general aim of this chapter is to analyse the tendencies connected with the phenomenon of 'local food' in Poland. The results of the research carried out as a part of the CORASON project show the ways in which the food-chains built around 'local food' are created or changed and what the relation is of such tendencies to the sustainable development of the countryside.

The issues related to local food production are analysed in the context of the Malopolska region (the southern province of Poland). This particular region has perhaps the richest local food tradition in Poland. This fact, together with a tourist boom, drives a real renaissance of traditional cuisine in this area. One can find many examples of successful food products embedded in the local tradition, with oscypek cheese being the most famous. The authors try to identify factors that allowed Malopolska to revive its food heritage. That includes the need to mention the unique culture of Podhale highlanders and an examination of the regional government's policy towards promotion of local products.

The case study presented in this chapter focuses on oscypek cheese from the Podhale area in the Malopolska region. oscypek – a smoked cheese made from sheep's milk or a mixture of cows' and sheep's milk – has become the best known example of regional food in Poland. This cheese is of great cultural significance, as it is a part of the shepherd's tradition and is at the same time a key tool used in preserving the shepherding heritage. On the other hand, oscypek cheese also represents a high economic value, constituting the base of a large and diversified production and retail sector. Its significance for the regional economy is hard to overestimate and can be illustrated by the involvement of the regional government in this question. Finally, oscypek is present in three different, coexisting food-chains. For the purposes of this paper we have labelled them as follows: (1) 'oscypek as a souvenir for mass tourism'; (2) 'oscypek as a conqueror of the food market'; (3) 'neo-traditional oscypek in European Union (EU) realities'. In each of these networks a different form and perception of oscypek can be identified. What is more, these chains have undergone dynamic changes over the last decade.

For the above-mentioned reasons the case of oscypek cheese provides a unique opportunity for identifying and analysing how different types of knowledge (expert, managerial and local) and their combinations can generate various different currents in local food production.

'oscypek' cheese as food product seems to be one of the very first beneficiaries of the initial period of Poland's membership in the EU. But it should be stressed that its success story did not begin on 1 May 2004, when Poland, together with nine other countries, became a member of the EU. Quite to the contrary, oscypek has been the longitudinal element of the history of the Podhale region and has lately become one of its icons. The Podhale is recognised in Poland as the centre of the very peculiar highlander tradition and shepherd's culture. That is why consideration of major characteristics of the region seems to be a necessary element in any attempt to analyse the situation with oscypek. Currently one can buy oscypek not only in its place of origin, that is the Podhale area. Oscypek cheese can be bought from the local street vendors selling their products along the main roads to the Tatra Mountains. Oscypek can also be purchased on the streets of several of the bigger cities in the south of Poland as well as in shops with organic and health food. Quite recently it has become possible to eat specially prepared and served oscypek cheese even in elegant restaurants, and all over Poland. These listed distribution channels comprise a complicated system, with oscypek circulating inside several alternative food chains.

An investigation of oscypek cheese in its social and economic context must begin with a brief description of the cheese itself. oscypek is a smoked cheese made of sheep's milk or a mixture of cow's and sheep's milk. It is an important part of the shepherding tradition, with a history going back to the fifteenth century (Cichocki and Kozak 2003). For hundreds of years it was produced in the mountains by local shepherds. It is of crucial importance that the cheese should be hand made in a mountain shed using non-pasteurised milk. The recipe is passed on from generation to generation by oral tradition, which makes the final product unique for each shepherd who makes it. As local highlanders say 'there is one technique of production and many recipes'. What can differ is the proportion of ingredients, temperature and duration of smoking, as well as its final size and consistency (Rybak 2001). What is common for all oscypek cheeses is the fusiform shape. In its original form, oscypek is produced and eaten in the summer season (from May to September/October).

As this short sketch shows, oscypek cheese was an intrinsic part of shepherd's culture. It was an important source of food in remote areas, with the mountain shed being a self-supply base for teams of shepherds. At the same time, however, it was also a product made for sale. Herdsmen provided it to local villages, in this way acquiring an additional source of income. Oscypek naturally became a popular tourist commodity when tourism appeared in the Tatra Mountains. This is how the situation remained until the fall of communism: oscypek was traditionally produced as an auxiliary activity to sheep breeding, and eaten by locals or sold to tourists. That was the starting point for the significant changes that occurred in the 1990s and which continue today.

To focus on the issues considered above, oscypek arguably highlights three basic dimensions of local food production, distribution and consumption, namely: the relation to regionality/territoriality, the social/political construction or definition of the product and its characteristics, and, last but not least, various forms of production-consumption chains.

Food production-consumption debates and rural development: some theoretical issues

Theoretical debates over food production and consumption seem to involve at least three basic streams of discussion and thought. The first is connected to de- and re-territorialisation of food production and consumption and links to a more general debate on industrialisation and globalisation of the agro-food system. Industrialised agriculture produces standardised mass products which, thanks to globalised processing and distribution networks, become de-territorialised commodities sold all over the world. Some producers in turn try to counter such tendencies, marketing territorially-characterised products under various labels that stress their non-standard character. As Tregear says, such products are offered as typical, origin labeled, traditional, regional speciality, artisanal, special quality or special farm products. All evidently have a common feature: 'special characteristics relating to territory' (Tregear 2003: 91). The use of 'special' and/or 'traditional' in the label denotes, in our opinion, a countertendency to de-traditionalisation in agro-food networks.

Another important issue should be mentioned in this context, namely 'quality'. As Goodman (2004: 4) has pointed out: 'In broad terms, the catalyst and fundamental theme of the Western European AAFN[1] literature is the perception of a "turn" by consumers away from industrial food provisioning towards quality'. The growing concern about quality makes the topic of food production methods a significant subject of social debate. Generally speaking, the notions of territoriality, quality and tradition introduce a new conception of food that seems to be an alternative or a challenge to the standardised products perceived as commodities produced in the industrial agricultural sector.

Such a contradiction between alternative 'models' of food seems to have a few important consequences. It leads to a peculiar duality in the agro-food system built upon two different types of networks. As Holloway et al. (2007: 2) claim, the debate on food economies and food production-consumption relationships is focused on 'a comparison between 'deterritorialised' conventional food networks and 'reterritorialised' alternative food networks'. This duality has been conceptualised in a slightly different way in European and North American rural

---

1 Alternative Agriculture Food Networks.

sociological literature. These differences have been perceived and evaluated by Holloway et al. in their seminal paper, as they stress:

> In the European tradition, alternative food networks tend to be discussed in terms of their potential to contribute to the survival of small businesses (particularly farms), and more widely to processes of rural development through processes of adding value in various ways to farm outputs, [while], in the North American literature, ... 'alternative' has tended to be used in a rather more politicised discourse of oppositional activism (Holloway et al. 2007: 4).

However, in our opinion, these two options should not be treated as disjunctive. In both perspectives, that is the European as well as the North American, the emphasis is being put on the impact of food production on various economic, social and political processes observed in the area of agriculture and rural communities. The only difference is the stress on questions of rural development (in the case of the European literature) and the issues of political discourse (in the North American case). But even bearing in mind this general distinction, it is arguable that the alternative food networks, connected as they are in the European tradition to the issues of rural development, are in some cases framed in the new social and political language of opposition to the agro-industrial and post-productivist regime. Their discourse is that of the 'contested countryside' (Marsden 2003). Moreover, it seems that in both approaches new/'alternative' food networks are treated as embedded in their social and political contexts, quite contrary to the 'conventional' food networks, which are seen as an integral part of the Fordist agro-industrial system.

Taking into account the Fordist character of conventional food networks, alternative networks can be treated as examples of post-Fordist regimes in agricultural production, food distribution and consumption. But is such a sharp distinction an appropriate conceptualisation for local food? From the analysis of the oscypek case, which will be further elaborated below, we can hypothesise that actually a kind of multi-level food network is under construction. As oscypek is offered by local producers at local markets in the Podhale region and at the same time can be bought in shops or in conventional supermarkets in other regions of Poland, what is involved is really a matter of different circulation networks for one, single product within both alternative and conventional food production-consumption systems.

Some authors try to conceptualise the same issue in a slightly different way. Let us note for example, the interesting consideration presented by Fonte and Grando (2006). They distinguish two general models of food chains, namely: 'local production for local market' and 'local production for global markets'. The first model, based on the 'local production-consumers net-chains', is characterised by a strong link with the territory. In this case, territory functions as 'the common ground of specificity, culinary culture and knowledge' (Fonte and Grando 2006: 7). As a result, relations between producers and consumers are non-formalised.

This applies both for production techniques (based mostly on tacit and lay local knowledge) and for the approach of consumers (key role of trust, personal contact, tacit knowledge and no need for official certificates confirming quality). In turn, the second model is formed by the 'local production-consumers net-chains'. Its distinctive feature is the significance of certification systems, which allow distant consumers to recognise the 'locality' of food products. The links between food and the territory of its origin are translated into the language of formal certificates that are understandable by extra-local actors. This tendency leads to the involvement of managerial and expert knowledge. Because the producer-consumer relations have become de-personalised one can observe the process of institutionalisation of trust. In the organisational dimension this model is characterised by the extension of food chains. 'The product can "travel" to distant markets, and still maintain its local identity' (Fonte and Grando 2006: 8).

In the previous part of this chapter we mentioned the notion of territorialisation as a key characteristic of local food products. However, this is merely one of many dimensions of this topic. One other dimension that we want to explore has to do with the consumption rather than the production side. Such a theoretical perspective focused on consumption opens up several interesting areas of analysis. Let us point to at least some of them, especially those that are mentioned in the literature. The first pertains to evaluation of social position. Consuming a particular type of food can be seen as an activity that strengthens or weakens social status. As Jordan (2007: 24) put it: 'what we put in our mouth is reinforced by and reinforces our social standing relative to others'. The same author develops this consumption motive further, stating that: 'these cases demonstrate … how the class and status entrance requirements to the spaces of consumption enforce distinction' (Jordan 2007: 25) and, elsewhere: 'the centrality of the spatiality of consumption becomes clear – where an object is consumed matters, and the creation of markets and tastes for such goods has a distinctly spatial aspect' (Jordan 2007: 37). To sum up, what we eat and where we eat seems to have an important impact on the way we are perceived and evaluated in society. However, we would argue that such an approach does not seem to be comprehensive. It is concentrated mostly on the spatial characteristic of food reconstructed on the basis of the attributes of the place where it is eaten (for example, a fancy or exotic restaurant). We argue that the underestimated characteristic here is that of the local context (community, region) where the particular product has been created (we purposely use the term creation instead of talking about 'production' in order to emphasise the social aspect). In other words, coming to the traditional leisure region (for example Podhale) and eating the local food (for example oscypek cheese) bought directly from the shepherd or from the street vendor seems to have less to do with the distinction of social status or class (Bourdieu 1984) and much more with the 'totality of experience'. This is linked to the popular idea of the tourist adventure, according to which visiting a holiday destination should be combined with taking part in the widest possible range of local attractions, for example skiing, trekking, climbing, taking a bath in hot geothermal lakes and – last but not least – eating local food.

This theoretical perspective has an important consequence for the analysis of the oscypek case in the following parts of this chapter. It leads us to the careful analysis of spatial and temporal context of oscypek cheese. We agree with Jordan when she stresses: 'that the study of taste, food and culture can, in some cases, benefit from the systematic inclusion of space (and time) into the analysis' (2007: 38). Therefore, the characteristics of the Podhale region, as well as the story of oscypek itself, will be taken into consideration below.

Temporal and spatial contexts of the case

In the broadest terms, the situation of oscypek cheese is to be seen in the context of food culture transformations at the national level. Investigating this issue, one is forced to acknowledge that Poland has a rich and diversified culinary culture. Many regions, having been influenced by different historical factors, are characterised by distinctive forms of cuisine. This heritage of traditional and regional products was endangered during the communist period. In the realities of post-war food supply shortages, national authorities promoted large scale production of cheap and accessible food. In the state-controlled economy there was no place for a variety of local cuisines. The principles of 'socialist equality' were also applied also in the realm of food. Luckily, despite the state-driven homogenisation, traditional food specialities survived in Polish families. What needs to be underlined is the fact that they were able to survive mainly due to rural communities. The Polish countryside succeeded in resisting the attempts at collectivisation, and the traditional mode of agriculture (family farms) survived and helped preserve the rural heritage. The situation changed after the transition to the market economy in 1989. Following decades of isolation and empty shelves in the shops, the national market was flooded by large amounts of foreign and relatively cheap food. The patterns of consumption in Polish society shifted towards 'Western' foodstuffs, which were a symbol of the new realities. The appearance of the international supermarket chains that have conquered the retail sector has additionally strengthened this tendency. The rapid 'Westernisation' of food habits turned out to have a negative influence on local food production. The market was dominated by the corporate grocery chains, for which quality was a secondary consideration to low price.

Such a state of affairs has begun to change in the last few years. Firstly, consumers are becoming more concerned about food safety (visible influence of foot and mouth disease, mad cow crisis and so on). Secondly, with the increasing wealth in society more money can be spent on a variety of more sophisticated comestibles. The appearance of shops with healthy, organic food, the proliferation of vegetarian restaurants, are good illustrations of this tendency. The third factor is linked to the development of tourism. In the popular tourist regions (for example the Podhale) a significant increase can be observed in the number of restaurants/ eateries serving regional cuisine in the folk style. It would even be possible to describe this development as a fashion for local food, as in almost every Polish city there is a place where the local version of 'peasant food' is being served. Short food-supply chains are becoming increasingly common as more and more consumers are prepared to pay for the 'original taste with local identity'. The popularity of regional and traditional products is being supported by a number of projects. These include both governmental projects (at the national and regional level: for example 'Agro-smak'; 'The Taste of Malopolska') and NGO (non-governmental organisations) initiatives (for example Slow Food Poland). It needs to be stressed, however, that Poland is still only at the beginning of the process of preserving and promoting its culinary heritage.

The Malopolska region has perhaps the richest local food tradition in the country. Along with a tourist boom, this region is witnessing a real renaissance in traditional cuisine. oscypek cheese is just one of many examples of successful local traditional food products. There are a number of factors that can help to explain why Malopolska has managed to revive its food heritage. The first has to do with the structure of agriculture: one distinctive feature of the region is the very small farm size – on average, farms cover only 2.6 hectares (the average in Poland is about 7 ha). Over 90 per cent of farms are smaller than 5 hectares, with farms of the relatively optimal size (15–20 ha) comprising only 0.2 per cent. Another factor seems to be the large number of plots within farms that extend over a large area. This arrangement is rather static since only one to two per cent of total farms annually change in size. In many areas plots are even further subdivided into smaller parts (Województwo Małopolskie 2004, available at www.malopolskie.pl, accessed: 25 November 2009)

Paradoxically, such a state of affairs favours the development of distinctive local food products. Because Malopolska has such a disadvantageous agrarian structure, the majority of farmers have been forced to adopt a strategy of self-sufficiency in food and to specialise in various types of niche production. Extensive local agriculture thus provides a variety of strongly individualised foodstuffs. Other positive factors are the well-preserved cultural heritage of the countryside and the marked regional identity, both of which support a wealth of culinary traditions. These elements are additionally strengthened by a vigorous tourist sector, which is a driving force behind local food valorisation in Malopolska. It should be mentioned however that the influence of mass tourism is not only positive as it also creates conditions favoring 'falsification' of the original local food products.

Podhale, the specific area of Malopolska examined in the study, is situated at the foot of the Tatra Mountains in the southern part of the province. Due to its natural features and cultural heritage, the Podhale is significantly different from the rest of the country. The region's mountainous character sets it very much apart from the plains of central and northern Poland. The region has been settled since the twelfth century, being a melting pot of the Polish, Romanian, Hungarian, Slovak and Rusyn cultures. These elements gave rise to a distinctive Podhale Highland culture that remains vital in language, regional costumes, shepherd traditions, literature, folk music, glass-painting and handicraft (see: http://www.podhale.z-ne.pl).

In the social description of the region, mention must also be made of the important role of religion (the Roman Catholic Church) and the spirit of independence that is such a significant feature of the local mentality. Traditionally the economy of the Podhale was based on agriculture and animal husbandry (above all cattle and sheep breeding). But due to the harsh conditions of life in the mountain areas a significant part of the local population was living in poverty, which became a driving force for mass emigration (for the most part to the United States). From the second half of the nineteenth century onwards, the area began to attract a growing number of tourists. This particular branch of the local economy continued to develop in the interwar period and under the communist regime, but the real boom occurred in the 1990s. Nowadays the Podhale region is among the country's most popular tourist destinations. It is estimated that around two million tourists a year come to visit one of the several national parks, to hike or to ski in the winter. The main centers of the region (Zakopane, Nowy Targ, Białka Tatrzańska, Bukowina Tatrzańska, Biały Dunajec) are also Poland's top skiing areas.

Discussion: three oscypek networks

In trying to understand why oscypek became a local food icon in Poland one has to consider two key factors. Firstly, with the transition towards a free market economy, many sectors of Polish agriculture experienced a deep crisis in the 1990s. This was especially visible in sheep husbandry. Losing their state subsidies, breeders faced a rapid rise in the price of inputs and the means of production, while at the same time the price of wool fell dramatically. It is estimated that as a result of this there was a fivefold decrease in the number of sheep in the Podhale region, from about 250,000 to only 50,000 ewes (1990–2005). In terms of oscypek cheese, this meant a huge reduction in supply, as only a limited amount of the primary raw material, namely sheep's milk, was now available.

At the same time, the Podhale was experiencing a real tourist boom. The number of visitors to the Tatra Mountains increased significantly through the 1990s. The whole infrastructure of the tourist sector has also undergone extensive development, largely in parallel with the expansion of skiing. This, in turn, led to a transformation in patterns of tourism. The winter holiday season has been extended and has become the busiest period of the year. As for oscypek, these changes opened a great potential market, but there was a complication – the window of sales opportunity did not coincide with the production season (May-September). As a result, only a very limited supply of oscypek cheese was available to meet the growing demand.

This triggered a revolution in the oscypek cheese market. It involved 'commercialisation of tradition', driven by economic realities. 'Commercialisation' in this context meant improvements in recipe. The crucial shift was from sheep's milk to cows', a modification that enabled sellers to overcome the two main structural obstacles. The limited number of sheep was no longer a problem.

'Improved' oscypek could be produced all year round and therefore in winter also. Furthermore, since oscypek had been decoupled from shepherding, it could now be made on farms without sheep. This led to further departures from the original technique, with the traditionally lengthy process of smoking often being reduced to just several hours, or even replaced by the procedure of steeping cheese in a tea-based blend. By implementing all these 'improvements' the local community has created a large and decentralised sector for home oscypek production, located almost entirely within the sphere of the black economy.

A somewhat similar situation prevails on the retail side. A large and decentralised small-scale retail sector is visibly growing in the main traffic streets. Street stalls and food markets have become the predominant sales points. Local people selling oscypek on the roadside, on fold-up tables or directly from their cars have become an integral part of the Podhale landscape. So many people wanting their own stall on the street has caused serious traffic problems, so much so that local authorities have introduced a method of drawing lots as a basic way of assigning space to the oscypek stalls on the main street of Zakopane (the biggest resort of the Tatra Mountains). Each year the national media report social conflicts following the announcement of the annual stall-space lottery results in Zakopane. Oscypek sellers who are denied a space for a stall seek 'justice' by complaining to the press, politicians or even to a local church.

This 'fake' oscypek network is in fact something like a self-sustaining economic system. It is a special case of local production for local markets but it is targeted to non-local consumers. In this food-chain, oscypek cheese has become a kind of souvenir for mass tourism. The product marketing is targeting 'distant' consumers who buy the commodity while visiting the Podhale region. The production-consumption chain is short but the ties between seller and consumer are rather weak. It is normally a single transaction, after which the two sides will not meet again. As far as power relations are concerned, it seems to be a producer-driven process. With demand still exceeding supply, producers are not under any great pressure. The main factor that shapes these relations is price. Being sold on the street, oscypek needs to be reasonably cheap so that the average tourist can afford it. There can therefore be no grounds for assuming any process of quality negotiation. It is frequently only in name and shape, therefore, that there is any resemblance between the 'fake' popular oscypek cheese of the street and the original. The 'fake' version is moreover pushing the traditional oscypek out of the market. According to estimates, only about 10 per cent of cheese being sold under the name oscypek in the Podhale region is actually made from sheep's milk, in line with the traditional technique. Tourists are being cheated, but they do not seem to complain as long as the commodity is affordable and does not threaten their health. Many experts, local politicians and, above all, herdsmen, are trying to reduce the power of this black market. This is very difficult, though, because both the production and the retailing are a significant source of income for a large number of local people. Given that the conditions for agriculture in the Podhale region are unfavourable and plots are very small, production of the 'popular' type of oscypek cheese is one of the very few remaining economic alternatives for local farmers. In some villages, especially those of lesser tourist potential, around half of the village inhabitants are implicated in this network.

Several tendencies seem recently to have been leading in the direction of contraction of the market for the 'fake' oscypek cheese. On the one hand local authorities are trying to bring both producers and retailers under greater control and on the other a gradual rise is taking place in consumer awareness. These two trends should eliminate the most blatant examples of falsification and force improvements in the quality of the oscypek.

The second distinctive food network was created in reaction to the above-mentioned 'oscypek boom'. The idea was to take advantage of the popularity of oscypek on the wider that is non-local market. Thousands of tourists, visiting the Podhale region and sampling oscypek there created demand for this cheese in the rest of the country also. Many people simply wanted to eat oscypek after their return from holiday. The key strategy of this second network was to satisfy the new demand. It should however be pointed out that two alternative ways of achieving this goal are to be distinguished within the same analytical category. Both target distant consumers, but in different ways. The first element in the network is oscypek cheese as a product of dairy factories. Several small and medium-sized dairies in Podhale have introduced into the market a machine-made cheese that resembles the original oscypek in colour, shape and name. As a result, one can buy such oscypek-like cheeses in supermarkets all over Poland. But, the use of machines and vacuum-sealed packages represents a drastic departure from the traditional character of oscypek. Above all it involves the use of cows' milk, and pasteurised cows' milk at that. Although highlanders say that 'such a product is anything but oscypek' (Tomczak 2003), more and more consumers buy it. One of the reasons for this success is the fact that dairies can guarantee food safety. Products are labelled, hermetically packed and tested regularly. The name of the producer assures traceability. And, last but not least, supermarket-type oscypek is very accessible. It is being distributed through the system of grocery shops and supermarkets. This, then, is a distinctive oscypek network, not only in terms of the product form (machinery-made, pasteurised milk) but also in terms of food chain construction. These features make it almost inevitable that this version of oscypek will generate controversy. The view can be heard that the commercial success of this cheese is achieved on false pretences. This is a critique made mostly by highlanders opposed to local businessmen having the right to use the name oscypek for this cheese of theirs that is produced in a dairy factory. They regard this practice as amounting to theft. The name is crucial as it imparts to the cheese in the shop all the charm of mountain food. But it is possible to adopt a more creative approach to the name issue. The biggest producer of factory-made smoked cheeses from Podhale uses the name 'scypek' to market its products. Although the difference in the name amounts to only one letter, this is enough to avert accusations of illegal competition.

The food-chain underlying this second network is not short but long. Power is shared between private (producer) companies belonging to the region and retailing units from outside. Quality is secured through extensive innovations in the recipe, with the cheese's local identity being underwritten by the name oscypek (or a similar word). This variety of oscypek has almost nothing in common with its shepherd roots. But there is no real conflict of interests between this network and the one previously mentioned. The two food chains simply embrace alternative market niches. They do not compete with each other.

As 'market conqueror', oscypek in fact targets another group of customers. Again, the customer to whom it is being offered is from a faraway place, but this time it is an offer in the 'luxury' mode. Two main distribution channels are involved: restaurants serving various kinds of traditional Polish cuisine and organic food shops. Oscypek here assumes the character of an exclusive product for selected customers who are willing to pay for regional delicacies. The main objective is to sell cheese that can be associated with a traditional mode of production. It should be stressed however that a specialised food supply chain is by its nature something quite different from the production process that starts from the local shepherd's shanty. Several difficult challenges are therefore involved in this particular marketing project. The most important has to do with national regulations on food markets. For many years it was illegal to sell traditional oscypek in shops because it is made from unpasteurised milk. The requirement to use only pasteurised milk in dairy products remained a fundamental principle of food hygiene regulations. Since, as the Highlanders claim, oscypek cannot be made in this way, the product, of necessity, had to move to the 'food underground'. The conditions changed with legislation introduced by the Ministry of Agriculture in 2002. Since then it has been possible to produce and sell several kinds of dairy products containing non-pasteurised milk. The so-called Kalinowski decree (Jaroslaw Kalinowski was the head of the Ministry of Agriculture at that time) opened a new era in the oscypek retail system. Nevertheless, obstacles continued to be erected. The products were allowed into the market, but producers were obliged to obey very strict hygiene regulations. Given the realities of Podhale they were difficult to enforce, for a number of reasons. First, it was often impossible to ensure hygienic conditions in a mountain shed (because of lack of infrastructure, the temporary character of the building and so on). Additionally, social factors played an important role. Known for their independence and distrustful attitude towards the communist regime, the Highlanders were reluctant to accept something they saw as enforced regulations. Moreover, based on knowledge derived from a long tradition of oscypek production, they treated the new requirements as superfluous. The shepherds' argument was as follows: since oscypek had been made in a 'primitive' way for centuries and 'no one ever got sick', this proves that 'natural conditions' for the cheese production are quite safe for potential consumers.

As a result, only a small number of local oscypek makers are involved in the above-mentioned food supply chain. They form kind of 'elite' of the most entrepreneurial breeders. They have not only made investments enabling them to fulfil all the hygiene requirements; they have also introduced certain marketing strategies. None of this could be successful without the promotion of regional products. 'Modern herdsmen' are thus engaged in publishing promotional leaflets, presenting their products at expositions or fairs, carrying out business negotiations with shops and restaurant chains. In analytical terms this activity corresponds to the logic of the modern, consumer-driven food supply chain of local production for distant clients. It has a decentralised character based on individual contracts between producers and retailers. Given that the product is marketed to a special group of consumers (with a high level of food safety awareness) the quality is the main object of attention. The retailers are co-operating only with selected oscypek makers who can guarantee the proper features of the final product. The value of regional food is also important in this arrangement, so cheeses are specially labelled to ensure both traceability and conformity to the traditional method of production.

The third network described in this chapter and built around oscypek cheese is also the most recent one. Its current profile has emerged from the practice of certifying traditional local food and is also dependent on the role of oscypek as a symbol of regional tourism. New categories of actors are involved in it, something that serves to distinguish it from the other previously described arrangements. The dominant role of local authorities and the collective representation achieved by the participating breeders has given this network a more organised character.

Let us commence our account by examining the phenomenon of tourism. The tourist boom in the Podhale mentioned before has been to a large extent a spontaneous process. Nevertheless, local authorities did finally begin to adopt a strategy of supporting and strengthening the development of the tourist sector. The region was promoted both through professional marketing campaigns and through creating new attractions for visitors. Oscypek cheese was chosen as a central motif in this strategy. A special event called 'the oscypek festival' was introduced into the calendar of annual events organised in Zakopane by the local government. During this festival, visitors can witness the traditional method of cheese making, listen to folk music and buy the oscypek cheese. Apart from that, oscypek is also an obligatory element in almost every larger public event in the Podhale during the tourist season. It has become an attraction in its own right. In cooperation with a number of herdsmen the local authorities have created the so-called 'oscypek route'. It is a tourist 'product' that includes visiting selected places where the cheese is made and getting to know its history and production technique. The possibility of staying overnight is also offered. The oscypek farms are specially marked and the whole network is promoted in informational leaflets (see *Oscypkowy szlak*, available at http://www.tatri.pl, accessed: 25 November 2009).

The famous smoked cheese is present on postcards and other materials promoting the Podhale region. It is without a doubt one of the most recognisable symbols of the Tatra Mountains and the town of Zakopane. The situation with the network is interesting in the sense that while on the one hand it is the oscypek cheese that is being promoted, on the other that promotion is a vehicle for advertising the whole area. This dual relation can be observed in the case of the contest organised by local authorities to develop a suitable packaging for oscypek. The primary aim was to get the cheese packaged as a kind of sophisticated gimmick to be used in a promotion campaign for the region. But in the longer perspective, when the new packaging is introduced on a larger scale it will serve as another factor in the process of increasing the popularity of oscypek. In the network described, the involvement of institutional actors is not limited only to co-organisation of the retail system. It is also visible in the production aspect. Given that local authorities are taking part in the promotion of the oscypek cheese, they have also decided to take action that might ensure the proper quality of the product. This is partly a reaction to the phenomenon of the 'fake' oscypek. In order to prevent potential misuse, the government launched a special commission in the late 1990s to regulate the herdsmen's production (the measure did not include the 'grey sector' of small-scale production). The commission includes representatives of authorities, veterinary experts and the association of sheep breeders, but it has no legal mandate to punish the refractory cheese makers. The only instruments it has at its disposal are advice (suggesting improvements) and awards to the best producers. The influence of this 'social' commission is thus rather limited.

The above may give some idea of the variety of forms in which oscypek cheese is present on the market. The fact that one and the same product circulates in networks representing opposite sides of the food system seems especially interesting. On the one hand oscypek can be found within the conventional food network (machine-made and vacuum packed cheese in supermarkets); on the other hand it is also being sold in the alternative chains. The question arises of whether these different networks can co-exist in a longer-term perspective and whether the status of multifaceted local food can be retained permanently by oscypek or whether is it just a transitory phase. We shall endeavour to address these questions in the following sections of the chapter, introducing the concept of knowledge dynamics.

Knowledge dynamics in local food production

When theorising on local food products one should recognise the variety of functions that food plays in social life. Arguably, food can sometimes be nothing more than a cultural object (Jordan 2007). Durkheimian totemic qualities could likewise be attributed to it (Goodman and DuPuis 2002). But to develop such a wide-ranging theoretical framework leads to further discussion covering various dimensions of the production-consumption debate in agro-food studies. Quite recently Holloway et al. (2007) listed several 'analytical fields' in food studies: site for food production, food production methods, supply chain, arena of exchange, producer-consumer interaction, motivations for participation, constitution of individual and group identities. Such a cluster of fields of interest has an impact on the changing perception of food. As Marsden (2000: 28) stresses: 'we need better models of *food governance* which build upon a more asymmetrical and differentiated understanding of food as a natural, social and political construction'.

Such a theoretical position implies that we need to focus on the various discourses and different types of knowledge used by various actors in the process of constructing food products. As a result of the plurality of analytical fields, the problem of integration emerges as one of the most crucial in the whole debate. Analysis of struggle over the various types of knowledge in different types of social and political food construction processes seems to offer such a tool for integration. In other words, a bridging approach of this type can be built on the theorising about power relations that might be observed in the various areas of agro-food studies. As Goodman and DuPuis put it, in their endeavour to illuminate this key issue: 'By integrating these arenas of struggle over knowledges, we can begin to bridge the divide separating perspectives which "know" food only as either Marxian fetish or Durkheimian totem' (2002: 6).

The problem of interactions between various social actors involved in food production/consumption processes is linked to the issue of different perspectives corresponding to different experiences, value systems, logics of acting and so on. In other words, it can be described as different kinds of knowledge brought by particular social groups (Bruckmeier 2004).

For analytical purposes one can distinguish three general types of knowledge: expert, managerial and local, this last one differentiated into lay and tacit (see the Introduction to this volume). The first conclusion resulting from our analysis of the oscypek case is that the three networks built around oscypek cheese are different in terms of knowledge dynamics. There is no doubt that oscypek in its original form is embedded in traditional local knowledge. Being a part of shepherd's culture, the production of the cheese has for centuries been based on lay knowledge, meaning unwritten recipes, production techniques and so on. But the first network seems to exemplify change that took place in recent decades and was especially visible in the 1990s. Oscypek turned into a mass tourist souvenir product that had very little in common with its original form. Both the recipe and the production methods have been 'improved' in order to bring more income to the local population. This involved a transformation of traditional local knowledge under the pressure of economic realities into a form that can be identified as adaptive local knowledge. This adaptive version of local knowledge in fact represents a misuse of tradition. It amounts to a forgery of the legacy left by previous generations which, in the short term, allows some groups to obtain substantial economic benefits. The dynamics of the transformation is illustrated in Figure 7.1.

Such modification of the knowledge basis allowed a number of new actors to come into the oscypek network. Local people who are not involved in sheep breeding are now producing and retailing 'modified' oscypek which helps them to supplement their families' incomes. Despite the opinion of many people that this is an example of the 'commercialisation of tradition' it could also be regarded as an example of adjustment strategy, a reaction by the local community reaction to

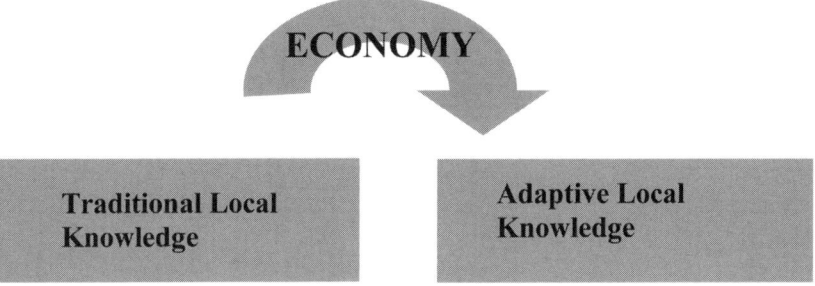

Figure 7.1   From local to adaptive knowledge

some of the challenges and opportunities that have resulted from the development of mass tourism.

The second network described in the analysis is characterised by a different knowledge dynamics, with so-called 'external' (scientific, managerial) knowledge now playing an important role. As far as mass production of oscypek is concerned, expert managerial knowledge is required to manage the dairy factory or market the products. This is a job for large and medium-sized business companies, so that objectively the position of local actors in the network is weakened vis à vis non-local ones. Moreover in its new 'reconstructed' form, the oscypek looses its local character. The Highlanders say that the oscypek has been hijacked by the 'big' producers but at the same time they add that it is not the 'real' oscypek but an 'run-of-the-mill imitation'.

In an alternative view, the transformation amounts to oscypek becoming a delicacy for a special group of customers. External managerial knowledge is also visible here, mainly taking the form of promotion of this particular food product. Only an 'elite of herdsmen' are involved in the organisation of this particular food supply-chain, modern shepherds who represent a mixture of local know-how and expert business knowledge. They belong mostly to the younger generation, open to market opportunities and not afraid of making changes in the tradition of their 'fathers'.

The last network has been built on yet another type of knowledge dynamics, with managerial/administrative knowledge playing a dominant role. Intensive involvement by institutional actors has led to centralisation of the food chain and to a shift of power away from individual producers. This is a real organisational revolution, as production and sale of oscypek cheese has heretofore been the domain of independent Highlanders. Along with the certification process, the crucial position in the network was assigned to the institutional actors, in particular the breeders association. It is worth mentioning that the Association of Sheep

Breeders (ASB)[2] does not represent the whole community of shepherds producing oscypek cheese. It is a voluntary organisation that was created by a group of market-oriented shepherds involved in large-scale production. Many of the individual producers have not joined the association and function independently. They dislike the idea of supervision by authorities or organisations. Some sheep breeders, especially of the older generation, oppose any changes in methods of production. They disregard suggestions that in the case of oscypek some of these changes might increase food safety. In this third network involvement of institutional actors is not limited to co-organisation of the retail system. It is also visible in the production aspect. Local authorities are taking part in promotion of the oscypek cheese, and they have understandably also decided to undertake actions that would ensure the quality of the product. The agency that has responsibility for this policing of the newly introduced system of EU approved control is the 'social' control commission mentioned earlier in the text. The dominant role of institutional actors derives from the issue of rights to use the oscypek name. Previously it was a regional common good and every producer could use it (hence the existence of three oscypek networks). Since official registration, access to the oscypek trademark has been strictly limited and possible only by permission of the Association of Sheep Breeders. As the name is the key to market success, this bestows great power on the institutional actors.

Several other problems and difficulties have arisen in connection with the registration/certification procedures. These are described in detail below as this is perhaps the most significant issue with a bearing on the immediate future of oscypek cheese.

The idea of certification was launched in the 1990s. The ASB decided to seek legal permission to retail oscypek and had formulated a set of rules to be adhered to during the production process. The application was unsuccessful at the time so the regulations remained unfinalised. The problem returned in 2002 when the media revealed that the name 'oscypek' was in the trademark registration process and the application had been submitted in secret by an individual who was a local politician. In this atmosphere of scandal a number of local institutions began arguing about the right to register oscypek as a trademark. The upshot of this 'oscypek war' was a decision by local authorities that a key role should be played by ASB. Given that the registration procedure involved significant costs, the association asked the local government bodies of three Podhale districts for help and all these entities submitted a joint application to the Polish Patent Office. The situation changed again in May 2004 when Poland formally became a member of the EU. Since European rules on registration of regional products were now replacing the national ones, the whole oscypek trademark registration procedure

---

2 Members of the association call themselves breeders as they own flocks of sheep. In most cases they are at the same time shepherds because they are involved in shepherding. However there are also shepherds who are hired by breeders to take care of sheep during the season. These temporarily employed shepherds do not belong to the ASB.

had to start again from scratch. ASB, together with local government, again played a leading role in this process.

Two important issues should be mentioned here. Firstly, the idea of registration is a symptom of local food globalisation. Some formal actions have been undertaken with a view to extending the market for oscypek cheese into the entire territory of the EU. This spatial extension required the involvement of institutional actors, so that what was involved amounted to a kind of centralisation of the food supply chain. Secondly, oscypek's entry into international markets is accompanied by the reinforcement of its traditional features. The regulations concerning the production process, ingredients, shape and size of the product have been formulated after comprehensive study of the regional herdsmen tradition. We would however argue that this 'return to the origins' should be described as creation of 'modernised' or 'reinvented' tradition.[3] Because in order to 'conquer Europe' oscypek cheese has to comply with sanitary regulations, the suggested production technique in many ways represents an 'improvement' over particular traditional practices. This 'neo-traditional' oscypek is both, embedded in local culture and well enough regulated as not to cause concern about food safety.

The registration procedure required a 'common norm' for oscypek cheese. As each herdsman has his or her own recipe supported by family tradition, the development of a single norm seemed to be very problematic. Finally ASB adopted general regulations stipulating the proportion of general ingredients, size and shape of the final product as well as the conditions for its production. It should also be mentioned that the 'registered oscypek' pattern imposes strict requirements in terms of the ingredients the cheese can be made from. According to the norm, the name oscypek can only be used for products made exclusively from sheep's milk or from a mixture of sheep's and cow's milk. For the latter option, important regulations have been introduced. They allow for only two kinds of milk to be blended in a certain proportion and the cow's milk has to be from the Polish Red Cow (the rare local breed of indigenous Polish cow).

There is also an important territorial aspect to the newly created oscypek norm. Registration at the European level is possible only if the territory of origin of a given product is defined. In the case of oscypek cheese this was a challenging task. The problem is that oscypek-like cheeses are produced in the greater Carpathian mountain region, that is an area not limited to Podhale and environs. It is for this reason that attempts to register the oscypek name in accordance with EU standards triggered protests from the Slovakian producers on the other side of

---

3   The notion of 'invented tradition' was popularised in the social sciences by Hobsbawm and Ranger (1983). They describe examples of particular local traditions which are presented as ancient, whereas in reality they have a very recent history or have been deliberately invented as a response to novel situations, which take the form of reference to the past, even when the continuity with past is largely fictitious. In case of the third oscypek network the 'reference to the past is based on real connections to old traditions. That is why we decided to use the term 're-invention', rather than 'invented', tradition.

the Tatra Mountains. They claimed that registration could lead to the market marginalisation of their own similar cheese called 'ostiepok'. The result of this dispute was that the registration procedure was blocked for more than a year. Registration proceeded after Polish and Slovak highlanders managed to negotiate a compromise solution. There was, however, also conflict in the domestic arena over the range of territorial applicability of the oscypek norm. The original application for the oscypek registration included only the Podhale region. But cheese of this kind is also produced in other parts of the Polish mountains. Wanting to retain the right to use the name, the producers who had been excluded protested against such discrimination. Leading institutional actors from the Podhale region took advantage of their dominant position as the authors of the application and negotiated only with selected partners. Consequently in the final version of the registration documents the area of the traditional oscypek production has been limited to the two administrative districts of the Podhale region (Tarzanski and Nowotarski poviats) and some parts of other districts to the west and north. This means that the EU-approved area of origin for oscypek is greater than the Podhale region but smaller than the actual area of production (see Figure 7.2). The agreement

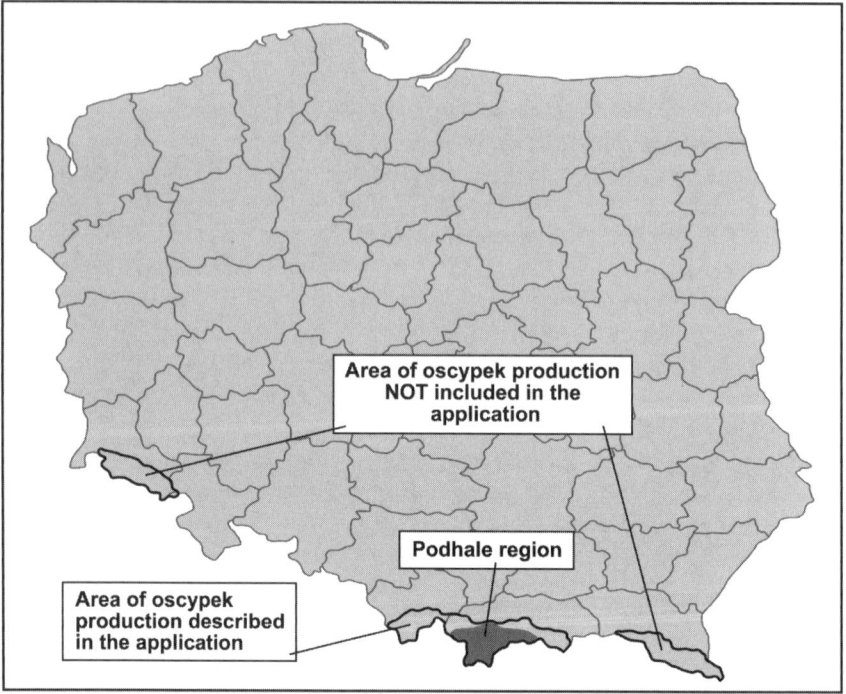

Figure 7.2   Map of the 'traditional' area of origin of oscypek cheese as social construction

left out of account the whole area of eastern Beskidy. Shepherds from that region cannot use the name oscypek even though their cheese meets all the standards in terms of production procedure and ingredients. The above described situation is an indication of how the 'area of origin' of oscypek was more the outcome of a social construction than of geographical research. Clearly the process of local food certification involves not only codification of tradition but also identification of the frameworks within which the tradition finds expression. The decision about which elements belong to the local tradition can be negotiated or taken arbitrarily by the strongest social actors. In the latter case the less powerful groups are deprived of their right to make use of a tradition, a loss which often has negative economic consequences for them.

Conclusions

Discussion of local food production is inseparable from the problem of rural development. Once it is hypothesised that consumption of oscypek is part of the 'totality of regional experience' then 'expansion of the oscypek complex' can be perceived as part of a regional development strategy. Such a relationship has been stressed in literature on agro-food networks. As Goodman (2004: 3) says: 'Recently AAFN (alternative agro-food networks) have been seen as the forerunners of a paradigm change, a territorialised, ecologically-embedded successor to the modernisation paradigm of rural development'. Also Tregear stresses that the analysis of the role of typical food 'needs to be based on a recognition that all manifestations of food-territory links are of potential importance and relevance to rural development' (2003: 104). Both authors mentioned here recognise the role of local products in the context of more complex and systemic change in rural areas. Therefore the analysis of such a role for oscypek cheese will form the conclusion of our chapter.

The linkage between local food production and rural development seems to be mainly a matter of economics. Production of local food is seen as a promising opportunity for rural inhabitants to acquire income. The first oscypek network provides a good illustration of this potential. A significant part of the local community is involved in production and retailing of the oscypek, which remains an important source of income for many families. It is especially important because the natural conditions are disadvantageous for agriculture in Podhale, meaning that an average farm can hardly ever be profitable. But the analysis also reveals some less encouraging aspects of this network. In the case of many producers, use of the oscypek name has not enabled them to establish a viable economic situation for themselves. A study of the first network indicates that the market success of this food chain has been made possible first and foremost by low prices and accessibility. So the economic contribution of this network is indeed important on account of its scope and the number of people involved, but at the same time it is a waste of the potential offered by local food. When local food turns into a cheap mass product it loses its main advantages over conventional food and could easily be pushed out the market by the latter. Moreover the impact of this network on other important dimensions of rural development seems to be quite ambiguous. One of the more desirable aspects of local food products is the way they can contribute to preservation of the cultural heritage of the countryside. With oscypek such an effect is rather marginal, as in its popular version it has almost nothing in common with the original Highlanders' food with its roots in the shepherd tradition. It is also difficult to represent the social dimension of the network as a positive, given that food quality is not a priority for producers and consumers.

The fact that oscypek cheese has the potential to play an important role in the Podhale's development has been recognised by local authorities and is evidently a central motive for their involvement in re-organisation of production/consumption patterns (network 3). The primary motivation behind the process of registration and certification of oscypek was that it should pave the way for entry into international markets. With legal protection from the EU, oscypek cheese is expected to reach new groups of wealthy 'Western' consumers, in turn generating some profits for associated producers. The whole project is thus supported by local governments and the ASB primarily on the basis of economic calculations. But there is also the prospect of achieving other objectives. One subsidiary goal is to counteract the invasion of 'fake' oscypek and in this way protect the original local food tradition. The introduction of oscypek into the wider European market is expected to have beneficial effects for promotion of the Podhale region generally, contributing to further development of tourism and helping to foster an overall valorisation of local resources. In addition, the rising profitability of oscypek production will help to preserve the heritage of shepherding. Given that the price of wool remains rather low, the supplementary income from oscypek sales plays a crucial role in maintaining sheep breeding in the Podhale. The next element that needs to be pointed out is the ecological aspect of the third network. As mentioned earlier, registration of oscypek at the EU level entailed elaborating a strict set of rules that must be observed by producers. These rules open international markets only for cheese made from a particular type of sheep's and cow's milk. The breeds involved are, indigenous and in the case of the cows (Polish Red Cow) rare. This means that the development of oscypek production will have a positive influence for preservation of biodiversity in domestic animals.

The conclusion we draw from the above is that local food production can take the form both of spontaneous actions by resourceful individuals and more organised, institutionally supported initiatives. It is moreover interesting that with oscypek both these alternatives can be observed in a single product. Such structural factors should be taken into account when analysing the contribution of local products to rural development. On the basis of the materials collected one cannot, however, draw the simple conclusion that project-type initiatives (network 3) are more beneficial than the spontaneous actions of rural inhabitants (network 1). The research shows that along with the positive effects of emergence of the third network there can also be some negative ones. Registration and certification lead to centralisation of power in the oscypek cheese networks, in effect triggering a revolution on the oscypek market. The coexistence of the three types of networks described in this chapter appears no longer possible as the cheese not produced in accordance with registered rules is not granted a certificate. This problem is particularly serious for the first network as the bulk of home production and street retailing centres on the oscypek made of cow's milk. People involved in the network are no longer able to use the name oscypek for their products. They cannot shift towards the authorised standard for the product as they have no sheep or mountain sheds. Moreover, if oscypek is to be produced in accordance with the registered rules it will become available primarily during the summer season. When taken in conjunction with an increase in exports, such a factor can affect the supply side and in turn lead to a rise in prices on the national market. This poses a real threat to the existence of first network. The large-scale producers will also find themselves in a difficult situation. They too are no longer authorised to use the name oscypek, though it is a key element in the marketing strategy for their products. The process of registration and certification of oscypek cheese thus opens up new market opportunities for some groups of producers but at the same has the potential to marginalise several other categories of social actors. Crises in the 'popular oscypek' sector may well have negative social and economic consequences, especially for farmers not engaged in sheep breeding. Without the right to use the name 'oscypek' they can be pushed out of the market and lose a significant source of alternative income. As a result, instead of empowerment of rural inhabitants, what could occur is social exclusion. It is also worthwhile emphasising that even though the production protocols finally adopted are quite general, standardisation of the cheese norm can still lead to negative homogenisation and deterioration of a variety of oscypek that is a part of our cultural heritage.

In terms of knowledge dynamics, the certification and registration of oscypek cheese can be regarded as an attempt to protect the 'true' tradition by protecting local knowledge. The attempt has been made by institutional bodies employing EU regulations as their tool. One interesting aspect of the oscypek cheese situation is that both external actors (network 2) and local actors (network 1) were seen as a threat to the regional heritage. Monopolisation of the right to use the oscypek name is likely to eliminate the other networks from the market. On the one hand it will support development practices based on local knowledge but on the other it could well cause serious social and economic problems. The production of oscypek in accordance with approved protocols is a highly exclusive activity, reserved for sheep breeders and shepherds. The rest of the local farmers will have to move to other products. The question remains as to whether this is the only solution. Is it possible to implement certification without destroying the fragile balance in the oscypek market? Some highlanders believe that the three alternative oscypek networks can still exist if the ASB and local government are prepared to concentrate on conquering the European market while at the same time being less strict in enforcing new regulations on the home front.

The three main theoretical conclusions emerging from our case study should be stressed once more. Firstly, the example of oscypek cheese confirms that the phenomenon called local food is very diversified in its nature. This is true not only because different kinds of local food are different from each other. It is true above all because a single local food product can be the base on which a complex food system is created consisting of alternative networks. Each of these networks may involve different actors and a variety of interests, as well as utilising different types of food chains. Direct sale and supermarket distribution can be involved at the same time. This is one reason why we should avoid theoretical generalisations constructed around sharp distinctions between the local and the conventional food, portrayed as polar opposites corresponding to post-Fordist and Fordist logic respectively.

The second conclusion is partly related to the previous statement. As outlined earlier, there is a strong tradition in academic literature according to which local food is the outcome of a 'quality turn'. It involves an assumption that local food and quality are two notions that are intrinsically interlinked, with quality always being a distinctive feature of the 're-territorialised' food. In reality this is not necessarily the truth. As the oscypek case (network 1) shows, local food products do not always meet general standards of quality. This may be particularly the case when it comes to foodstuffs targeted at tourists in popular leisure regions. When local food becomes an attraction in the context of mass tourism, accessibility and price become the first priority. In case of mass tourism, the desire for 'total experience' makes travelling consumers less quality sensitive. This aspect of the reality should be taken into consideration in the interests of theoretical reflexivity. There should perhaps be less wishful thinking in portrayals of the relation between local food and quality.

The last conclusion is linked to the dimension of knowledge dynamics. There is no doubt that the introduction of such an analytical concept into research can be of assistance in studying the complex character of local food systems. As the oscypek case has shown, the alternative food networks might have different knowledge dynamics behind them. This, in turn, determines their dissimilar form and shape. The analysis of knowledge types that have been involved in a particular network could reveal hidden processes or internal conflicts within a given local food system. There is, for example, a very interesting question of ownership in relation to local food. To whom does local food belong? The seemingly obvious answer is that the local community and some food-related authorities/agencies might possess such rights. But in reality there can be several distinct groups within the local population who have somewhat contradictory interests *vis à vis* the same food product. There are also important external and quasi-local actors involved in the system. This raises many significant questions in terms of local sustainable development: on what basis should it be decided who is more entitled to use the original name? Can several versions of the same product be allowed on the market? Or is it perhaps more beneficial if only one dominant type of local food product is imposed? These issues remain a challenge not only for policy-makers but also for further theoretical discussion.

References

Bourdieu, P. 1984. *Distinction: A Social Critique of the Judgment of Taste*. London: Routledge.
Bruckmeier, K. 2004. Theoretical and Methodological Framework, CORASON, Working Paper no 2. Available at: http://corason.hu/download/wp2.pdf [accessed: 13 November 2009].
Cichocki, W. and Kozak, A. 2003. *Tło historyczne tradycji wytwarzania oscypka, bryndzy i bundzu na Podhalu*. Zakopane: Muzeum Tatrzańskie im. dra Tytusa Chałubińskiego Zakopanem.
Fonte, M. and Grando, S. 2006. Local food production and knowledge dynamics in the rural sustainable development. Input paper for the Working Package 6 (Local food production) of the CORASON Project. Unpublished manuscript.
Goodman, D. 2004. Rural Europe redux? Reflections on alternative food networks and paradigm change. *Sociologia Ruralis*, 44(1), 3–16.
Goodman, D. and DuPuis, E.M. 2002. Knowing food and growing food: beyond the production-consumption debate in the sociology of agriculture. *Sociologia Ruralis*, 42(1), 5–22.
Hobsbawm, E. and Ranger, T. 1983. *The Invention of Tradition*. Cambridge: Cambridge University Press.
Holloway, L., Kneafsey, M., Venn, L., Cox, R., Dowler, E. and Tuomainen, H. 2007. Possible food economies: a methodological framework for exploring food production-consumption relationships. *Sociologia Ruralis*, 47(1), 1–19.
Jordan, J.A. 2007. The heirloom tomato as cultural object: investigating taste and space. *Sociologia Ruralis*, 47(1), 20–41.
Marsden, T. 2000. Food matters and the matter of food: towards a new food governance? *Sociologia Ruralis*, 40(1), 20–29.
Marsden, T. 2003. *The Condition of Rural Sustainability*. Assen: Royal Van Gorcum.
Rybak, J. 2001. Oscypek, czyli góralska tajemnica. [Online]. Available at: http://www.nsik.com.pl/archiwum/74/a20.html [accessed: 16 July 2008].
Tomczak, A. 2003. Oscypek. [Online: British Studies Web Pages]. Available at: www.elt.britishcouncil.pl/elt/r_Oscypek.htm [accessed: 16 July 2008].
Tregear, A. 2001. What is a typical local food? An examination of territorial identity in foods based on development initiatives in the agrifood and rural sectors. Centre for Rural Economy Working Paper Series, Working Paper 58, University of Newcastle. Available at: http://www.origin-food.org/cherch/ukat.htm [accessed: 13 November 2009].
Tregear, A. 2003. From Stilton to Vimto: using food history to re-think typical products in rural development. *Sociologia Ruralis*, 43(2), 91–107.

Chapter 8
Traditional Food as a Strategy in Regional Development: The Need for Knowledge Diversity

Gunn-Turid Kvam

Introduction

The development of local food and agro-food systems has been a strategy for revitalising rural communities in Norway since the early 1990s. One important obstacle to accomplishment of this strategy has been lack of knowledge about traditional processing methods, such knowledge having been lost in many parts of the country. This chapter takes a closer look at one region, Valdres, where traditional food production is still practised and where a strong emphasis is placed on the re-localisation of food to increase local added value and enhance rural development. This study focuses upon the establishment of local food systems in Valdres based on two of their traditional products: semi-fermented fish and dried cured salami. The focus is on developing local food systems from the perspective of knowledge forms and dynamics. Important questions in the two cases presented are: what knowledge forms – local, managerial and scientific – are activated in the process of establishing a local food system and what the status and power are of these different forms of knowledge.

This chapter starts with a description of the context of Norwegian agriculture, local food and rural development. This is followed by a presentation of the theoretical approach that informed the study and a description of the method of data collection. A brief overview of the socio-economic characteristics of Valdres is followed by a presentation of each case study of traditional food production. This leads to a discussion of the kinds of knowledge forms that are commonly deployed, including the complex interplay between these different forms of knowledge. The chapter concludes with a discussion of the main findings from the study.[1]

---

1 The chapter is based on a report prepared for the EU CORASON project on the topic 'Local food production and knowledge dynamics in rural sustainable development'.

The context: agriculture, local food and rural development

Although only four per cent of the labour force in Norway is directly employed in agriculture, many rural communities are highly dependent on agriculture. The authorities' relatively strong support for agriculture and rural areas has been a crucial factor in the dispersion of settlement all over the country. At the same time, however, the drastic restructuring of agriculture following a gradual liberalisation of agricultural policies and adaptation to international regulations has led to farms closing down. In 1979 the number of active farmers was 125,300 and in 2006 there were about 51,200 active farmers left (Statistics Norway 2005/2008; available at: http://www.ssb.no/emmer/).

The agro-food industry represents around 12 per cent of total employment in manufacturing, with about 33,000 employees (Borch and Stræte 1999). During the 1990s the Norwegian agro-food industry was exposed to increased threats from newcomers to the market, increased cross-border trade and increased consumer negotiating power. The increasing concentration and strengthening of links between wholesalers and retailers parallels the emergence of an agro-food complex where four wholesaler-retailer companies account for about 99 per cent of all food products sold (Dulsrud 1999). For the food-processing industry, the result of structural changes has been an intensification of pressure, necessitating market adaptation through cost efficiency and a further concentration of production. The focus for the processing industry has been on achieving economies of scale and minimising the cost of production (Almås et al. 1997).

Prior to the 1990s there was little activity to promote local food production in Norway, and farm-based food production was nearly nonexistent. One important reason for this situation was that the cooperatives within the agriculture sector had for decades been responsible for processing and marketing food products on behalf of the farmers. Nearly all farmers in Norway were members of the agriculture cooperatives and thus had to deliver their products to the cooperative (Kvam et al. 2003). This meant there was little opportunity for the farmers to add value to their products or tap into niche areas to market their differentiated products.

The early 1990s saw the emergence of greater consumer demand for niche food. This necessitates food production in which qualities such as local aspects, traditions and history, are embodied into the product and for which customers are willing to pay a premium price (Stræte 2008). Some Norwegian farmers recognised the potential in farm-based quality production, in many cases based on success stories from the south of Europe. The desire for greater variety and less vulnerability in the rural economy combined with a sharper consumer focus on traditional food, or food with specific qualities, were important reasons why the Norwegian Government encouraged the establishment of farm-based food production at the beginning of the 1990s. Since then support systems have been developed, both at national and regional levels, to assist farmers and other small-scale food producers and manufacturers in establishing local quality food firms (Kvam et al. 2003).

Local food production was in a way 'reinvented' in the mid 1990s. A considerable number of farmers have established such niche production in recent years. In a few areas of the country newly established producers have based their production mainly on traditional local products. In other areas local food products – in particular cheese – have been developed by drawing upon knowledge from cheese makers in other European countries, such as France and The Netherlands. Because Norwegian local food production is generally a farm-based emerging industry, most firms are still very small and the economic impacts are currently minimal (Magnus and Kvam 2008).

Theoretical approach and method

The quality turn in agriculture and the importance of 'the local' (Goodman 2003) are important elements in this study. They are associated with a development that can be interpreted as a reaction to large-scale industrialised food production involving little variety or distinctiveness. Local aspects are embodied in the quality food product and competition is based on quality rather than on price (Ilbery and Kneafsey 2000, Murdoch et al. 2000, Goodman 2003). In this context, quality may be related to a geographical origin, tradition and/or a specific characteristic of the product. Quality is a socially constructed phenomenon shaped in the relationship between producers and consumers; it is not only a matter of what the producer does, but also of what the consumer perceives (Stræte 2008).

In this case study on the re-emergence of traditional foods we are considering two local products with strong, long-standing traditions in Valdres. In such products, local forms of embodied knowledge are generated in practice and usually can only be weakly codified (Tovey and Mooney 2007: 9–12). Knowledge exchange has mainly been informal and based on trust, and is therefore embedded in local networks.

In the CORASON project, knowledge was divided into the categories of local knowledge, scientific knowledge and managerial knowledge (see the Introduction to this volume). While local knowledge is unexplicated and 'prediscursive', scientific or expert knowledge typically denotes highly articulated, elaborated and codified scientific and technical knowledge (Giddens 1976). Scientific knowledge also differs from local knowledge in that it is both standardised and formal. Managerial knowledge is also, as scientific knowledge, a form of expert knowledge. It involves a blend of scientific and local knowledge relating especially to organisational and bureaucratic matters (for example, hygiene norms, certification, marketing, cooperation) and may be shaped by past experiences in management practice and/or formal education. In the CORASON final report local knowledge was also differentiated into local lay knowledge and local tacit knowledge (Tovey and Mooney 2007: 103; see also the Introduction to this volume).

In the Norwegian study the significance of tacit knowledge as part of local lay knowledge is apparent in the cases studied. Tacit knowledge is defined as knowledge that is not codified. According to Polanyi (1967) it is not possible to transmit tacit knowledge through writing or speech. Polanyi characterises tacitness as the elements of knowledge and insight embodied in each individual which are poorly defined, not codified or published and therefore cannot be expressed. Tacit knowledge thus remains something individual, not susceptible of transmission orally or in writing. From our point of view, however, it is possible to transmit tacit knowledge to some extent. For example, a scientist may be able to codify parts of a local producers' tacit know-how about production. But when connections are too complex it is some times too difficult for experts to accomplish scientific calculations, and not possible to codify the local tacit knowledge (Kvam 1995).

In our study we also observed that tacit knowledge does not necessarily exist only at the individual level. Traditional food products embody tacit knowledge about production and processing held by the group of local producers. The tacit part of their knowledge may be more or less difficult to codify for external actors, but usually it is only weakly codified by the producers. The extent to which knowledge is tacit among the producers may vary, depending on each person's ability to articulate and share knowledge about a specific case (Molander 2000). Local know-how about social and relational matters employed in managing interaction with other people will also be partly tacit. This is a kind of knowledge local people apply more or less unconsciously (Tovey and Mooney 2007: 102).

As indicated at the beginning of this chapter, tradition and/or specific characteristics of particular food products are construed as being aspects of 'quality food'. In the case of traditional food products from Valdres, such qualities are arguably correlated with local lay knowledge about production. Because a part of local knowledge is tacit and thus not easy to transmit, it may constitute a competitive advantage from the producer's point of view. At the local level, distinctive geographical conditions such as climate and soil type can affect the potential for developing products with special qualities that cannot be replicated elsewhere, thus providing a possible longer term market advantage. Such factors and conditions as differentiate the product from other products may come to impart to the goods on sale a quality of distinctiveness, especially when the item is being marketed outside its own region of origin. For such qualities or competitive advantages to endure on a long-term basis, however, it is important for firms to develop specific resources and capabilities to provide the necessary support for this (Wernerfelt 1984, Barney 1991, Grant 1991). Certification may be one way of distinguishing the product.

Norway has established a certification for local food products called *Spesialitet* (Speciality). This certification requires that food products have special qualities that distinguish them from others. Products certified for *Spesialitet* are usually products with a local or a regional connection. As in most other European countries, Norway has established a national regulation to protect the designations of agricultural products, fish and seafood. There are three forms of designation in Norway: Protected Designation of Origin (PDO), Protected Geographical Indication (PGI), and Traditional Speciality Guaranteed (TSG), based on EU regulations 2081/92 and 2082/92. These certification schemes have been developed to protect against replication and encourage the marketing of products with specific regional and traditional qualities, thus contributing to an increase in the value of the products in question. Certification is regarded as a way of displaying the quality and uniqueness of the product to customers. Methods of production are defined by a code of practice that guarantees certain standards or quality, which customers value and for which they are willing to pay a premium (Tregear et al. 2007). The requirements for being accorded certification for protected designations are more stringent than those for certification with *Spesialitet*.

Certification processes mean codifying at least some of the local tacit knowledge about production that is embodied in regional, traditional products. Scientists and technicians are brought into the process, with what we define as scientific knowledge, to control complex processes of transformation and, in the case of the two case study products, fermentation. Managers and experts on regulations and certification schemes are also necessary for accomplishment of these aims, bringing into the process managerial knowledge (Fonte 2008). While this certification process is meant to give the product a market advantage over others, it has a negative aspect: the codification of what was previously local tacit knowledge might make it easier for competitors to imitate the product (Goodman 2004). Thus the process of certification may make the products more vulnerable to competition.

Data collection for our study took two main forms. During the first stage we tried to define the background and context through the desk-top research we conducted. This involved analysing the websites of the individual producers and their cooperative societies, websites of the regional authority and local press articles. We also reviewed previous studies and written material relevant to these cases. Secondly, in-depth interviews were conducted with the producers involved with the niche food items, fermented fish and salami, as well as with the head of the board of each of the two cooperatives. Other key operatives such as representatives from local development agencies and from external support agencies, both national and local, were also interviewed. As part of the fieldwork study we visited producers and were introduced to the production methods for these niche products. We also visited shops and outlets where the products were sold.

Valdres and its strategy for local food development

The Valdres Region

Valdres is a region in Norway located in the county of Oppland. It is a valley comprising six municipalities. In 2000 there were 18,367 inhabitants in Valdres; as in many rural regions in Norway, the population had declined between 1980 and 2000 (SSB 2005).

Each region in the county of Oppland has established a Regional Council which is responsible for developing a strategic plan for its particular region. One of the main challenges that is being mentioned has to do with the restructuring of the agricultural sector. There are about 950 farmers in Valdres. Most of the farms are small in size and the farmers often have other kinds of paid work in addition to farming. The decline in the farming economy and demands for more efficient production has led to farm closures (Kvam 2006).

Valdres occupies a unique position in the culinary culture of Norway because the people there have maintained their traditional forms of food production. Valdres is well known particularly for its production of semi-fermented fish and salami (dried fermented sausage) – a tradition kept up by local people. The valley is also well known for summer farming, which is still part of a strong tradition, particularly in Valdres. Summer farming means that farmers move their cattle from the farm to high mountain pastures for a few months during the high summer. The mountain pastures are still an important resource for farmers in Valdres to ensure enough food for their animals through the winter. The valley also enjoys a rich cultural tradition which is kept very much alive and is closely connected with the inhabitants' identity (Kvam 2006).

Activities to strengthen local food production

As early as the beginning of the 1990s regional authorities were developing a strategy for strengthening local quality food production to increase the volume of local, value added produce in the agricultural sector. The authorities established a regional development agency called Valdres Næringsutvikling in 1990 for the purpose of working towards this goal, among others. Ownership of the company was in the hands of the six municipalities of Valdres and the County of Oppland (Kvam et al. 2003).

Valdres Næringsutvikling contributed to the promotion and development of its regional produce through organising a number of events and fora aimed at increasing the quality and the appeal of local food. One such event was the 'Norsk *Rakfisk* Festival' (Norwegian Fermented Trout Festival) in Valdres, which was organised for the 15th year in succession in 2005. It is a well-known commercial festival that attracts a large number of people from Valdres and elsewhere (Kvam 2006).

The Valdres Matforum (Valdres Food Forum), a cooperative association, was also established by Valdres Næringsutvikling in 2002. The express aim of this cooperative was to promote among local food producers awareness of new developments in the food sector, and to connect the food and tourism sectors in a more integrated way. The Valdres Matforum is financed by public support (75 per cent) and local food producers (25 per cent). The goal of the cooperative society is to construct the economic foundations for production, processing and sale of food from Valdres through quality production based on distinctiveness and local traditions. There are about 40 members of the Valdres Matforum (Kvam 2006).

The regional authorities have been very conscientious in extending support to local food initiatives and in particular to those that focus upon traditions and local culture as a means of strengthening agriculture and hopefully maintaining small family farms in the region (Kvam 2006). Valdres Næringsutvikling was important for establishing the Valdres Rakfisk cooperative and the Valdres Matforum for establishing the Valdres Kurv cooperative, which will be described in more detail below.

The situation with semi-fermented fish: a traditional product

Valdres is the area in Norway where most of the production of semi-fermented fish takes place. '*Rakfisk*' from Valdres is something like a household name in Norway. The tradition of fermenting fish is known all over the world, but the term *rakfisk* seems to be peculiar to Norway. *Rakfisk*, or 'fermented fish', is a processed product. In the case of '*rakfisk* from Valdres' the reference is to fermented trout. The fish is first cleaned and salted, perhaps with the addition of a little sugar, and then placed in a bucket and covered with a marinade for six to 12 weeks at 4–8 °C. Traditionally the fish was kept in wooden barrels and stored in a dark, cool place (Riddervold and Heuch 1999), in earthen cellars under the house or in specially constructed hillside storage rooms.

Preparing semi-fermented trout has been a food tradition since the sixteenth century or earlier, with the producers sourcing trout from local lakes. This method of conservation was once important for preserving food for longer periods. The fish was prepared in the autumn and ready to be eaten around Christmas. Fermented fish was an important source of food during winter – a product that was produced locally for local consumption. Producers in Valdres have their own method of preparation which distinguishes this product from its counterpart in other areas of the country. Today fermented fish from Valdres is food for special occasions. It is usually eaten with crisp bread, potatoes and onion and in some cases other accompaniments. Sales of *rakfisk* peak around Christmas, but nowadays it is possible to buy the fish all year round.

The process of certification   The manager of Valdres Næringsutvikling initiated the process of establishing closer cooperation among the producers of fermented fish in Valdres. According to him the producers were not much inclined to cooperate. When a local competitor started to import trout from Denmark and sell it as *rakfisk* from Valdres, the six producers saw the need for cooperation if they were to profile themselves as the 'real producers' of *rakfisk* from Valdres.

Collaboration centred on the task of having the product certified as part of a strategy for increasing production and sales and expanding from a local to a national market. The manager of Valdres Næringsutvikling played an important part in this process, not least because of the managerial knowledge he brought to the project. He was a key driving force behind the certification process, helping to build a necessary common platform for cooperation between the six producers and connecting them with supportive professionals. He was also involved in coordinating the financial aspects of the process.

By the time 12 months had passed, the *rakfisk* producers had succeeded in establishing contact with many institutions that supported them in the certification process. Matforsk (the Norwegian Food Research Institute) handled the first part of the process, conducting a scientific study on the food safety aspects of fish fermentation. It is well-known that many bacteria can develop during the fermentation process, with botulism being the most dangerous of them. The aim of the chemical research study was to optimise food safety and hygiene. This science-based research had significant influence on the outcome of the application for certification, both for Speciality and for Protected Geographical Indication (PGI) (Kvam 2006).

The *Senter for produktutvikling i næringsmiddelindustrien* (SPIN) (Centre for Product Development in the Food Processing Industry), and Matmerk (The Norwegian Agricultural Quality System and Food Branding Foundation) were also important contributors to the certification process, particularly in respect of image building and branding. The six producers first participated in a pilot project where the aim was to establish the certification label Speciality (*Spesialitet*). Later, the protagonists in the certification project also undertook to help with elaborating the application for PGI (Protected Geographical Indication). This involved working out common requirements for processing (the production protocol), necessitating a codification of part of the local tacit knowledge and compilation of historical documentation on the links between *rakfisk* and Valdres. There was also close cooperation with the local food authorities and veterinary authorities, who applied their scientific knowledge and expertise to monitoring the production process so as to ensure observance of hygiene and safety requirements.

A design firm was recommended by Norsk Designråd (the Norwegian Design Council) to develop a common logo for 'Valdres Rakfisk BA'. The goal was to develop a logo symbolising something both stylish and delicious, whose important attributes were reliability and quality, cultural pedigree and prestigious origins. The design firm's main contribution was to provide managerial knowledge for promoting and branding the product.

The intention behind the process of guaranteeing the quality of the product was that the fermented fish had to be produced the same way as they were 200 years ago, but in compliance with today's safety requirements. Many elements have changed: the slaughter site on the shore of the mountain lake has been replaced with modern slaughter houses; the wooden tub has been replaced with tubs of plastic, the storage cellar has been replaced by modern fermentation installations which maintain a regular, stable temperature. But even so, the recipes used today by the producers are much the same as the earlier ones, and the *rakfisk* has kept much of its original consistency and taste (Amilien and Hegnes 2004).

The specific character of the six manufacturers' fish products is primarily determined by a combination of the geophysical peculiarities and the area's production knowledge (that is, local lay and traditional knowledge): the fish comes exclusively from Valdres, either from fish farms or from mountain lakes, and the product is prepared locally in a traditional way, with local products (as far as possible). *Rakfisk* from Valdres was the first locally branded product for which Norwegian PGI status was sought. In terms of identity and territory, *rakfisk* fits these concepts with its natural origins and its traditional fermenting process. But because the salt is imported and the fish food is bought from outside the region, it is not possible to obtain the Protected Designation of Origin certification.

According to the producers, the process of certification influenced their production practices. They have established some shared routines so as to achieve a stable, consistent and high-quality product. They have also, for instance, agreed not to use a starter culture to initiate the fermentation process. Despite the fact that the use of a starter culture contributes to a more even quality, the producers decided to maintain the original methods of processing so as to secure traditional distinctiveness. The producers have also agreed on when to start processing so that their products can be sold during the same time period, from a certain date. But maintaining the differentiation between producers and products has also been an important aspect of the process. Each of the six producers in Valdres Rakfisk production still uses his own traditional recipes. They have coordinated some activities and processes, whilst maintaining the product distinctiveness of each of the producers involved. In this way they have upheld and reproduced local traditional knowledge. The fact that producers use their own distinct recipes that have been based on trial-and-error processes for generations implies that individual tacit knowledge in production still exists.

Knowledge accumulation in production: one producer's experience Both in the production of trout for fermented fish and in processing there are, according to one producer, many factors that influence the quality of the end product. One factor of great importance is the quality of the water. The producer has laid pipes from the mountains to his fish farm to secure clean water. The skin of the fish harbours different bacteria depending on the quality of the water. Different bacteria influence the fermentation process in different ways, and clean water is therefore of major importance for the quality of the end product.[2] The producer has developed his own system for keeping the water clean inside the fish dam as well for securing healthy trout. This system was based on scientific knowledge from the Norwegian University of Life Sciences, but put into practice by combining external scientific knowledge and own experience-based knowledge which in some instances is tacit.

The slaughtering conditions represent another key quality factor. According to the producer there are at least 60 to 70 factors and combinations of factors that influence the quality of the product. One important factor is not to stress the fish prior to slaughtering. The same producer has developed his own system to

---

2 According to him: 'For fermented fish, the water means what the soil and the light means for grapes' (Personal interview).

avoid stressing the fish. He has also developed his own system for cleaning the equipment and the processing room and supplying clean air to avoid unwanted bacteria inside the processing building. In establishing these methods and systems, he has also acquired knowledge from outside, both through research via scientific readings and through expert advice from the local Food Security Office. Scientific knowledge from outside has in many cases confirmed and codified his own experience-based tacit knowledge.

The importance of differentiation with producers' merchandise   It is clear from the situations outlined above that individual tacit knowledge among the different producers still plays a role in production, despite the certification process that has codified part of the tacit knowledge in the possession of individuals and producers as a group. There are various reasons why producers wanted to preserve this differentiation between the different items they manufacture and not standardise every single element in production. The most important reason for keeping differentiation was to enable producers to keep their own 'brand identity' and their pride in their own distinctive products. The experience of the producers is that the same customers tend to come back year after year to buy products from their favourite producer. To reach even more customers some of the producers have developed different grades of matured fish, thus differentiating products even further. In this way producers' knowledge about customers' preferences also exerts an influence on the product and its development.

Another reason to keep the distinction between the individual items of merchandise marketed by the different producers is the Norwegian Fermented Trout Festival, where one important activity is to vote for the producer of the year's best *rakfisk* from Valdres. There are two prizes, one awarded by professional judges (for 'expert knowledge'), and one awarded on popularity (for 'local knowledge'). Different producers win the prizes in different years, something that highlights the differential evaluation, depending on whether the criteria are local knowledge or expert knowledge.

Results from the process of certification   Each Valdres Rakfisk producer has his/her own land-based fish-farm and buildings where he produces and processes his own trout for sale. One producer has a hatchery from which the others buy their trout. Valdres Rakfisk was established with a common brand, logo and design on sale materials, but each producer takes care of his own distribution, sales and marketing, with his own refrigerated vehicle for bringing the fish to customers. One of the producers sells nearly all of his products from his own farm shop. He also sells products to Smak av Valdres (Taste of Valdres), a local shop in the main town of Valdres which specialises in local food products, and to two other specialised food shops in Oslo. This producer emphasises the necessity of producers' controlling distribution and sales themselves, because *rakfisk* is a perishable product and it is important in particular to maintain a steady, low temperature so as not to lower the quality of the fish. The other producers sell their products through grocery shops in the region and in the southern part of Norway.

The establishment of Valdres Rakfisk has been a success story for the six producers. They were the first producers in Norway to be awarded the Speciality certification in 2001 and they have applied for PGI. By any standards this has been a flourishing rural development project, with a turnover of about 20 million NOK for Valdres Rakfisk in 2005. Cooperation between the producers has resulted in higher production volume, more efficient production, higher prices for products and increased sales. The six producers turned out a total of about 150 tonnes of fermented fish in 2007, with around 70 people receiving a salary through employment in this new industry. But many of these people are seasonal workers. About 15 people work full-time in the six fermented fish enterprises. Since establishing their links of collaboration, the producers have experienced growth both in product turnover and in the number of employees. Company decision makers from Valdres say that the certification process for Valdres Rakfisk has increased the status and sale of the product locally and contributed to strengthening of the local identity. For the Valdres community, this development has 'spearheaded the development of local food' (Interview to the project manager, Kvam 2006).

Outside competitors One competitor who imports trout from Denmark and some other countries is the only processor of fermented fish in Valdres not to be a member of Valdres Rakfisk. He is a producer with a long tradition in making *rakfisk* in Valdres. Tradition is also an important element of his marketing strategy. There are two main reasons why he is not included in the cooperative. He imports trout and he adds a starter culture to the fermentation process, unlike members of Valdres Rakfisk, for whom this is not permitted. The competitor has set up production in a former dairy and runs a more industrialised production process than members of Valdres Rakfisk. He is the largest producer of fermented fish in Norway. His product is marketed all over the country in a range of grocery shops. Because of the Valdres Rakfisk certification process he is not allowed to use Valdres in the name of his product. His product is market as *rakfisk* combined with *traditions* and the *name* of the producer.

This producer is seen as a competitor, but also as an important positive factor in keeping the production volume of *rakfisk* at a certain level. The demand for *rakfisk* is greater than the six producers of Valdres Rakfisk can fulfil. The publicity from the establishment of Valdres Rakfisk BA has been positive for the competitor and has increased demand for his product (Kvam 2006). Even though the competitor uses different production methods and imports his trout, he can take satisfaction in the high score his products obtain at the Fermented Trout Festival.

Barriers to further development There is one major barrier to further expansion for the Valdres Rakfisk cooperative. Due to the possibility of expanded development compromising water quality, the County Governor's Department of Agriculture is very sceptical about allowing the established producers in Valdres Rakfisk to increase the volume of production through expanding fish-farming. Moreover no new actors will be permitted to establish fish-farming in Valdres. Because of this obstacle to growth and a sustainable economic development, the producers have joined a research project to examine ways of finding a solution to this problem. It is not a problem for the competitor because he imports the fish he uses in his production.

The case of dried cured salami: Another traditional product Valdres also has a long-standing tradition in producing dried, cured salami. People use their own local name for the salami, which is kurv. One important reason for the diversity of salami production in Valdres is that the local abattoir Helle Slakteri has for many years processed the salami from local peoples' own recipes. Nowadays much of the salami is produced for sale, through farm-based micro firms and a few small companies. There is a great variety in the salami, which can be made from a whole range of meat sources including beef, elk, reindeer, sheep and goat and various combinations of these and others. It is also common to add different spices and herbs to the salami. The salami is food for everyday consumption, for example on open sandwiches for breakfast or in the evening. It is quite common for salami to be eaten with crisp bread and beer as an evening snack.

The recipes vary as between producers located in the north and south of the region as well as between individual producers, which makes for wide variety. Most of the relevant businesses sell their products on the local market, but the largest also sell part of their product outside the region through specialised food outlets. The main market for the product within the region has been with a particular group of people from outside the area who own cabins in Valdres: 'They are better able to see the qualities in the products than local people' (Interview to the project manager, Kvam 2006). There are nevertheless also local people who buy the salami. It is a niche product whose contribution to the local economy is by no means negligible, with a turnover of around NOK 25 million in 2005. Producers believe there is vast potential for increased sales, particularly in the southern part of Norway.

The inland region where Valdres is located is very dry. This means that the salami can be dried in the natural way, hanging in buildings at temperatures more or less the same as those outdoors. Sugar in some form is added to start the process of fermentation. The salami is hung for two to five months depending on the temperature and the weather. At least one producer in Valdres uses a bacterial culture to start the process of fermentation. This producer also takes the salami into a climate-controlled room to shorten the process of maturing and drying. This way of producing is the same as in the industrial production of salami.

The certification process In 2001 Valdres Mat forum initiated the process of certification for salami producers. The main goals have been to establish a distribution and sales system, to obtain a higher price for the products and to add value by establishing a brand and a common logo. Ancillary goals are to increase production and widen the market, benefiting both the salami producers and the farmers delivering meat to this commercial enterprise.

As part of the process Valdres Matforum contacted Matmerk (The Norwegian Agricultural Quality System and Food Branding Foundation) in 2001. Matmerk recommended an advisor to provide an educational course on salami production. This person was a sausage maker and food technologist. His course was therefore based on scientific knowledge. About 25 producers from Valdres participated in the course, during which the advisor recommended that the they should add bacterial cultures to start the fermentation process and ensure a consistent product quality. The producers, however, refused the expert advice, asserting their local knowledge and choosing to continue with the traditional methods in order to guarantee that the product quality remained more nuanced, rather than standardised. As in the case of Valdres *rakfisk*, the salami producers also deem it important to maintain the traditional way of processing so as to safeguard the uniqueness and the distinctive quality of the product.

The project manager from Valdres Matforum believes it would be unfortunate if all salami producers followed the experts' advice. The products would be more homogenous and it would be more difficult to differentiate them from industrial production. This is something they learned from the development process in Valdres Rakfisk: the importance of maintaining distinctiveness, differentiating each producer's merchandise. The local Food Security Authority has supported the producers in establishing an appropriate system of 'internal control' and thus has contributed with scientific and management knowledge.

In Valdres 15 salami producers, out of 30, established a cooperative society called 'Valdres Kurv BA' in 2006. The Helle Slakteri abbatoir is the largest, with 14 employees or thereabouts. The others are micro-firms with just one or two employees, or else salami production for them is part of more general farming activity. The three largest companies have their own processing equipment, but the same is true for only one of the smaller ones. The other micro-firms have their animals slaughtered and the meat processed at the abattoir.

At their first meeting the members of Valdres Kurv explored their common ground and discussed what the rules for membership should be. They decided, as already indicated, that the salami should be processed in the traditional way, without use of a bacterial culture to start the fermentation process. They also decided that all producers would be required to add fresh or dried blood to the salami. Adding blood is a tradition in Valdres. It makes the sausage black outside and helps to distinguish Valdres salami from other similar products. The salami must be processed in the colder part of the year so that the products develop similar levels of fermentation. Another decision taken is that livestock must be slaughtered and processed within a radius of 50 kilometres from Fagernes, which is at the centre of Valdres. 'Valdres Kurv' is the name shared by all the products, but each producer may also have their own brand name and the producer's name on them. This will effectually enable them to remain independent producers selling their own products. One marketing goal is that merchandises from the different producers are presented together in food shops.

The salami producers want to apply for PGI status. For this certification it is acceptable to use raw materials in small quantities from outside the district as long as the main ingredients are locally produced. Traditionally, producers in Valdres have used pork fat to achieve an acceptable consistency in the salami. Today there is too little pork production in the region for them to use local pork fat, so it has to be brought in from outside. The traditional method of processing also includes the use of horse intestines, which today are imported from countries such as Poland and Brazil. The intestines of horses dry very conveniently and evenly and can be hung outside in different kinds of weather. Producers also use various spices in the production and in many cases these too are imported.

To apply for and be granted PGI certification it is necessary to provide documentation that the product is traditional to Valdres. Such documentation already exists. But before a group can apply for certification there are some other important points to be agreed upon. It has been recommended that they should employ a professional designer to develop a common logo and sales material. According to the project manager, they are very much alert to the need to control the development process themselves, consulting others as required. Because they do not know what the certification will be worth in future, the relevance of this process is that it enables them to establish a positive attitude towards quality, distinctiveness and cooperation (Kvam 2006).

Knowledge dynamics in the valorisation of local traditional food

In both of the cases presented, all of the types of initially defined knowledge forms have arguably been important in the process of revitalising traditional products and developing the associated agro-food systems. Management knowledge was, in both cases, brought in through a local development agency, which saw the potential of cooperation among local producers and so initiated the process of establishing the cooperatives. Specifically, in the case of the semi-fermented fish, this agency fulfilled an important mediating function between the different producers and in this way contributed to establishment of a common platform for cooperation among producers. In both cases the input of management knowledge from people working with local agencies provided important support for those involved in the certification process. Expert knowledge could be drawn upon whenever necessary. It was obviously crucial for the task of establishing a cooperative and for financing and running it that management knowledge could be enlisted. Although local producers act as custodians of local knowledge, they are not in possession of managerial knowledge. Another important factor in the deployment of the managerial knowledge was that the key people in the local development agencies were themselves locals, who knew the people, the culture and the society very well. This made for an effective combination of managerial knowledge with some aspects of local lay and tacit knowledge.

Expert knowledge in the form of scientific and management knowledge was important at some stages of the certification process, where the scientific element was contributed primarily from external sources. These forms of knowledge were used as a basis for ensuring hygiene and product quality; both necessary for obtaining certification. The certification process implied that part of the knowledge that had previously been of a more local and experience-based nature, and in some cases tacit, had to be codified and to some extent standardised. But notwithstanding this utilisation of outside expert knowledge, the local knowledge of producers remained fundamental in maintaining the distinctiveness of the products.

Another interesting point to be noted from the case descriptions is the way one producer of fermented fish combined together local and scientific knowledge as a way of increasing the quality of his product. He is the only producer to have laid pipes from the mountains to get cleaner water into his fish farm. He has also applied his own experience-based methods to the task of ensuring the good health of the trout, and securing hygienic and safe food processing. The producer has combined scientific, local traditional knowledge and own experience to develop optimal solutions. When expert knowledge is introduced it is mostly for the purpose of improving the quality of the water and the fish and to ensuring hygiene and food safety. Local traditional knowledge, on the other hand, is still essential for retaining the distinctiveness of the fermented fish from Valdres. The fact that each individual participant uses his/her own recipes and develops his own processing techniques demonstrates how traditional knowledge and a considerable element of tacit knowledge are still embodied in the products.

The relatively strong position of local knowledge in the process can be explained by the fact that locals both initiated and controlled the process. In both the instances cited above, Valdres Rakfisk and Valdres Kurv, all producers joined the certification process. Along with the close links with local consumers and local food culture, this degree of self-determination might be one important reason why local knowledge remained a key element in the development processes.

Conclusion

Valdres society has experienced economic and social marginalisation. One strategy they have used to try to reverse it is a focusing on quality food production based on traditional practices and distinctive food products. Building on long standing cultural traditions and product distinctiveness, the central protagonists sought to develop quality food production and in this way defend agricultural production and the institution of the small family farm. As indicated, the strategy has been based on the employment of local forms of knowledge and the ability to work collectively to transform the essential aspects of this knowledge into commercially viable products. Local food networks in Valdres have thus arguably prompted re-evaluation of local lay and traditional forms of knowledge and techniques as a specific and distinctive resource for the revitalisation of agriculture and for food relocalisation.

The local people described above are very conscious of their desire to control their own development. They want to maintain local traditions and culture as well as an embodied form of distinctiveness. There is an awareness that local – and in particular traditional – knowledge is important for preserving variety and distinctiveness, which are in turn important for differentiating local products and retaining competitive advantages. But there is a similar awareness of the need to enlist managerial and expert knowledge in developing products and activating networks to secure the distinctive local knowledge when necessary.

References

Almås, R., Kvam, G.T. and Stræte, E.P. 1997. From productivism to flexible specialisation? Experiences from a restructuring process in the Norwegian dairy industry. *Journal of Rural Cooperation*, 25(2), 65–82.

Amilien, V. and Hegnes, A.W. 2004. The cultural smell of fermented fish, about the development of a local product in Norway. *Journal of Food, Agriculture and Environment (JFAE)*, 2(1), 141–147.

Barney, J. 1991. Firms resources and sustainable competitive advantage. *Journal of Management*, 17(1), 99–120.

Borch, O.J. and Stræte, E.P. 1999. *Matvareindustrien – mellom næring og politikk*. Oslo: Tano Aschehoug.

Dulsrud, A. 1999. Markedstrender og utvikling i distribusjonsmønsteret, in *Matvareindustrien mellom nærings og politikk*, edited by Borch, O.J. and Stræte, E.P. Oslo: Tano Ascehoug.

Fonte, M. 2008. Knowledge, food and place. A way of producing, a way of knowing. *Sociologia Ruralis*, 48(3), 200–222.

Giddens, A. 1976. *New Rules of Sociological Method*. London: Hutchinson.

Goodman, D. 2003. The quality 'turn' and alternative practices: reflections and agenda. *Journal of Rural Studies*, 19(1), 1–7.

Goodman, D. 2004. Rural Europe redux?: reflections on alternative agro-food networks and paradigm change. *Sociologia Ruralis*, 44(1), 3–16.

Grant, R.M. 1991. The resource-based theory of competitive advantage: implications for strategy formulation. *California Management Review*, 33(3), 114–135.

Ilbery, B. and Kneafsey, M. 2000. Producer constructions of quality in regional speciality food production: A case study from South West England. *Journal of Rural Studies*, 16(2), 217–230.

Kvam, G.T. 1995. *Teknologioverføring fra et FoU-miljø til små og mellomstore bedrifter*, Doctoral Thesis 1995: 41, Norges Tekniske Høgskole, Trondheim.

Kvam, G.T. 2006, Norway WP6 country report for the CORASON project. Available at: http://corason.hu/download/wp6/wp6-norway.pdf [accessed: 3 November 2009].

Kvam, G.T., Hålien, E. and Olsen, P.I. 2003. Utvikling av nisjematorientert næringsutvikling i distriktene – lokale organisasjonsbehov og – muligheter. *LØF*, (1), 27–38.

Magnus, T. and Kvam, G.T. 2008. *Vekststrategier for lokal mat*. Frekvensrapport nr. 8/08, Norsk senter for bygdeforskning, Trondheim.

Marsden, T., Banks, J. and Bristow, G. 2000. Food supply chain approach: exploring their role in rural development. *Sociologia Ruralis*, 40(4), 424–438.

Molander, B. 1996. *Kunskap i handling*, Göteborg: Bokförlaget Daidalos AB.

Murdoch, J., Marsden, T. and Banks, J. 2000. Quality, nature, and embeddedness: some theoretical considerations in the context of the food sector. *Economic Geography*, 76(2), 107–125.

Polanyi, M. 1967. *The Tacit Dimension*. Garden City, NY: Doubleday Anchor.

Riddervold, A. and Heuch, H. 1999. *Rakfisk*. Oslo: Teknologisk Forlag.

Stræte, E.P. 2008. Modes of qualities in development of speciality food. *British Food Journal*, 110(1), 62–75.

Tovey, H. and Mooney, R. 2007. CORASON final report. [Online] Available at: http://www.corason.hu/download/final_report.pdf [accessed: 3 November 2009].

Tregear, A., Arfini, F., Belletti, G. and Marescotti, A. 2007. Regional foods and rural development: The role of product qualification. *Journal of Rural Studies*, 23(1), 12–22.

Wernerfelt, B. 1984. A resource-based view of the firm. *Strategic Management Journal*, 5(2), 171–180.

Chapter 9

Traditional and Artisanal Versus Expert and Managerial Knowledge: Dissecting Two Local Food Networks in Valencia, Spain

Almudena Buciega Arévalo, Javier Esparcia Pérez
and Vicente Ferrer San Antonio

Introduction

In the last two decades we have seen a significant growth in both the availability of and the demand for quality food. On the demand side, consumer interest is increasing and people are ready to pay more, not only for products of higher quality in terms of raw material, characteristics of the production process or exclusiveness, but also for products embedded in local traditional knowledge and connected to the territory. On the supply side, in the absence of other productive activities, some rural areas have been able to reconvert existing local know-how into marketable commodities able to create employment and generate wealth, while contributing, at the same time, to the preservation of cultural identity. In fact, local foods have been conceptualised by many authors as a form of cultural capital, with the potential to generate wider social and economic benefits for rural areas. Several empirical studies have indicated that regional foods can indeed play this role (Tregear et al. 2007).

Many rural areas have become involved in important processes of transformation in their traditional food production systems, articulated around a few key objectives, such as improving product quality, enhancing traditional production processes through adoption of innovations, expanding into external markets, marketing and publicity. Behind many successful experiences in the agro-food industry we find small family firms that have been in the business for many generations. It is in fact striking how, being the agro-industry the most important Spanish manufacturing sector, 82 per cent of the companies in 1999 were small companies employing less than 50 employees and thus constituting what is close to being the most fragmented industrial fabric in the European Union (EU), surpassed only by Italy with 88 per cent in the same typology of enterprise (FIAB 2004). One rather interesting way of counteracting the potential disadvantages of the small size of these firms has been to build alliances, partnerships and other relationships of cooperation which, under the umbrella of a territorial or quality label, serve to promote local agro-food products. These alliances usually have an institutionalised profile; since 1992 the EU, along with national and regional government, has been promoting the protection of agro-food products under different quality labels. EEC Regulation 2081/92 launched the *Protected Designation of Origin* (PDO) and the *Protected Geographic Indication* (PGI) labels, which in the Spanish context are the equivalent of the *Designation of Origin* (DO) and the *Designation of Specificity* (DE). To these two, Spanish legislation has also added the *Generic Designation* (DG). The PDO and PGI labels derive from a wine qualification system that dates back to the nineteenth century, when protocols were developed to protect producers in French wine-growing regions from fraud, following the phylloxera outbreak. These labels imply that the character of the qualified products is linked to physical (for example, soils, climate) and cultural features (for example, traditions of production and processing) of a territory (Tregear et al. 2007). According to Ilbery et al. (2005) there are many complex reasons for the proliferation of such schemes. They include the need to increase revenue; to protect and enhance the environment; to defend local traditional products and the social and economic structures that sustain them; to find alternative and more socially just means of producing food.

The potential that local or regional food systems may have for rural development can vary depending on the strategy that the relevant actors adopt when promoting them. Pacciani et al. (2001) distinguish two types of strategy: *the supply chain strategy*, which implies the building of a strong network of actors around the production and processing of a regional product. Under this approach, the regional product contributes to socio-economic well-being, increasing revenues and employment opportunities (Tregear et al. 2007). In the second type of strategy, based on *territorial quality*, actors see local food interlinked to other environmental, cultural and socioeconomic resources. Thus, regional foods are perceived as contributing, potentially, to a wide range of initiatives that encourage diverse activities and novel interactions between actors (for example, tourist trails, markets, festivals, educational initiatives, community events).

In this chapter we present two case studies, both located in the same geographical area, the Utiel-Requena *comarca*,[1] (see Figure 9.1) that highlight the two different strategies referred to above. One case study is related to the construction of the label of origin for the Utiel-Requena wine DO: that initiative corresponds to the supply chain strategy, though in the last years it has also been developing in the direction of the territorial approach. The other concerns the Requena's cold meat and might fit more appropriately into the second type of strategy, albeit not without a mixture of some features of the first. Both cases are interesting in the way they combine tacit and expert knowledge, not to mention the equally interesting networks of cooperation established around them.

---

1 A *comarca* is a physically and historically defined territory, grouping together a number of municipalities.

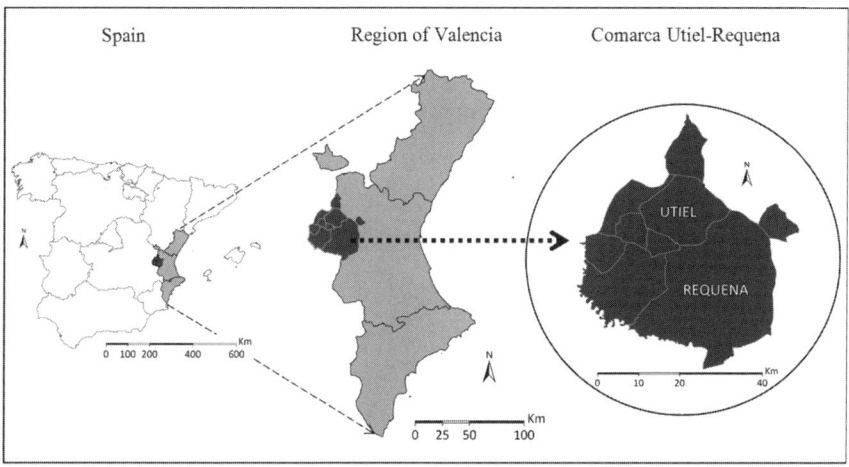

Figure 9.1 **Study area in the region of Valencia**

The context

The agro-food production system is of importance for the whole economy of the region of Valencia. In 2006, the agro-food industry accounted for 20.6 per cent of total Valencian exports, occupying a position in the ranking of export sectors second only to the automobile and automobile components (Conselleria de Agricultura, Pesca y Alimentación – Agriculture, Fisheries and Food Department, Regional Government – 2008). But these figures may serve to conceal the existing divide in the region between an intensive irrigated agriculture and a dry-crops agriculture which is not profitable, is highly subsidised, and is the predominant form of agriculture in the backward rural areas of the Valencia region. It is in these areas that strategies focused on the enhancement of product quality and establishment of linkages with specific localities have gained increasing importance as a way of adding value and differentiating products in a highly competitive market (Moreno Sánchez 2005).

Among the most important agro-food products manufactured in the Region of Valencia are olive oil and wine; products conspicuously present in the rural landscape, highly embedded in the culture and economy and traditionally an important source of income in the rural family economy. Other products (for example pastries) are widely diffused and produced in different types, each reflecting the differences in the know-how involved in producing them. Finally there are products whose identity is linked to specific places, where there are certain raw materials or physical conditions, around which a specific expertise has coalesced (for example, meat products, cheese, herbal products). In some cases these products have succeeded in overcoming the limitations of local markets, becoming differentiated products appreciated in wider markets.

In examining this process it is worth looking at the role played by institutions, including the existing context of qualification and labelling systems. The qualification process effectively transforms the local knowledge and the natural resources embodied in regional food (the basis of the added value) into the collective intellectual property of the relevant actors (Thiedig and Sylvander 2000). In the Region of Valencia's government, the *Conselleria de Agricultura, Pesca y Alimentacion* (Agriculture, Fisheries and Food Department) is the key institution for its competencies in agro-food products. Situated within this department is the Valencian Institute for Agro-food, a body that promotes 16 different agro-food products under two main quality labels: *Origin Designation* (DO) and the *Protected Geographic Indication* (PGI). Each product has its own regulatory board, which is a partnership of stakeholders with an interest in the particular product.

In a manner that is both compatible with and complementary to these two existing labels, in 1998 the Regional Government created what was at first called *Calidad Valenciana* Valencian Quality label, and later became known as the *Certificación y Validación*, Certification and Validation label. These labels respond to the consumers' increased concern for quality and protect products from possible fraud and misunderstandings (Moreno 2005). To date, labels have been made available for olive oil, honey (rosemary and orange blossom) and some types of cold meat.

The next sections of this chapter deal with two study cases exemplifying possible alternative trajectories for the evolution and consolidation of local agro-foods. Both converge on the same point albeit via very different routes. Information gathered for the purpose of documenting these cases is derived, on the one hand from secondary sources and, on the other from in-depth interviews with 10 relevant stakeholders and experts.

Case study 1: The Utiel-Requena Label of Origin (Utiel-Requena DO) for wine

Context and background

La Plana de Utiel-Requena is a *comarca* in the west of the region of Valencia. It includes nine municipalities, among them Requena and Utiel, the most important urban centres. In 2008 the area had a population of 39,970 people and contrary to what is typically the case in inland rural areas, the broad demographic trends for the whole *comarca* have been positive in the last 15 years. Nonetheless, population growth is mostly concentrated in the two main cites, Utiel and Requena with approximately 12,000 and 20,000 inhabitants respectively. Most of the other smaller municipalities have lost population in the last 10 years.

The Utiel-Requena Plain is a territory where viticulture is the main agrarian activity and indeed the base of the area's economy (Piqueras 1998). The service sector, the most important in terms of the number of jobs created, is very much linked to the production and marketing of wine. This, in other words, is a traditional wine-producing region, with one of largest, but at the same time most compact, vineyard areas in Spain. The production of wine here dates back to prehistoric times, but it is only in the latter half of the eighteenth century, a time of great expansion in the manufacturing of spirits, that it became a commercial activity. The year 1854 is of particular note, for it was then that Catalonians and French manifested a clear interest in the wine produced in this area, which was seen as ideal as a complementary ingredient to lend colour and strength to their own wines without altering the original flavour. The key element behind this attractive red colour was a variety of grape called *bobal*, an autochthonous variety extremely well adapted to the area's climate and soil. It is the most important type of grape in the area in terms of volume of production, followed by *tempranillo* and *garnacha*, introduced more recently.

At that time, the increase in the demand for this type of wine, known as 'doble pasta' – or double pomace, encouraged local investment in the planting of new vineyards, and the area became highly specialised in the production of this type of 'colouring wine'. It is important to note that the darkening produced through addition of *bobal* is very difficult to obtain using other varieties so this was, and still is, a much appreciated product in this particular market.

The commerce soon gave rise to the first growers' association: the Guild of Growers of Utiel (*Gremio de Cosecheros de Utiel*), founded in 1861. It brought together 35 large producers whose objectives were to improve cultivation and processing techniques while at the same time securing control over the local and foreign trade (Redacción 2008). Until 1977 farmers concentrated on the production of this type of wine, which was then sold in bulk to other parts of Spain and France. The priority was not quality, but quantity. The aim was to produce a large volume of wine of the darkest possible colour. No attention was paid to modern techniques or to oenologists' criteria for improving wine quality.

Priorities did not change until the 1980s, when there were two significant developments. On the one hand, agrarian overproduction encouraged changes in the Common Agriculture Policy (CAP) orientation, overproduction affecting not only wine, but also other products. The EU policy that, between 1985 and 1996, encouraged the progressive abandonment of vineyards was not totally negative for the area, because it contributed to renewal of the existing vineyard areas, that were old and unproductive. Moreover, in the wine sector, there was a gradual but fundamental re-orientation towards the production of quality wines, implying an important upgrading of technical standards in the sector. Farmers carried out an important restructuring of vineyards, replacing old *bobal* plantations with other 'classic' varieties such as *tempranillo* and *garnacha*, in order to improve the quality of the wine. The 1990s saw introduction of other internationally recognised varieties, such as Cabernet-Sauvignon, Merlot and Chardonnay.

Though even today wine in bulk for colouring or for cheap consumption accounts for a high proportion of the total wine produced in the area (more than two thirds of the total production), in recent years the Utiel-Requena area has been gradually increasing the proportion of quality wines in its production. These quality wines are properly bottled and labelled; 60 per cent of them are exported and the rest is sold mostly within the region of Valencia. Various factors lie behind this shift: the need to adapt to market demands; requirements of the EU; the professionalisation of the sector, with an increasing presence of technicians and experts in the production of wine; the introduction of external capital; and the establishment of the label-of-origin system for wine made in the Utiel-Requena area (Utiel-Requena DO). However, contrary to what one might think, the last factor is regarded by experts as probably the least relevant in all this process of change.

Relevant actors and their network: objectives and actions

Viticulture and winemaking in Utiel-Requena involve a complex of interlinking networks including farmers, wine cellars, cooperatives, large and small firms, technicians, public administration, the DO Board, intermediaries, consumers and so on.

As far as local actors are concerned, it is possible to establish a time-chart showing how historically they have been intervening in the processes of wine production in the area. Prior to the First World War subsistence wine production predominated. Farmers used to cultivate vineyards and make wine that was later consumed within the families. At this stage, when farming was the most significant economic activity, crop cultivation was of more importance than winemaking. Later, when wine became a commercial product, (with the production of doble pasta) farming became something different from wine-marking, which was subsequently undertaken by cellar-men. With the development of commercialisation, cellars, that originally had a common family profile, might become one of three different types: (i) There were still some family cellars which reproduced the old model, where farmer and winemaker coincided in the same person; (ii) There were commercial private cellars, some owned by foreign companies, which were attempting to control production *in situ* and to reduce costs; (iii) There were wine-cooperatives, large and small, grouping most farmers. These acquired considerable strength, and remain important to this day.

According to Piqueras (1998), in this area the most important 'actors' in setting the profile of today's wine production scene are: 1) the School of Wine-producers and Winemakers, established in the 1950s; 2) the School of Viniculture and Oenology, which, starting from 1961, became essentially a continuation of the aforementioned school; 3) the second-degree cooperative COVIÑAS, which in 1968 introduced onto the market the first quality wine produced in the area.

In the late 1970s a process of change in the direction of specialisation, greater reliance on technology and technical knowledge, and the search for quality was spearheaded by the Requena *School of Viticulture and Oenology*, and the Regulating Committee of the *Utiel-Requena Origin Designation (DO)*.

The 1980s saw the consolidation of commercialised wine production, with an increased importance of private firms. Wine cooperatives too began to adopt an entrepreneurial philosophy and practices, with a gradual increase in specialisation and a search for highly differentiated products (single-variety artisan wines, wines of complex structure and so on) for niche markets, where small and family-owned cellars could compete. The new structure consolidates the separation of functions between vineyard owner, farmer and winemaker. Farmers accordingly become grape producers while wine is produced by winemakers under the supervision of highly specialised technicians (oenologists, agricultural engineers and so on).

A number of different types of actors are involved in present-day wine production: big firms with foreign capital that have set up operations in the area in order to reduce costs (for example Murviedro or Hen); small family cellars; new cellars owned by entrepreneurs coming from other economic activities, attracted by this type of business. But agrarian cooperatives are still the most important and the most numerous structures in the Utiel-Requena viticulture sector. The area has one of the highest levels of participation in cooperatives by vineyard owners; 2,400 owners took part in the creation of the area's 36 cooperative cellars. By 1982 participants had risen to 10,000 (Redacción 2008). The cooperatives are associative structures involved in activities pertaining to production, commerce, provision of services and finances and aimed at improving conditions for farmers. Following Catholic Church '*one village, one cooperative*' doctrine on rural associationism, agrarian cooperatives used to be an inseparable element of life in rural areas. The Valencian agrarian cooperative movement originated as a protest movement historically linked to catholic institutions and political parties (Gómez 2004). Agrarian cooperatives were strongly embedded in the culture of every rural village and – because they used to include most of the important actors in rural political and cultural life – their importance went beyond the productive and economic functions. They were an instrument with the potential to generate social capital and a very specific form of local governance and democracy. In the Utiel-Requena area, cooperatives remain to this day key structures in the dynamics of the agrarian sector, and particularly in the production of wine.

Professionalisation and co-existence of new and traditional structures

In this area people have traditionally been dedicated to functions linked to the production of wine. Since the beginning of the twentieth century winemaking has been the most important economic activity. With wine production, however, contrary to what applies with other products also manufactured in the area, such as cold meat, traditional know-how acquired over years of experience has not been at the centre of the current production of quality wine. The reason for this is that the area has been highly specialised in producing *doble pasta* (that is, wine for adding colour to other wines), rather than in making quality wine for direct consumption. It has therefore been necessary for technical and specialised knowledge to be introduced from outside, which has affected everything in the winemaking process, from cultivation and harvesting techniques to marketing and sales.

Even in the 1960s, before the creation of Label of Origin, professionalisation and technification of wine production was well advanced. One key factor in this development has been the presence of training structures imparting technical knowledge for the production of quality wine. The story starts in 1920, with the emergence of the *Estación Enológica de Requena* (Requena Oenology Institute). A pioneering institution, it made a great contribution to defining the economic strategy for wine production that should be implemented to support the local economy. The *Escuela de Capataces Bodegueros y Viticultores* (School of Wine-producers and Winemakers)' started in 1961 and later became the *Escuela de Viticultura y Enología de Requena* (School of Viticulture and Oenology). The first-mentioned school employed local technicians who had been educated and trained in the old wine-producing regions of Spain and France. It remains a key point of reference in Spain and maintains the tradition of employing local technicians as teachers.

Very important in this process was the presence and demonstrative potential of small innovative cellars that have invested in quality and encouraged others to do so. Cooperatives have in fact jumped aboard this bandwagon since the 1980s, implementing modernisation and reorganisation of infrastructures, equipment and management. Spain's entry into the European Community in 1986 was marked by an expansion of second-degree cooperatives in response to the new framework of regulation, necessitating the creation of larger structures with greater technical, human and financial capacity and higher competitiveness. These processes of amalgamation have been encouraged by various sectors of government and, notwithstanding the criticism they have encountered from some producers, brought benefits to the sector including new organisational models more conducive to efficient management. One good example of restructuring and adaptation to market demands is the second-degree cooperative COVIÑAS. This cooperative has three directors (technical, commercial and financial) and a number of technicians (oenologist, agrarian engineer and so on) that make and guide decisions within the cooperative. Some interviewees recognise that this type of professionalisation is fundamental prerequisite for producing and marketing competitive quality products.

The case of COVIÑAS is a good one in the sense that it addresses one of the most important challenges facing the agricultural sector in general and the wine sector in particular, that is adaptation of traditional institutions such as the agrarian cooperatives – which in this region are so important – to the exigencies of the market economy and the entrepreneurial world. According to the interviewees there is no need of modernisation in terms of infrastructure and equipment, as important improvements have already been made in that respect, but first and foremost in terms of organisational structures and decision-making mechanisms. Decision-making processes in cooperatives were in the hands of the Directive Committee and in some cases even of the whole Assembly, the latter having evident advantages in terms of democracy, networking and good governance generally. Admittedly, some have put forward the criticism that, considering the profiles of the participants, these decision-making structures may become static, reluctant to change and an impediment to strategic decisions and economic efficiency. It is undoubtedly true that personal issues (disputes, personal relations, rivalries and so on) and the specific situations of individual farmers can acquire unwarranted ascendancy over common interests. Some interviewees made the point that '*this is a very traditionalist model in a highly competitive world*'. One important consideration is that for most farmers agriculture is a part-time occupation, with the result that the intensity of their involvement and their capacity to assume risks is much lower than in the case of full-time farmers who have viticulture as the main source of income.

In the case of COVIÑAS one of the key elements in its success has probably been harmonic conjunction between the interests of the second-degree cooperative structure and those of the smaller cooperatives. COVIÑAS controls the part of the local cooperatives' wine production that is bottled and marketed by them, but it is up to each local cooperative to decide what to do with the other part of the production. It is usually sold in bulk and the result is that while the local cooperatives are subordinated to COVIÑAS, at the same time they still maintain a considerable autonomy.

In terms of knowledge integration, this type of cooperative combines the highly professional knowledge and expertise provided by the management structures with the traditional lay knowledge embedded in the local cooperatives. The expectation is that in practical terms most decisions will be informed by that technical knowledge. Therefore, as some of the interviewees put it, even when the Directive Committee has been taken over by the presidents of the local cooperatives with a risk that personal tensions and rivalries are brought into the Committee, agreements are reached as long as clear economic benefits are visible.

The role of the Label of Origin system (DO)

The Utiel-Requena DO, a broad formal network integrating all relevant actors linked to wine production in the Requena area, warrants a special section to itself. The Label of Origin was created in the 1970s in consequence of a national law prescribing that each wine-production area in Spain should have an Origin Designation Label (DO). Utiel-Requena DO became fully operative in the 1980s, so it is really quite a young DO compared to others in Spain, and especially to Rioja and Ribera del Duero, which were established long before the passage of the relevant law.

Utiel-Requena DO groups together the whole *comarca*; its current registered membership comprises 108 owners of lands totalling almost 39,000 ha, where different varieties of red and white grape are cultivated.

The Label of Origin has four main functions: (i) Control of quality through the different stages of production and commercialisation; (ii) Protection and promotion of the varieties of grape that are grown in the area; (iii) Provision of informational and promotional material to farmers, wine producers, cellarmen and so on, to enable them to achieve required standards of quality; (iv) Implementation and management of promotional actions. Label of Origin is managed by a *Regulatory Board* (DORC), an entity under the jurisdiction of the Regional Department of Agriculture, Fisheries and Food. The Board's directive committee currently includes 19 elected members who are stakeholders from all over the area's wine-producing sector: wine-producers, members of agrarian groups and cooperatives, delegates from private firms and the regional administration. Responsibility for wines tasting is in the hands of the *Tasting or Qualification Committee*, which is independent from the DORC and currently includes a group of 26 technicians. It is these technical experts who have the job of deciding whether wines are entitled to the DO label that testifies to their quality.

In order to be sold under the DO label, wine production must go through a series of technical control processes starting in the vineyards and ending at the marketing stage. According to the DO publicity committee, the most important function of the label system is technical, involving control not only over the quality of the wine, but also over the crops. A farmer registering his crops with the DO is required to comply with some technical specifications covering such matters as fertiliser use, density of cultivation and so on.

Almost all of the region's vineyard area is today included in the DO registration. But not all of the wine production from these hectares is bottled and sold under a commercial label. Cooperatives, in particular, and big enterprises, leave most of their production to be sold in bulk. Smaller firms and cellars concentrate on the production of bottled wine with a label. Producers who have registered part of their production under the DO can also process non-DO wine as long as the two production processes are kept entirely separate.

Through their insistence on a strong link to the ecology and culture of specific places, label-of-origin systems re-embed a product in the natural processes and social context of its territory. But it is interesting to analyse how social systems of organisational coordination institutionalise and legitimate a given interpretation of the product. To be effective this legitimating process must not only be carried out within the territory of production but must be nested within multiple levels of coordination from the local to the global (Barham 2003).

It would seem reasonable to assume that, apart from encouraging and ensuring the production of high-quality wine, the DO board should be responsible for promoting these linkages. But some stakeholders criticise DO management for not being able properly to market the linkages between wine and the local area, with consumers not perceiving the wine as a territorial product coming from an important wine production area, as one sees, for example, with Rioja or Ribera del Duero. One of the interviewees associated with the DO stated, surprisingly, that: 'The DO does not have to sell a territory, it must sell a product. Some consumers may like to know where the products come from, but others may be interested only in consuming a high-quality product. It doesn't matter where it comes from'.

Effective promotion should instead be based on a territorial strategy, promoting the autochthonous variety, that is *bobal*, and the production of wine typically produced with it. The overall objective should be to achieve a distinctive product and change negative consumer perceptions of this variety.

To cite another failure of the Label of Origin objectives, as identified by Barham (2003), inclusion under a common label has not succeeded in building cohesion among producers and municipalities. Historical rivalries between the two main cities in the area, Requena and Utiel, are not easily overcome. Localisms and individualisms overshadow common interests. For instance, as pointed out by one of the interviewees, '… when we go to fairs each has its own stand … the DO has its own stand, but Requena (council) has also got one with Requena wines, and Utiel do the same, even when both villages are included in the DO'. The fact that the DO includes in the same forum a variety of very different producers (for example, big and small cooperatives, enterprises, limited companies, small artisan enterprises, family cellars and so on), with correspondingly varied strategies and objectives, adds further complications. Finally, we should bear in mind that the DO is not only a technical, but also a political structure, with power games between distinct types of actors within it. Decisions made by the DO are very much linked to political decisions, and this may translate into significant internal disputes.

In contrast to these criticisms, on the positive side, the DO structure has succeeded in moving beyond its merely formal functions to do with control and marketing of wine and, in collaboration with other public and private actors, has become involved in such actions as promotion of the Utiel-Requena DO Wine Route as a tourist product. An association was created to manage this initiative, including a variety of actors participating in the project. The Utiel-Requena DO Wine Route Association brings together a representative sample of the local agro-food system: 21 cellars/producers; one wine shop; four heritage assets (three museums and some historic old cellars); eight restaurants; six accommodation establishments; two artisan firms; along with some public and semi-public structures: the Utiel-Requena DO Regulatory Board, the Requena cold meat Protected Geographic Indication, Mancomunidad Lands of Wine, and the 'Utiel-Gastronomic' Quality Label. The initiative is of great interest owing to its capacity to generate synergies in the territory, strengthening the linkages between wine, place and local culture and involving a significant number of key protagonists. Despite all this, the initiative is not without its critics. Some think that not the DO but a different organisation should be leading it, providing the process with a more integrated profile. There does in fact appear to be a problem of legitimation of the DO structure, some of the mistrust being political in character. The DO Regulatory Board is after all in the jurisdiction of the Agriculture, Fisheries and Food Department of the regional government and this is something that cannot be overlooked.

Strategies

Scientific studies have proven the quality of the Requena's autochthonous grape (*bobal*) and its potential value in production of both young and matured wines. In their quest for differentiation some smaller cellars have started working on the production of good wines using this local type of grape. It is in fact acknowledged by a number of different producers, not only by small cellars but by larger cooperatives as well, that the production of *bobal* wine is probably the best way to find an outlet for a distinctive and recognisable product in a highly competitive market. In an area where quite a few new varieties have been introduced in response to market demands, to promote the manufacture and maturation of wines that use the local variety of grape can be a good course to follow. It is indeed the strategy that has recently been promoted by the DO. The objective is to have this variety of grape valued and known as linked to Utiel-Requena in the same way that *tempranillo* is linked to Rioja. It is a strategy in glaring contrast to DO's initial policy, which consisted in introduction of new varieties to the detriment of the local variety.

Agro-food firms choose one of two main options when designing their competitive strategies: they seek to achieve competitiveness either through cost leadership or through consumer value creation. While the former is to be found in commodities markets for undifferentiated agricultural products where price is the most important consideration for buyers, the latter is more appropriate where there is a differentiated final product with multiple attributes designed to cater for an increasingly segmented and personalised consumer demand (Sanz and Macías 2005).

The Utiel-Requena area seems to combine both strategies, with one prevailing at one time, the other at another. Indeed at the end of a long process during which numerous efforts have been made to ensure that the area becomes a landmark in the production of quality wines, a subterranean debate still continues on the expediency of this or that strategy. Some producers look back to past achievements in producing cheap wine (*doble pasta*) sold in bulk. Others are more impressed by the idea of the area finding its niche market through specialisation in production of single-variety wines made from the native *bobal*. Reasonable doubts have been articulated in relation to this last strategy, the most important relating to over-production. Despite the elimination of a considerable proportion of the former vineyards, the area continues to produce much more than recommended by the EU, so that producers cannot escape from having to sell part of the production in bulk. Secondly, there is competition from other very efficient and more consolidated wine producing regions (for example Rioja or Ribera del Duero), and from regions producing with lower production costs (for example Chile). Thirdly, the high number of intermediaries in the commercialisation channel has the effect of increasing final prices. The fact is that the market externalisation strategies followed by most producers (60 per cent of the DO production is sold abroad) and the long production-consumption chains impact negatively on benefits both for farmers and producers. That is why some firms, family cellars and cooperatives try to control the distribution process by selling the products, for instance, directly to restaurants and specialised shops.

Final remarks

Whatever the virtues of existing ancient traditions in winemaking, it is indisputable that expert knowledge has a most important role to play in shaping the wine production system in the Utiel-Requena area. But there are different aspects to this type of expert knowledge. As noted by Herbert-Cheshire and Higgins (2004), empowerment is largely a matter of experts providing the training and the technical wherewithal for individuals to exercise personal responsibility within their own particular cultural communities of ethics and lifestyle. Is this the way knowledge and structures of expertise have functioned in the Utiel-Requena area? In this connection it is worth pointing out the positive role of institutions such as the School of Viticulture and Oenology, which educates and trains technicians who then work in the area's cooperatives and other firms. There is also a clear connection between expert knowledge and the success of cellars that have concentrated on a well-defined strategy for production of quality wine. There is also evidently a concentration of expert knowledge in specific structures with an acknowledged capacity to define general strategies for the sector. One important question is the extent to which these structures are efficiently and legitimately integrating the views of all those involved in the agro-food wine system. To the extent that they fail to do this, contradictory messages will be sent to local actors, who will not be able to exercise their responsibilities.

Some successful linkages have clearly been developed between old and new forms of knowledge, as may be seen if one examines this particular productive system. One example is the renewal of traditional agrarian cooperatives introducing specialised knowledge into the different areas of production, management and marketing. But some opportunities for greater enhancement of local knowledge have also been missed. For example, the predominant strategy that was followed for many years was one of introducing new varieties of grapes while downplaying the potential of the native variety of grape (*bobal*) for the production of distinctive quality wine. It has not been until recent years that a different strategy began to be promoted, involving small firms, big cooperatives and the Viticulture school, where research on the use of this type of grape has been taking place. As previously indicated, the viticulture agro-food system is a broad and complex entity, encompassing many different actors, from farmers to final consumers, participating in it in different ways and at different levels. If they are to be successful, agro-food strategies must take into account the knowledge and perceptions coming from the demand side. Some small firms, for instance, are producing 100 per cent *bobal* wine and the final product is very much in demand among highly specialised consumers. However, the market this product addresses is very small, because the *bobal* grape does not have in general a good reputation among consumers. So it is not only that more research and technical controls are necessary in the winemaking process; consumer perception of the product need also to be improved.

Case study 2: Requena cold meat Protected Geographic Indication

The second case we would like to introduce is in relation to an initiative carried out in the same area, Utiel-Requena, and has been selected because it typifies an entirely separate dynamic, with an externalisation process and a pattern of interrelations between expert and local knowledge altogether different from the case presented above.

La matanza has been a traditional practice in the inland rural areas of Spain since medieval times, and still persist in some parts where climate conditions permit it (the climate must be very dry). It involves the killing of one or several pigs, on a certain day, in the course of which entire families of relatives come together to process the meat, preparing cold meat and other similar products to be stored and consumed during the course of the year. The practice was a key element in the rural family economy in the times when this was mainly a subsistence economy, but it had also important cultural and anthropological dimensions attached to it. The introduction of new legislation linked to sanitary controls and changes in personal habits and lifestyles have contributed to the contraction of this practice to the family context and, in some way, to its professionalisation.

The practice is at the origin of current production of different types of cold meat and other meat-based products manufactured in the village of Requena in the Utiel-Requena *comarca*. There is an authentic culture of meat consumption in the area and it may come as a surprise to find that in a village like Requena in the 1980s there were 40 butchers. Today the numbers are fewer, but still considerable. To illustrate the importance of butchers and of the meat-consumption tradition in the area, one interviewee spoke of how young children go into butcher shops and ask for small pieces of *fuet* (thin sausage), which they have become accustomed to getting for free and chewing as if it was toffee.

The evolution in recent years of artisan production of cold meat has been marked by three main events: (i) organisation of Requena's Cold Meat Fair; (ii) adoption of a label for the product, that is Embutido de Requena (Requena cold meat); and (iii) acquisition of a Protected Geographic Indication (PGI) quality distinction label. All these developments have involved mobilisation and a combination, to a greater or lesser degree, of traditional, experiential, and technical forms of knowledge.

The ideas of organising a fair and creating a label emerged at the same time in 1992, but it was easier to implement the first idea, while the creation of a label was a more protracted process that finally came to fruition in 1995. The quality distinction for the product was obtained from the regional government in 1999 and undoubtedly contributed to the success achieved by the annual Cold Meat Fair. The PGI label covers the process of manufacturing the product, but not the origin or characteristics of the raw material used, as happens with labels of origin (DO). The pigs used in making the cold meat do not need to be of a particular autochthonous breed, or indeed to be from the area at all. They need only meet the hygiene requirements prescribed by EU regulations.

Knowledge in local food production

As we have said, the point of departure is the existence of an artisan product highly embedded in the local gastronomic culture and which has also enjoyed some degree of popularity outside. People emigrating out of the area and the proximity to the city of Valencia have probably contributed the products' achieving such popularity as they have acquired. But it has been above all the marketing strategy linked to the organisation of the Fair that has given a tremendous boost to these products.

The producers have introduced a number of new practices, along with instruments that have enabled facilitation of the manufacturing process while ensuring compliance with the necessary standards of hygiene and health, but they also insist that the process should be kept strictly *artisan*. In fact this requirement limits producers' capacity to grow and to expand their markets, but this is something they have taken into account. The interviewees recognised that the key for success lies in the transmission of old recipes and manufacturing know-how from parents to children. The combination of spices and manipulation of the ingredients is one of the prime secrets that each family jealously guards as a precious treasure. These indeed are the features that distinguish one variety from another. As one of the interviewees put it:

> For the product to be manufactured there has not been a need for any additional new knowledge. It is based on transmission of traditional knowledge that has been handed on from parents to children ... Within the association we have held courses on food processing and other related topics. But when it comes to making our products, no-one can give us any advice. We've been doing this for many years.

Tacit lay knowledge therefore constitutes producers' main asset for ensuring manufacture of a distinctive product in a competitive market. Producers follow a dual strategy, including this tacit knowledge as a key component. On the one hand, lay knowledge is involved in the manufacture of a product that to some extent has been codified with the introduction of the PGI regulation. One might imagine that this codification or regulation could limit the potential for introduction of personal knowledge by producers. But in the production process a small margin of variation in the use and combination of ingredients gives the product its distinctive characteristics, and it is this small variation that each family of butchers guards as a precious treasure. On the other hand, Requena cold meat is sold and promoted as an artisan product, manufactured using traditional methods, and this is an important factor motivating consumers to buy it. Producers utilise, and indeed benefit from, the positive perception in the collective imagination of notions such as 'artisan', 'traditional', 'natural', 'hand-made'. These notions are identified (overtly or subliminally) with 'tacit knowledge'. Consumers believe that some venerable, traditional, secret knowledge that has been handed down from one generation to the next lies behind the product that they are buying. This is unquestionably a selling point for consumers: the uniqueness of a tacit knowledge embodied in a product which is based on a secret jealously guarded by each producer.

The granting of a formal label to a product implies that the product has satisfied the criteria established by the institution promoting the label. In the present case, the PGI for Requena cold meat is administered by the Institute of Agro-Food Quality, which is an agency of the regional government's Department of Agriculture, Food and Fisheries. The products must be subjected to analysis by the Institute to determine their composition (proportion of fats, salts and so on) quality, and various taste characteristics.

Though producers may insist that it has not been necessary to acquire additional knowledge in order to improve products quality, since 1992, and with the adoption of the label, they have been gradually acquiring new (technical) knowledge related to handling of specific machinery and specific aliments. Managerial knowledge for better marketing and selling of their products has also been upgraded, with local producers attending international food fairs and there encountering new ideas that may be put into practice at the next Requena Fair. Each new fair launches new products, including new cold meat recipes and new ingredients. Producers clearly favour innovation and adaptation to consumer expectations of new taste experiences. Here too we catch a glimpse of the new knowledge that comes with introduction of variations into the sound foundations of tacit knowledge possessed by local producers.

Social and ecological sustainability

The idea of organising a Fair and creating a label for promoting Requena cold meat originated within a group of local producers who had become uneasy over the impending construction of a motorway in 1992. They had concluded that this new item of infrastructure might have unpredictable effects on their businesses, attracting competitors from the metropolitan area while at the same time discouraging would-be customers to come into the village as before. A group of 12 producers formed the association which then established the Fair. The first edition, held in 1993 attracted 20,000 people, and the number has been growing ever since. More than 100,000 people attended the most recent Fair in 2009. The Fair is well-known in Spain today as the first fair dedicated entirely to the promotion of cold meat.

At the public level, two institutions have been fundamental to the organisation and promotion of the Fair and the marketing campaign for it, and in the creation of the PGI quality label: namely the regional government's Agriculture Department and the Requena City Council. The emergence of the Fair also provides impressive evidence of the way the initiative and the social capital of individuals – in this case the former president of the Requena Cold Meat Association, who maintained very good relations with external political actors – can play a catalytic role in the implementation and the success of new ideas.

The adoption of a Protected Geographic Indication (PGI) for Requena's cold meat necessarily implied the creation of an PGI Regulatory Board, which, when established, brought together the 12 producers that had initiated the PGI, along with representatives of other bodies, such as the local council and the Autonomous Community Government. There are currently more than 20 butchers in Requena. Of these only 11 adhere to the PGI. In fact, no new members have come into the association since the Fair started. Qualification processes are not always either easy or homogeneous, as Tregear et al. (2007) recognise. In practice the qualification process may be dominated by a single set of actors pursuing a single set of interests, so skewing the distribution of rent within a supply chain and/or a territory. If other actors happen to contest the approach of the dominant actors, conflicts may well emerge. In the case of Requena there has not been contestation but passivity; some local butchers are not ready to embark on the readaptation involved in promoting the label because they already have their own specific market segment. The label is not important in the local market because local consumers have other information on the basis of which they can make their choice. But for consumers elsewhere the label is important. It is also important to stress that this is a local initiative, that does not include other municipalities in the area as in the case of wine. The reason put forward in explanation of this is that it is necessary to uphold the quality associated with Requena. Producers are clearly pursuing a strategy of preserving and enhancing local knowledge, something that appears to be of even greater value for as long as it can be linked exclusively to one place.

The fact that an initiative of this type has not managed to involve all the relevant actors, that is local producers, introduces an important public goods' dilemma that can be observed in the tense relations between PGI members and non-members. The latter benefit from the increasing popularity that the product has gained as a result of the marketing campaigns developed in the context of the PGI, while at the same time they are not making any contribution to the development and consolidation of the label. Moreover, there have also been some problems of unfair competition, pointing to an evident lack in this group of entrepreneurs of the kind of social capital that is perhaps necessary for achievement of an effective synthesis of common interests.

Leaving aside these problems, the Fair has nevertheless become an appropriate instrument for the generation of synergies in the local economy and the involvement of other local actors. This involvement has taken place at different levels; firstly, at the institutional level, as we have seen; secondly, other economic actors have participated and have benefited from popularity of the Fair, for example bars and restaurants, shops selling local products, wine cellars. Among these economic actors wine producers and wine cellars are particularly relevant because wine is a key product complementary to the consumption of cold meat and so is very much in evidence at the Fair. Finally, there is the involvement of the local inhabitants who are active participants in the Fair.

Strategy

The typology of firms involved in this initiative – very small family enterprises – not to mention the strategy of preserving the artisan character of cold meat production, place limits on producers' capacity to grow and to reach new markets, and this is something they largely accept. It is, moreover, this objective of manufacturing quality products that makes it possible, in the opinion of the producers interviewed, for all these small businesses to survive. It is the focus on quality that distinguishes them from other industrial producers. People travel to Requena from the city of Valencia (about 70 km away) just to buy these products.

Another key feature of the initiative that also helps to explain some of its limitations are its endogamous and static character. This is attributable largely to structural conditionings. A large investment, for example, is necessary if the legal requirements covering hygiene that are a prerequisite for opening the business are to be fully complied with. There is also a very high element of risk. Competence is one relevant factor here, as are the difficulties in positioning when the market to be entered is a limited one.

With the expansion of the market to the metropolitan area of Valencia, with its 1.5 million potential consumers, some producers have introduced a corresponding shift in their marketing strategy. Currently three out of the 11 producers sell 90 per cent of their production outside the village, and specifically in the city of Valencia, at specialised delicatessens and in a big superstore. Producers not covered by the PGI follow a different strategy, selling their products in the local market, to locals and visitors. PGI producers employ a sales representative to distribute their products to their final sales points. For non-PGI producers the link between producer and consumers is direct. Two strategies thus converge: on the one hand the strategy of short food supply chains directly linking producers with local and external consumers; this strategy has been reinforced through institution such as the Cold Meat Fair, and by virtue of the increase in the numbers of visitors and tourists coming to the village. But on the other hand external food supply chains have also been promoted: producers dispose of their wares on external markets, now selling in shops located in the metropolitan area of Valencia; these products have a special characteristic: they are sold under a quality label that is clearly recognised by consumers.

Current objectives of PGI producers include securing of Designation of Origin (DO) status for Requena's cold meat and to this end the Association has accordingly initiated negotiations with local farmers with a view to tracing the origin of this meat.

Some issues for comparison

The two cases we have presented above represent two different examples of agro-food systems: the first pertains to wine, where the local/global aspect is clearly identified from the outset. The second involves cold meat, whose local/local profile in some cases expands to embrace an external market, but within clear limits established by the production capacity (in distinct contrast to the situation with wine, where there is overproduction).

The different cases also show how the 'quality' that is associated with some products can be differently understood, and how it can be created, transformed or commoditised by various actors. In the case of wine, the quality standards associated with the product seem to correspond to fairly homogeneous and generally-shared criteria operating at the international level, though often heavily influenced by the 'gurus' of wine, who have great influence on production parameters in the various wine production areas. There is a Designation of Origin (DO) Board that has traditionally established the parameters for wine production in accordance with more global trends. In the case of cold meat, the point of departure is the existence of some traditional products and a clear intention to preserve them. The producers' strategy involves enabling consumers to appreciate the singularity of the product, given that it has been manufactured in this specific area. It does not involve modifying the basic recipes so as to adapt the product to some particular demand, though some changes are in fact introduced when it comes to including new and innovative products. For that eventuality there is an organisation for implementation of standardised technical rules to cover product quality, first and foremost in relation to sanitation and safety, in effect codifying tacit knowledge already existing among local butchers.

Both products, wine and cold meat, originating in the same geographical area, share a common point of departure that is a long-standing local tradition in production. That said, the processes involved in the evolution of the wine sector are much more complex than for cold meat, as indeed are the implications for knowledge and its dynamics. We will try to summarise them.

In the case of wine, new knowledge has been gradually incorporated, transforming the traditional product in order to respond to the quality parameters imposed by external markets. Technical and managerial types of knowledge have been fundamental to the advances in quality achieved with wine, indeed to the overall evolution of the wine sector. With cold meat the point of departure was a traditional product embedded in the local food culture, which had already gained a high degree of acceptance in a certain sector of the market. In terms of the implication for knowledge forms and processes, two key features of this product's trajectory are worth emphasising: on the one hand the existence of a certain stock of traditional knowledge on the production of cold meat that has been transmitted from generation to generation as an important family secret; on the other the contribution of specialised marketing knowledge that has contributed to enormously increase the popularity of the product.

In the evolution of Utiel-Requena wine the input of technical knowledge has played a decisive part in effecting the transformation. But the technical knowledge was introduced in such a way as to preclude inter-communication between lay traditional knowledge and technical knowledge. To quote Piqueras Haba (personal interview):

> traditional knowledge in wine production disappeared when co-operatives were constituted and an oenologist was put in charge of the manufacturing process. Before that, old people used to make their own wine. For example, my father used to do it. But then the wine-growing process was separated totally from the manufacture of wine. Nowadays wine-growers know about vineyards but they do not know about winemaking. And wine-producers know how to make wine but they have no idea about vineyards.

In the second case study, technical knowledge was introduced in the form of new equipment that improved efficiency and hygiene in manufacturing and processing. But this innovation had no effect on the traditional knowledge that was at the basis of the products' success, for example traditional recipes, the use of natural tripe for preparing cold meat and so on. As far as preparations for the Cold Meat Fair are concerned, expert knowledge was contributed by technicians working on the local council, who played a facilitating role. There was also an input from a type of expert-experiential knowledge that has been acquired by organisers from the international and national food fairs that they attend.

Another important difference between the two products we have analysed has to do with the size and the characteristics of the networks involved in the related production and marketing processes. Requena cold meat originates with a very limited network of only 12 producers in the same village. Their marketing strategy consists in maintaining tradition and quality even when this means accepting a continued low market share rating. The network is quite homogeneous in terms of the characteristics of its members and in terms of its objectives. It collaborates closely with public institutions and other structures in the area such as the DO Regulatory Board for Utiel-Requena wine. It has not however managed to integrate the other half of the villages' butchers. With the wine the situation is totally different and much more complex: there are numerous different networks of actors linked to the product, because there are over 100 different enterprises in the area working in this field, and a diversity of product strategies. We have focused on the DO network because it includes most of the area's producers, embodying a formalised and standardised technical knowledge that is applied equally to all firms registered under the DO, the objective being to achieve a specific level of quality. Most interviewees agreed that the main obstacle to achieving a more efficient functioning of the DO is the heterogeneity of this group.

Wine production in the Utiel-Requena area is a clear example of a local agro-food local system that has shifted from being a family-based system to being a local-global system of local production and global consumption. Traditionally local wine was sold in bulk and there existed no strategy for the production of a clearly identified product, that is quality wine from Utiel-Requena. Current policy is for a dual local-global strategy, with benefits coming both from the selling of bulk wine and from the manufacture of quality wine bottled and labelled under the DO. By contrast, the strategy being followed by the Requena cold meat producers could be characterised as the kind of defensive localisation that, as Hinrichs (2003: 37) puts it, 'tends to stress the homogeneity and coherence of the "local", in patriotic opposition to heterogeneous and destabilising outside forces, perhaps a global "other"'. This is because the 'tradition' involved in the elaboration of the products is carefully guarded and preserved. The products' qualities are exclusively linked to the village of Requena, and entry into the producers' network is closed to other villages' producers.

Considering the variety of strategies that have been applied and the current situation of both cases analysed, the conclusions could be drawn that wine has been affected by standardisation and only some small cellars and firms are now trying to produce distinctive quality products (for example 100 per cent *bobal* wine). There has been no comparable standardisation with cold meat. It is still a distinctive product whose singularity derives precisely from the preservation of tacit knowledge.

Acknowledgment

The authors wish to thank all those that were interviewed and provided the information upon which this chapter is based.

References

Barham, E. 2003. Translating terroir: the global challenge of French AOC labelling. *Journal of Rural Studies*, 19(1), 127–138.
Conselleria d' Agricultura, Pesca i Alimentació (2008) Comercio Exterior Agroalimentario de la Comunitat Valenciana, in *Informe del Sector Agrario Valenciano 2006*, Valencia [Online]. Avalilable at: http://www.agricultura.gva.es/ publicaciones/revistasint.php?id=4 [accessed 13 November 2009].
FIAB (Federación Española de Industrias de la Alimentación y Bebidas) 2004. *Ganar Dimensión. Una necesidad para la industria agroalimentaria española*. Available at: http://www.mapa.es/alimentacion/pags/Industria/informacion_economica/ganar_dimension.pdf [accessed: 13 November 2009].
Gómez López, J.D. 2004. Las cooperativas agrarias de la Comunidad Valenciana frente al proceso de globalización. *Cuadernos de Geografía*, 75, 1–16, Universitat de València: Facultat de Geografía e Història.
Herbert-Cheshire, L. and Higgins, V. 2004. From risky to responsible: expert knowledge and the governing of community-led rural development. *Journal of Rural Studies*, 20(3), 289–302.
Hinrichs, C.C. 2003. The practice and politics of food system localization. *Journal of Rural Studies*, 19(1), 33–45.

Ilbery, B., Morris, C., Buller, H., Maye, D. and Kneafsey, M. 2005. Product, process and place: an examination of food marketing and labelling schemes in Europe and North America. *European Urban and Regional Studies*, 12(2), 116–132.

Moreno Sánchez, M.T. 2005. *Redes y Estrategias de Comercialización de Productos y Servicios Turísticos de Calidad en Áreas Rurales de la Comunidad Valenciana*. Tesis doctoral. Universitat de València. Depatament de Geografia.

Pacciani, A., Belletti, G., Marescotti, A. and Scaramuzzi, S. 2001. *The role of typical products in fostering rural development and the effects of Regulation (EEC) 2081/92*. Paper to the 73rd EAAE Seminar *Policy Experiences with Rural Development in a Diversified Europe* (Ancona, Italy, 28–30 June). Available at: http://www.origin-food.org/cherch/itss.htm [accessed 13 November 2009].

Piqueras Haba, J. 1998. Cambios recientes en el sector vitivinícola valenciano. 1977–1997 *Cuadernos de Geografía*, 63, 177–194. Universitat de València: Facultat de Geografía e Història. Available at: http://dialnet.unirioja.es/servlet/oaiart?codigo =715267 [accessed 13 November 2009].

Piqueras Haba, J. 2000. Expansión vitícola y reparto de la propiedad, *Cuadernos de Geografía*, 67–68, 351–380, Universitat de València: Facultat de Geografía e Història. Available at: http://www.uv.es/ cuadernosgeo/CG67_68_351_380.pdf [accessed: 13 November 2009]

Redacción. 2008. Los vinos de Utiel-Requena, *Enología y Viticultura*, 2 January 2008. Available at: http://www.noticiasdelvino.com/history.php?dmes=1&anio=2008&pag =5 [accessed: 13 November 2009]

Rose, N. 1996. The death of the social? Re-figuring the territory of government. *Economy and Society*, 25(3), 327–356.

Sanz, C.J. and Macías, V.A. 2005. Quality certification, institutions and innovation in local agro-food systems: Protected designations of origin of olive oil in Spain. *Journal of Rural Studies*, 21(4), 475–486.

Thiedig, F. and Sylvander, B. 2000. Welcome to the club?: An economical approach to geographical indications in the European Union. *Agrarwirtschaft*, 49(12), 428–437.

Tregear, A., Arfini, F., Belletti, G. and Marescotti, A. 2007. Regional foods and rural development: the role of product qualification. *Journal of Rural Studies*, 23(1), 12–22.

Chapter 10
Reclaiming Local Food Production and the Local-Expert Knowledge Nexus in Two Wine-producing Areas in Greece

Apostolos G. Papadopoulos

Introduction

In many of the discussions about local food systems the place of food and the 'localness' of production and/or consumption are set against the modern agro-food system. The socio-spatial embeddedness of local food production is contrasted with the distance and complexity of agro-food chains, which are blamed for loss of agricultural diversity, degradation of the environment, dislocation of community, loss of identity and sense of place (Feagan 2007).

The global agro-food system has delocalised production in order to service the rapidly changing large-scale global food market. From a political-economy perspective the global agro-food system seems invincible, indeed almost impossible to challenge, on many levels. Its major strengths are: its ability to cater for mass consumption by virtue of its mass production capacity, its facility of access to capital, and its profitability. At the same time the system faces many challenges, such as the difficulty of serving smaller, more differentiated markets, reorienting large global food system clusters, developing relationships of trust with consumers and providing a solution to the social and environmental problems it creates (Hendrickson and Heffernan 2002).

In this context there is room for alternatives based on personalised and sustainable visions that activate authentic social, economic and ecological relationships between actors in the food system. 'Local food' may be considered a shortcut formula for seemingly converging movements such as sustainable agriculture and food, food sheds, community food security, civic agriculture, short food chains and fair trade (Feagan 2007, Guthman 2007, Pratt 2007). Local food production thus coincides with 'alternative' food production, implying changes in the practice and politics of food production in the advanced capitalist economies (Niles and Roff 2008). Two main versions of local food production may be discerned: the first is based on the specificities of the locale and is considered 'defensive localism' (Winter 2003b); the second emphasises the local production process, criticising the agro-food system as a whole and prioritising a 'reflexive localism' (DuPuis and Goodman 2005).

This chapter examines two specific networks of local production, and specifically of wine, in order to explore developments, constraints and potentialities in local food production and processing. On the basis of specific guidelines provided in the context of the CORASON project (Fonte and Grando 2006), two study areas were selected. The first, which could be described as the site of a 'local to local' food production network, is the area of Lake Plastiras, located in the Prefecture of Karditsa, Region of Thessaly. The second, which corresponds to the model of the 'local to global' food production network, is the area of Nemea, located in the Prefecture of Corinthia, Peloponnese Region.

Qualitative research was carried out by means of semi-structured interviews with important stakeholders (representatives of cooperatives, officials from the local departments of agriculture, the local administration, producers of wine grapes, winery owners, technicians-oenologists) in both study areas. The field work was carried out in 2005 to 2006 but field visits to Nemea were also organised in May 2008. The research aimed at evaluating the prospects for locally-adapted food production and processing systems relying on available knowledge bases, in the context of ongoing economic globalisation.

The remainder of the chapter is divided into six parts. The first part explores certain theoretical considerations pertinent to the theme of local food production. Next there is a brief exposition of the Greek context. Then follow descriptions of the two case studies, containing a profile of the study area and an analysis of the local production network. After that there is a discussion of the local-expert knowledge nexus in the two areas. Finally, the findings from the two case studies are compared and some concluding thoughts put forward.

Theoretical considerations

Globalisation of the agro-food system, which provides the context for any discussion of local food production, has been a major theme encompassing four interconnected issues: (a) global processes of capital accumulation reshape food production, (b) industrialisation of the food sector requires favourable local conditions, meaning that food chains must always be adjusted to the local/regional socioeconomic contexts in which they are activated, (c) the place specificity of food production is being dissolved and (d) globalisation of the agro-food sector should be treated as a 'contested process' in which the corporate actors 'condition' the actions of local actors and the latter may challenge the dominance of the former (Marsden et al. 1996, McMichael 2000, Murdoch et al. 2000). The agro-food system is a configuration constructed by, and situated within, wider networks and can be analysed by using actor-based notions (Goodman and DuPuis 2002, Goodman 2004).

The emerging discipline of agro-food geography involves the study of new phenomena along the food chain and re-establishment of the relationship between food production and food consumption (McMichael 2000, Winter 2003a, Watts et al. 2005). It includes as one of its elements an increased attention to food governance, deriving from an assessment that there is now 'a more asymmetrical and differentiated understanding of food as a natural, social and political construction' (Marsden 2000: 28). The influence of the state, civil society and the local stakeholders on how food supply and consumption is organised has a significant impact upon the articulation of an 'industrial mode' and/or 'alternative modes' of food provision.

An argumentation has developed concerning the need to establish reconnections in agro-food geography. Four sets of reconnections are sketched out: farming and food, food and politics, food and nature and farmers and agency (Winter 2003a). The ultimate purpose of the call is to rebuild the lost trust of the public in food chain actors and consumers' confidence in food (Smith et al. 2004, Fonte 2008). Despite the note of caution over the implicit neo-liberal utopianism of similar calls (Harvey 2002, Guthman 2007), there is a clear trend in rural research and policy towards proposing a reconnection of product, process and place in the food production system (Parrott et al. 2002, Ilbery et al. 2005, Watts et al. 2005).

It certainly cannot be denied that global agro-food chains are spatially embedded due to the strong interconnections between different localities and regions, rural and non-rural. This does not however mean that the local specificities of food production are taken into account. The way the 'localisation' of the agro-food chains occurs is not moreover conducive to 'localness' of food production. Food and agriculture firms follow a number of different strategies over time. First they horizontally integrate through expansion at the same stage in the commodity system. Secondly, they follow a process of vertical integration, which necessitates expanding upstream and downstream in the agriculture and food commodity chains. Thirdly, they may globalise to reduce uncertainty or to expand business (Hendrickson and Heffernan 2002: 350).

In general, two spatial processes of food production can be discerned. One is the de-territorialisation process linked to the rise and dominance of the conventional agro-food chains; the second is the re-territorialisation process associated with the so-called alternative agro-food chains. As regards the first process Morgan et al. claim that de-territorialisation does not imply that 'it comes without any actual geography; rather its geographies are the result of corporate capitals' attempts to continue to intensify and to appropriate some of the functions of agriculture in ways that stretch the links, networks, and chains between production and consumption spheres' (2006: 53). As for the second, which is conceptually juxtaposed and/or interconnected with the first, they argue that it is 'a process whereby local and regional geographies come back again to play a central role in reshaping food production and consumption systems' (Morgan et al. 2006: 53).

Many researchers emphasise the permeability and multiplicity of global production systems while at the same time recognising their power and influence (McMichael 2000, Goodman and DuPuis 2002, Niles and Roff 2008). It is moreover likely that both the conventional and the alternative agro-food chains are competing or interlocking since both processes may coincide in the same region and at the same time (Whatmore and Thorne 1997, Hendrickson and Heffernan 2002, Morgan et al. 2006). The so-called 'conventionalisation' of organic farming that resulted from industrial attempts to meet the increasing demand for organic produce and healthy food also imposed a productive logic antagonistic to the historic agro-ecological ideals of organic production (Guthman 2004).

The 'food relocalisation' thesis, basically an outgrowth from one of the critiques of the conventional agro-food system, also stresses the importance of product, process and place against the unreliability of the mass agro-food system (Fonte 2008). The 'relocalisation' thesis derives from a discussion of the (re)embeddedness of food production. It implies the necessity for drawing another map of the food sector, highlighting 'those regions and localities that have not been fully incorporated into the industrial model of production and that have retained the ecological conditions necessary for quality production' (Murdoch et al. 2000: 111). The basic assumption is that food quality is intrinsically linked to the 'localness' of food production and that there is therefore a profound relationship between local production relations, local culture, local knowledge and food quality. Some researchers moreover indicate that the 'embeddedness' of local food systems has a bearing on a variety of consumer motivations for supporting them, ranging from defensive localistic attitudes to direct marketing of food products (Hinrichs 2000, Winter 2003b).

The growth of agro-food networks and more particularly the trend towards food relocalisation has been encouraged within the EU, partly as a result of the attempts to pursue sustainable rural development in the countryside (Fonte and Grando 2006). Food relocalisation is an aspect of difference, three types of which might be identified: a) difference in 'quality' between locally specific and mass-produced products, b) difference between geographical anonymity of food provenance and territorial specificity and c) difference in the way and processes that foods are produced (Ilbery et al. 2005). The way to verify such differences and communicate them to consumers is through labelling and certification schemes that attach locality to food products. The best-known certification scheme is the French *Appellation d'Origine Contrôllée* (AOC) classification, which was created in 1935 to protect the reputation, authenticity and quality of wines (Moran 1993a, Moran 1993b, Gade 2004). The central idea behind this scheme is the notion of *terroir*, implying a close linkage between particular territorial, natural and cultural characteristics and quality attributes of food products (Barham 2003). Many writers consider that the *terroir* concept serves to justify differences among agricultural products and establish intellectual property rights over the use of a certain territory (Moran 1993a, Moran 1993b, Broude 2005, Josling 2006).

The role of knowledge forms and structures is considered important for local food systems. Critics have challenged the technocratic optimism of the conventional food system, which has been based on the top-down application of science to predict and control agricultural conditions and food security. The traditional and local knowledge that is inscribed in the notion of '*terroir*' is pivotal for local production. What has been called 'local ecological knowledge' (a subset of local knowledge in general) has been defined as a 'wide array of practical skills and acquired intelligence in responding to a constantly changing environment' (Scott 1998: 313). The knowledge form that is generated in specific *terroirs* is place-specific, situation-related, detailed, and often contrasted to expert/scientific or managerial knowledge. A number of local communities have developed extensive expertise based on observation and interaction with their local natural environment. The utilisation of such place-based knowledge can improve the understanding of local conditions. For viticulturalists and oenologists alike not only viticultural and oenological techniques but also human factors such as history and socioeconomics are part of *terroir* (van Leeuwen and Seguin 2006).

Local wine production, which is considered a typical commodity chain, depends on cultural taste (consider Bourdieu's concept of a 'cultural capital' that creates distinctions); its distinctiveness is certified by a name. Knowledge of how wine is made and how it can be consumed is often a sign of class and can be analysed as a form of 'cultural' capital. Style of wine is related to regional cultures and thereby embedded in those practices that turn regionality into a way of life marked by distinctive structures of feeling (Harvey 2002: 9).

The distinction between different kinds of wine as between the old and the new wine-producing countries is largely based on knowledge forms and their characteristics. The advocates (both producers and consumers) of the '*terroir*' wines stress the place particularities embodied in wine, that are the result both of the natural characteristics of the soil and of the traditional and expert knowledge incorporated in the treatment of vines and in vinification techniques long experimented with by local producers. On the other hand the new wine countries (for example Argentina, Australia, Chile, New Zealand, South Africa) attach more importance to expert vinification techniques that bring about a transformation in wines, making them popular among a wider consumer population not habituated to older wine tastes.[1] There are two generic wine tastes: one is the 'cosmopolitan taste' for a standardised, global product that has developed over the past decades; the other is a 'locally differentiated taste' that is linked to lived experience of *terroir*.

The writings of wine experts[2] have moreover been instrumental in promoting awareness of quality in wine and so recognition of the importance to consumers of subtle differentiations in the characteristics of different wines (Harvey 2002, Gade 2004). Together with large retailers, these writers play a mediating role between the worlds of production and consumption. Their knowledge systems contribute in

---

1 Expert knowledge dispensed by 'flying winemakers', that is frequently travelling oenologists who are consultants for winemaking companies, is a recent powerful factor exerting significant influence on the global wine industry (Lagendijk 2004). A good illustration of this development can be seen in the documentary film *Mondovino* (2004) directed by Jonathan Nossiter.

2 Wine critics such as Robert Parker in the USA, Jancis Robinson in the UK and James Halliday in Australia comment on differences between brands and on various labels.

various ways to upholding the education of consumer taste in the direction of more varied and more differentiated wines (Gwynne 2006).

Finally, despite the apparent gap between 'industrial wines' and '*terroir* wines' there has been a certain convergence of the two patterns in recent decades. The French notion of *terroir* was initially adopted by the 'new' wine producing countries, but in the process more attention was paid to brands and wine varieties as an indication of quality in wine production. On the other hand, the notion of *terroir* has slowly adapted to the changing conditions of the wine market and cannot be considered solely in terms of soil, climate and environment (van Leeuwen and Seguin 2006). The term should be regarded as a shorthand designation for an ensemble of local social and cultural characteristics that allow for construction of the pattern of local governance conducive to the establishment of connections between 'human' and 'natural' factors (Jones and Clark 2000, Gade 2004). *Terroir* is thus used to account for the hierarchy that prevails among high quality wines, focusing specifically on the relationship between the quality, taste and style characteristics of wine and its geographical origin, the implication being that the latter might influence the former (van Leeuwen and Seguin 2006).

The Greek context

In the European context, Greece has the longest history of vine growing and winemaking. Vines are very common in Greece and are often cultivated alongside other crops. The depopulation of rural areas has led to a decrease in the amount of land in which vines are planted. This trend is especially true of mountainous and arid regions. Greek vineyards are often made up of little plots, in some cases few of them owned or maintained by one farmer. The result of this is small-scale local production on the one hand and pluri-activity on the part of the farmer on the other. Vine growers are not able to make a good living out of wine. Since 1981, when Greece entered the EU, there has been an overall improvement in the quality of Greek wines. During this time farmers received support for restructuring and modernising vine growing. The EU policy supported the eradication of vines (which led to the loss of local varieties) and placed a limit on planting new ones.

In the first half of the twentieth century a number of events occurred that are important for the evolution of Greek vineyards and winemaking. The outbreak of phylloxera disease at the beginning of the century, the consecutive wine crises, the Balkan wars, the Asia Minor war and the two World Wars had an important impact on vines and wines. In 1928 the first national law on oenological practices and additives, Law 3501, was passed. Greece implemented its appellation laws in 1971 and 1972. The technical aspects of the legislation, which utilised criteria similar to those of France and indeed of most European countries, were established by the members of the Wine Institute. A number of factors play a role in determining the areas and products that qualify. They include: the suitability of the grape varieties; their pedigree and historical role; soil composition; the elevation of the vineyard;

yield per hectare; sugar levels; the effect of oenological practices such as barrel ageing; any additional factors likely to affect the quality of wine in the regions under consideration (Nollas 2005).

Wine production in Greece is divisible into two broad categories: the *quality wines produced in specific regions* (VQPRD) and the *wines without a geographical indication*. In what follows the first two types are assigned to the VQPRD category and the remaining two to the second, more ordinary, category: The quality appellations are a) OPAP,[3] which currently includes 25 products, almost all of them dry red and white wines, and b) OPE,[4] which includes nine regions and nine products, all of them sweet wine. The other two types are c) *Topikos Oenos* (local wine), the Greek equivalent of *vins de pays*, including 139 products, and d) *Epitrapezios Oenos* (table wine), which is the equivalent of the French *vin de table*.

One of the distinct advantages of the Greek wine sector is the number of unique indigenous grape varieties grown in the country. Greece offers a large variety of interesting native white and red varieties, yielding a broad range of styles and exciting new flavours. Over 300 indigenous varieties of grape have been discovered in Greece, and they are the grapes grown in the large majority of vineyards.

Between 1951 and 2000 there was a drastic (43 per cent) decline in the area covered by vineyards, and a 23 per cent decline in the area covered by winemaking vineyards (Nollas 2005). The official data reveal a downward trend in the number of farm holdings including areas under cultivation with grapevines.[5] From 259,166 ha in 1989 the vineyard area dropped to 142,399 ha in 2003. What this amounts to is a 30 per cent decline in the total vineyard area, notwithstanding the fact that the area of vineyard used for producing quality wines during this period has remained stable at 13,358 ha. Total wine production decreased by 11.5 per cent, but quality wine production followed an upward trend, reaching 1,070,000 hl in 2003. Notably, 25.8 per cent of vineyards are now used for quality wine production and quality wine production now accounts for 23.6 per cent of overall wine production (NSSG 2005).

In the last 25 years, the Greek wine sector has undergone complete transformation. There has been huge investment in modern winemaking technology, with state-of-the-art wineries being built throughout Greece. The current generation of Greek winemakers has been trained at the world's best wine schools, including, Bordeaux, Dijon, Adelaide and UC Davis. This trend coincides

---

3   OPAP stands for *Onomasia Proelefseos Anoteras Piotitos* (Protected Designation of Origin of Superior Quality)

4   OPE stands for *Onomasia Proelefseos Eleghomeni* (Protected Designation of Origin).

5   This figure includes two categories of vines: a) vines for grapes and b) vines for wine. The second category is subdivided into vines for ordinary wines and vines for VQPRD wines.

with a professionalisation of knowledge in viticulture and winemaking that has not received any significant support from the Greek state.[6]

The unprecedented improvement of the quality of Greek wines has been due to a number of factors such as: the institutionalisation of a specific modernised legislative framework for the enhancement and protection of wine production; the establishment of detailed rules for production, distribution and trading of geographical origin wines, local wines and traditional wines; the research work of Wine and Vine Institutes; the systematic restructuring of Greek vineyards; the modernisation of wine processing installations; and the human resources (that is oenologists, technicians, viticulturalists, wine journalists, sommeliers) (Lazarakis 2005).

Although in Greece during its recent wine transformation much attention was centred on foreign varieties, few producers, large or small, ever aimed at abandoning what remained of the traditional Greek vineyard. In many cases the orientation towards foreign varieties was regarded as merely instrumental: a way of acquiring the prestige necessary for attracting international attention to wines derived from local varieties. Many successful new wineries in the 1980s and 1990s gained a significant part of their prestige from the way they handled common western varieties. Some leading winemakers have argued that foreign varieties are a passport, a standard against which the capabilities of Greek winemakers can be measured. The assumption has been that the outside world could come in this way to have a basis for accepting the competence of Greek winemakers, and so be in a position to judge and appreciate vinifications of indigenous varieties.

The two wine-producing areas

The two wine-producing areas chosen as locations for the study of local food production are: a) the Lake Plastiras area, where the Mavro Messenikola VQPRD is discussed as an example of a local-to-local food network, and b) the Nemea area, where the VQPRD Nemea (Ayiorgitiko) is analysed as an example of a local-to-global food network.

The first study area: Lake Plastiras

This area includes the catchment basin of the Plastiras reservoir. Lake Plastiras, located 18 km to the west of the city of Karditsa, is an artificial lake formed as a result of the construction in the period 1956–1962 of an arched dam on the Tavropos (Megdova) river, a tributary of the Acheloos. This reservoir serves a large

---

6   The Greek case differs from that of Spain, where there has been state support for the development of professional knowledge in winemaking (see Buciega et al. in this volume).

number of different purposes and is intimately interlinked with the socioeconomic activities developed in the wider area.

Given that the study area is located in the general vicinity of Karditsa, it is the Prefecture of Karditsa that provides the administrative framework within which the research is conducted. The region is characterised by geomorphologic polarisation between the low-elevation plains of its eastern part and the mountainous Agrafa areas in the West. The study area is situated right on the edge of the mountainous areas a few kilometres away from the high-productivity farmland of Karditsa. The study area used to comprise a compact socioeconomic unit consisting of a small number of adjacent municipalities and communities, but in the late 1950s when the work commenced for the construction of the new dam a fragmentation process was initiated: a water barrier was introduced that disrupted old communication and trade routes between villages and communities.

The lake covers the previously fertile mountainous plateau of Nevropolis, an area of 24 km^2. The rural settlements around the lake cover an area of 31,400 ha, that has been classified as a Less Favoured Area (LFA). Though the reservoir was designed as a hydroelectric project, in recent years it has functioned as irrigation infrastructure. The existing operational status of the reservoir is one of irrational management of water resources. The reservoir does not cover specific demand for water but is awarded water resource status on the basis of its annual water capacity.

Due to the natural beauty of the landscape around the lake, in the 1990s tourist development got under way, with tourism supplementing the two other uses the lake was now serving (that is irrigation and water supply). There was already antagonism between these two uses and the addition of the third, the urban-centred function of tourism, only intensified it. But it did introduce pressure for better management of the reservoir/lake. In addition, the water supply needs of the city of Karditsa, albeit limited, do tend to raise demands for higher water quality.

The part of the Thessaly plain that surrounds Karditsa specialises in the production of industrial crops (cotton, sugar beet and tobacco). The mountainous zone, by contrast, places relatively more emphasis on food production due to the non-intensive character of its farming activity. The farm size in the plains is very much larger than the farm size in the mountainous zone, and especially on the slopes, where the wine growing area of Messenikola is located. The hegemony of the agro-industrial model in the most productive zone of the wider study area has overshadowed the defensive, marginal rural development model of the mountains. Food production has therefore been regarded as a complementary farming sub-sector in Karditsa. Three products produced in the mountainous zone of Karditsa are protected by the EU designation of origin: a) PDO Feta Agrafon (white cheese), (b) PDO Graviera Agrafon (cheese) and (c) Mavro Messenikola VQPRD (a type of red-black wine).[7]

---

7 It should be mentioned that the whole mountainous zone of Pindos, to which the Lake Plastiras study area belongs, includes a significant number of PDO products: nine

Mavro Messenikola VQPRD is obtained from the Mavro Messenikola variety of grape, traditionally cultivated only in the villages of Messenikola, Moschato and Morfovouni which are designated as a wine producing zone. Nearly six hectares of Mavro Messenikola VQPRD are cultivated on the slopes of the mountainous zone to the east of Lake Plastiras (ANEM 2003).

Mavro Messenikola VQPRD   The Mavro Messenikola VQPRD was recognised as a wine appellation in 1984, but the production protocol took its present form in 1994 when it was decided that three varieties of grapes could be used to make the wine, in different percentages, of course: 70 per cent Mavro Messenikola and 30 per cent Carignane and Syrah (two French varieties).

The local wine production network in the Lake Plastiras study area is a limited one and locally bounded. Though a long-established local wine, the Mavro Messenikola VQPRD has only recently become known outside its area as a result of improved marketing. Located in the steep hills and on the slopes of the southern Pindos mountainous area, the wine zone has been a traditional wine growing area since the Byzantine period. The growing interest in VQPRD wines in the study area, as well as in the wider Thessaly region, is attested by a recent diagnostic study on the economic impact of vines and wine (ANEM 2003).[8]

Only two wineries produce Mavro Messenikola VQPRD.[9] One of them is the winery of the Karditsa Union of Agricultural Cooperatives, an organisation with a long history, but producing VQPRD wine, not one of the quality varieties. Though located outside the wine region the Cooperative winery has been granted permission to produce VQPRD wine legally until such time as it establishes a new winery inside the zone. This winery absorbs a large proportion of the farmers' produce, both because vine growers as members of the Cooperative are under the obligation to sell their produce to it and because there are not many alternative solutions.

The second winery is a private winery called Karamitros Winery S.A. It is a family business, and a small-scale one, producing around 80 tons annually. It was built and modernised through LEADER II funding by means of a local investment plan: it processes the grapes from its own four-hectare vineyard supplemented by the production of a number of other local vine growers.

Figure 10.1 represents the local production network of Mavro Messenikola VQPRD. A central feature of this network is the Local Quality Convention (LQC), which was conceived and set in operation by the Local Development Agency

---

cheeses, four types of olives, three extra-virgin olive oils, two varieties of fruits and three other PDO products.

8   Vines and wines are considered both by local economic actors and local and regional administrative bodies to be among the region's six strategic agricultural products (ANEM 2003).

9   A recent account raises the number of local winemakers in the area to 15. However, they do not produce VQPRD but wine from other varieties grown in the area.

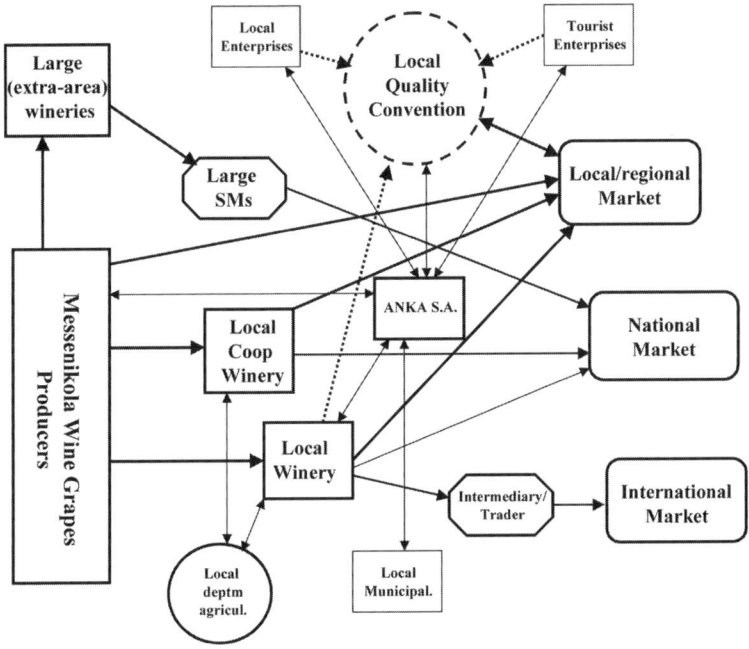

Figure 10.1   The local production network of Mavro Messenikola VQPRD

(ANKA S.A.) for the whole of the Lake Plastiras area. ANKA was in charge of the local LEADER programme and was able, by virtue of its ability to approve investment plans in the area, to promote LQC as a local agro-touristic quality certification scheme. The scheme has not yet been activated and needs to be integrated into a common framework to be arranged by the Hellenic Organization for Standardization (ELOT). The Karamitros Winery S.A. is part of this network with the explicit mission of supplying the local tourist market with its VQPRD wine.

Moreover two different market orientations are being pursued by this local production network in the Lake Plastiras area. The first orientation is that of the cooperative winery mostly targeting the cheap local market, as befits the quality of its wine, which is standardised only to a limited extent. The second orientation is that of the private winery supplying not only the local market but also other local markets in the country, on the basis of its (mainly informal) marketing network, as well as exporting a significant proportion of its VQPRD wine to Germany. The opening of the export channel came about as a result of an exhibition in which the private winery participated and where a company of brokers (who also publish a periodical about food and wine) showed an interest in Mavro Messenikola VQPRD. Nevertheless, the majority of customers in Germany are still Greek migrants and/or Greek restaurants and the volume of wine exported is only a small fraction of the quantity of the wine that is consumed locally.

Apart from the local wineries there are also the large wine companies (that is Tsantalis, Boutaris and Kourtakis) which buy wine from the local farmers but do not produce VQPRD wine. They simply take advantage of the local wine to produce cheap anonymous wine for the wider mass market. The marketing schemes created and supported by the large wine companies cannot easily be penetrated by the smaller cooperative or private wineries. The main reason for this is that large companies sell to supermarket chains which aim at low prices and buy a wide range of products. The small wineries cannot handle the large quantities, low prices and wide varieties of products demanded by the retail companies that supply the large supermarket chains.

The vine-growers in this area are mostly small to medium farmers, primarily concerned with producing as much as possible to maximise their returns. Their professional identity is clear to them: 'we [sic] are good vineyard cultivators but we are not winemakers. We don't want to become winemakers because we are professional farmers'.[10]

Many of them have licenses to make wine, but they make poor quality wine which is sold either in plastic bottles or in five-litre containers. Their customers are local people who either live in Karditsa or once lived there and have moved outside the region and are familiar with the distinctive taste of Mavro Messenikola. The informal networks of internal migrants who return frequently to the area on vacation, for pleasure or after retirement, seem to play an important role in the distribution and marketing of local food and wine products. They consider these products more trustworthy than the anonymous products bought at the supermarket. These migrants like their local food and by buying it and consuming it retain, in a way, their links with the place where they originated and to which they feel they belong.

In the case of Mavro Messenikola VQPRD, there is no explicit local wine production network and therefore no 'project' definition or network animator. What does exist is a weakly organised actor network which is, effectively controlled by the more powerful conventional food producers, in this case the large wine companies. The latter are the main actors because of their mobility inside and outside the specific wine zone and because of their capacity to introduce significant differentiation into their production on the basis of both mass and niche consumer demand. They moreover have at their disposal a high level of technical, scientific and managerial expertise and they occupy a privileged position in the formal marketing networks. The result of all these factors is that the gradual shift of the wider wine market in favour of quality wine is not evidently paralleled by any corresponding shift in the wine production of Lake Plastiras.

---

10 This is a quote from an interview to a group of farmers, in the village of Messenikola, in winter 2005.

This area has a long tradition of involvement with vine growing and winemaking, but local knowledge of winemaking has been significantly modified since the 1980s, due to farmers' concern with higher yields, as demanded by wine companies. This has implied the loss of the local knowledge that has been linked to lower yields and the use of less sophisticated machinery. Expertise has been channelled into the area above all by the agronomist-extentionists whose objective has been to secure adequate wine production. Modern oenological techniques have been introduced by the private winery, which has a family history of making wine, and by some small wineries that recently started to use modern oenological technology. It should be stressed that wine production in the area has been modernised only very recently, that is in the last decade or so.

There is no particular demand from local consumers that wine quality should be improved. The majority of traditional local consumers who know Mavro Messenikola are primarily interested in being able to buy it cheaply. It is the private winery that raises, and wants to raise, wine quality standards in the area. But the higher costs entailed by higher standards can only be paid when the product is traded abroad or in niche markets over a wider area. One marketing option pursued by the private winery has been to produce small bottles (0.33lt) for people who want to consume small quantities of quality wine in a local bar or in a local restaurant. This new local quality market means that modern local customers are offered a new experience and are able to reassess the flavour, and the other qualities, of the Mavro Messenikola VQPRD.

The Karditsa Union of Cooperatives and the private winery (Karamitros Winery) are thus pursuing differentiated strategies. The private winery has started to implement a policy of improving the quality of Mavro Messenikola VQPRD wine.

The Union of Cooperatives explains the low quality of the wine by pointing to the low quality of the grapes produced by the farmers, whose only interest is in gaining a higher income from selling larger quantities. And farmers argue that there is in any case no real differentiation in prices corresponding to differentiation in quality. As members of the cooperative, farmers also think that the latter is obliged to buy their produce irrespective of the quality of their grapes. The discourse on wine quality is thus undeveloped in the Lake Plastiras area.

Finally, local policies and above all the LEADER programme, are important for establishing and/or strengthening the links between Mavro Messenikola VQPRD and territorial development of the wine production zone. Both the private company's investment plan and the LQC that aimed at upgrading and sustaining high-quality products and services in the area have been the result of local policy implemented by ANKA.

The second study area: Nemea

The Nemea study area includes the Nemea Municipality, which is located in the Prefecture of Corinthia in the north-eastern Peloponnese. Nemea is located 25km from the city of Corinth and 110km from Athens. The Municipality of Nemea consists of 10 departments, the town of Nemea itself having 4,249 inhabitants. Nemea covers an area of 216.9km^2 or 21,690ha and is characterised as semi-mountainous, ranging from 280m to 800m above sea level. The quality of its soil resources, combined with the characteristics of its micro-climate, is especially favourable for the cultivation of vines. The local climate is dry semi-humid. Nearly 54 per cent of the area is covered by forest, more than five per cent is pasture and 39 per cent is agricultural land (CSPP 2005a). In terms of geomorphology the whole area is regarded as less favoured.

In the 1971–2001 period the area's population declined by four per cent. Despite the significant attempts to maintain the size of the local population there is a general decrease in population, ageing and deterioration in all demographic indices. Nearly 65 per cent of the working population are employed in agriculture, 8.5 per cent in the secondary sector and the rest, 26.5 per cent, in the tertiary sector. The area faces severe structural problems, typical of any marginal economy, due to the ageing of the local population, the absence of entrepreneurial capacities, the lack of educated and/or skilled human capital, the over-dependence of the local economy on viticulture, and the limited diversification of local economic activity.

The average farm size is above the national average (5.7ha as against 4.4ha) and specialised vine growing occupies nearly 55 per cent of total agricultural land in the area. Olive trees occupy 31 per cent of total land. The rest is taken up by fruit trees and by cereals and vegetables.

Viticulture thus constitutes the main agricultural activity in Nemea. Most local income is generated from the wine economy. The Nemea VQPRD is one of Greece's major wine zones. It comprises 17 rural communities,[11] and is based on the wine variety named Ayiorghitiko or Mavro Nemeas, derived from an indigenous cultivar producing dry red wines. 'Ay-Ghiorgis' (Saint George) is the former name of the city of Nemea. Nemea VQPRD is 100 per cent Ayiorghitiko. It is a wine appellation with a long history, embodying a similarly long tradition in vineyard cultivation and winemaking. In our analysis we recognise Ayiorghitiko as an enormously dynamic wine variety which has had a significant impact on the economic development of Nemea.

Nemea VQPRD Nemea VQPRD became a wine appellation in 1971 when most wine appellations were registered, long before the implementation of the European Council Regulation 2081/92. Between 10,000 and 12,000 tons of wine are produced in the Nemea area each year, of which 3,000 tons are Nemea VQPRD. The Nemea Wine Cooperative, which owns the area's largest winery, absorbs more than 50

---

11   The villages in the Nemea wine appellation are the following: a) in the Prefecture of Corinthia Kefalari (750m), Bogikas, Titani, Kastraki, Asprokampos (750m), Dafni (450m), Psari (650m), Petri, Aidonia, Koutsi (450m), Galatas, Leontio, Nemea (300m), Archaies Kleones (300m), and b) in the Prefecture of Argolida Gymno (300m) and Malandreni (300m).

per cent of the local production. The Cooperative dates back to the 1930s and has around 1,700 members. It provides extensive wine processing and storage facilities, but members are also free to sell their grapes to other wineries. There are also many private wineries, as well as individual winegrowers who make their own (non-certified) wine and sell it on the local market at a low price.

There are vineyards and wineries all over the Nemea area, many of which, within the area, have been relocated in the course of the last decade or so. In recent years, Nemean wines have been steadily gaining in market share both nationally and internationally. As well as the over 30 wineries in the area there is a firm packaging currants and a firm packaging table grapes. Most of the wineries are modern and fully equipped and produce certified wine.

Specialists see Ayiorghitiko as a promising wine of significant flexibility that is able to produce four or five different types of wine, all the while maintaining a very bright red colour with high (13–13.5 per cent) alcohol level. Taste and alcohol content depend on producer know-how, the altitude of the vineyard, and personal preference. The elevation considered ideal for dry Ayiorghitiko red wines is 350–600m and the most favourable sub-zones within the wine appellation are Asprokampos and Koutsi. Ayiorghitiko vineyards extend over an area of 2,100ha. A number of other varieties are still cultivated in the area but they are not considered part of Nemea VQPRD. The Greek wine variety Savatiano, which used to be the only white cultivar in the area and to a large extent covered the needs of local wineries for white wine, has been in decline since the expansion of Ayiorghitiko. Another local wine produced in small quantities is Roditis. Lately two well known foreign red wine varieties, namely Cabernet and Merlot, have been introduced by winemakers to experiment with them in blends with Ayiorghitiko. Finally, local farmers still cultivate Corinthian Stafida (Corinthian currants), the second most important cultivation in the area, covering 1,500ha. Some vine-growers also produce dried currants or sultanas, mainly for export and mainly for use in the confectionery industry. The vinification of Corinthian currants produces low quality and low price rosé wine. The main problem with this kind of wine is said to be that its low quality damages the wider area's reputation as a centre of high quality wine production (CSPP 2005b: 17–18).

The viticultural practices employed for many years by the local vine-growers have been largely replaced by the latest international viticultural practices. This is a process that can be traced back, in some respects, to the beginning of the twentieth century. Due to the phylloxera epidemic, American cultivars were grafted with Ayiorghitiko vines, which were then planted in lines, with a distance of 2.2 to 2.4m between the rows to allow for mechanical cultivation with tractors. The vines were tied together with wire to assist with normal growth of the plant. This replaced a previous system that aimed at shaping vines into a 'cup' form with a relatively limited distance (1.2m) between the vines. The new system supports fewer vines per hectare, with a number of interrelated consequences. The vine-grapes have more space to grow and the amount produced is significantly reduced, making for fewer, but higher quality, grapes. Third, there is a short pruning of the vine. Fourth, although legislation prohibits irrigation of the Ayiorghitiko, recent changes in the climate with attendant significant decreases in precipitation, necessitate managed irrigation of the vines. The winemakers who want to secure a high quality Ayiorghitiko wine have established networks of trust with vine farmers who follow the proper viticultural practices. Fifth, as regards fertilisers there are huge differences from sub-zone to sub-zone within the wider wine appellation area due to the differences in the soil. In general Ayiorghitiko requires little in the way of fertiliser. It depends on the needs of the vines. Sixth, in recent years Ayiorghitiko vines have responded well to the techniques of organic farming. The EU financial support that has been made available to assist conversion to organic farming has played a significant role, as has the interest that some winemakers have shown in wine made from organic grapes. Finally, new technologies have been used in vine-growing. Since the end of the 1980s modern machinery (small tractors, sprayers and so on) has been introduced which has decreased the cost of cultivation and increased both the quality of grapes and the income of farmers. In parallel with this, new measures have been adopted that have helped farmers use modern phytosanitary methods.

The grape harvest is carried out by hand using scissors. Nearly 4,000 labourers (seasonal and permanent) are employed at harvest time, the majority of whom are economic migrants (CSPP 2005b: 31). The low cost of immigrant labour and the fact that higher-quality wine production requires attentiveness and care have slowed up the introduction of harvesting machinery to the vineyards.

In terms of productive capacity and quality of wine produced, there are four types of winery in the Nemea wine region (CSPP 2005b: 62–64). The *first type* consists of wineries with a capacity up to 7,000hl which do not bottle wine (for example Zafiris Winery, Liakos Winery). Wineries of this type sell bulk wine through informal networks or undertake subcontracting for outside companies with which they have been working for many years. The *second type* comprises wineries of limited capacity (less than 1,000hl) (for example Gofa Domain, Raptis Domain) mainly producing for the domestic market. This second type of winery also utilises informal networks to sell their produce. They also work with large wine stores, being paid a set percentage from sales. The *third type* are wineries with a capacity of between 1,000 and 4,000hl (for example Palyvos Winery, Gaia Estate, Semeli Wines, Papaioannou Domain), which are more dynamic in advertising and promoting their quality wine for export. These are modern winemakers who participate in wine fairs and exhibitions, both in Greece and abroad, invest in advertising and account for nearly 20 per cent of Nemea's wine production. The *final type* consists of wineries of large capacity (7,000 to 100,000hl) (for example Koutsodimos S.A., Oenotechniki Ltd, Wine Cooperative of Nemea) that sell their wine to the large Greek wine companies (for example Kourtakis, Boutaris, Achaia Clauss).

Figure 10.2, which was compiled on the basis of information collected primarily through interviews with the social and economic actors involved, represents the Nemea VQPRD wine production network. The majority of wineries

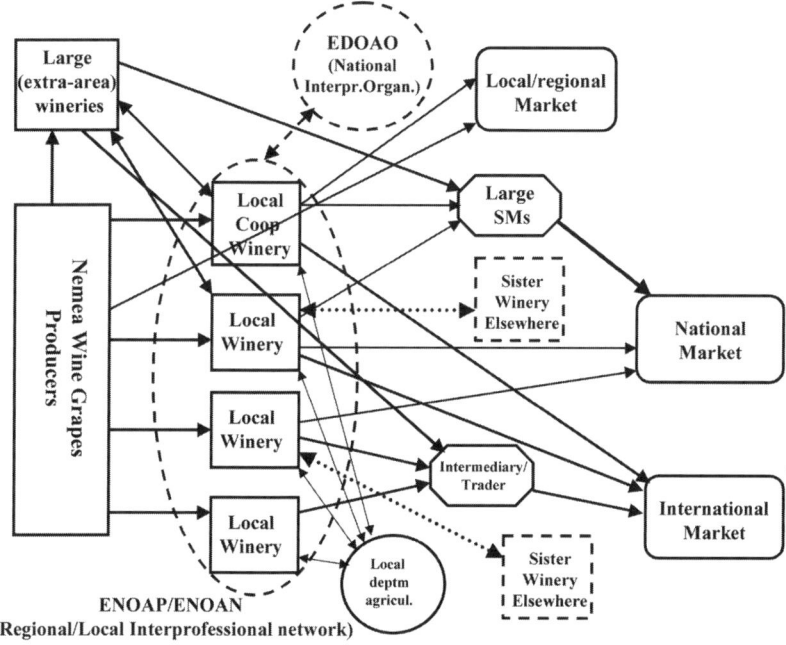

Figure 10.2   The local production network of Nemea VQPRD

were established in the 1990s, with some established just in the last five years. A number of them belong to local entrepreneurs (Papaioannou, Lafkiotis, Mitravelas, Peppas, Zafiris, Zaxarias). Many of the winery owners are themselves oenologists or have acquired formal technical knowledge as agronomists.

Recently, with the help of financing and consulting provided by an INTERREG IIIB programme entitled COHESION, a more formal, interprofessional network of the local economic actors was established. Three social actors: the Nemea Wine Cooperative, the majority of the private wineries and the Municipality of Nemea together found common ground in Ayiorghitiko to further develop agro-tourism in the area. By 2005 a new local interprofessional wine network had been established entitled 'Nemea Union of Winemakers and Viticulture' (ENOAN). Over 20 winemakers have joined the network, and a number of others indicate interest in participation. Some of the preliminary actions of the initiative were: the construction of a congress centre in Nemea, the establishment of a Centre for the Promotion of Wine, the designation of the wider area, and the introduction of vocational training for people working with wine. ENOAN's next steps included elaboration of a total quality scheme, coordinated promotion and marketing of the product abroad, eco-tourist and agro-tourist activity and so on.

ENOAN has by now become a formal wine production network,[12] part of the corresponding regional organisation (ENOAP) that has set itself similar tasks at the regional level, that is promotion and improvement of the image of the wines of the Peloponnese; opening of the Peloponnesian vineyards to agro-tourism. The ENOAN network is in fact the animator of the regional interprofessional organisation, which in turn is a member of the national body entitled EDOAO: 'The National Interprofessional Association of Vines and Wines'. EDOAO was, established in the year 2000 out of a merger between two central organisations of the sector: the 'Central Union of Vine and Wine Producing Cooperative Organisations of Greece' (KEOSOE) and the 'Greek Wine Association' (SEO). The national interprofessional organisation links together the cooperatives and represents both farmers engaged in vineyard cultivation, and winemakers.[13]

It is important to underline the fact that a large proportion of Nemea VQPRD is exported: to the rest of Europe, to the USA, Canada, Asia and Australia. The smaller wineries export to the EU (notably Germany) and the larger ones to America, Asia and Australia. The exporting wineries have established permanent cooperation with trading companies based abroad. Many Nemea wineries face problems when they try to establish domestic marketing networks because of the intense competition with the large wine companies and with imports from other wine-producing countries in Europe or beyond Europe.

The policy of the large wine companies in Nemea is to remain mobile and maintain cooperation with local wineries that allow them to vinify in the area. Another option is for them to buy grapes from the area, produce Nemea wine without the appellation label and then sell it at a lower price. In the domestic market there is significant demand for good-quality cheap wine that is not necessarily certified. It is considerations of this kind that also justify the strategies of wineries producing certified wine to target it mostly at the international market.

The wine production network in Nemea pays more attention to quality than does its counterpart in the Lake Plastiras area. This is primarily due to the export orientation of most certified wine produce, testifying to the fact that competition is harsher in the international market than in the domestic, where there is only limited demand for certified wine. But quality strongly correlates with identity and with recognition of the wine, which is in turn linked to flavour and to familiarity. The wineries made strenuous endeavours to familiarise consumers abroad with Greek wines, especially Ayiorghitiko, which is closely linked with Nemea, both in its identity as an ancient site and in the Mediterranean landscape features of

---

12   See ENOAN's *Memorandum of Association* of ENOAN. The president of ENOAN is also president of the Nemea Wine Cooperative, whose vice-president is a private winery owner.

13   The distinctions between the different levels of the interprofessional organisations reflect an organisational hierarchy that sends out a strong reminder of the social drive to scale and/or rescale institutions (see Marston 2000).

its mountainous Peloponnesian setting. The connections are easy to make. The problem has been one of establishing a reputation for this type of wine.

The wine production network in Nemea is a consumer-driven network: wineries have relocated to this area in response to the expanding demand for quality wine. The export orientation of many of the area's wineries is an index of the importance of the international consumer market in sustaining the local wine production network. Yet it is arguable that it is conversely the export orientation of the wineries that has made possible the strategy of quality valorisation.

The pattern is for wineries to own a significant area of vineyards and/or to have taken out long-term leases on large areas of land. Within the wine production network a certain amount of power accrues to those who have scientific knowledge and are able to utilise modern machinery and equipment to carry out delicate vinifications and introduce differentiations into the wines they produce.

The establishment of an interprofessional association in the Nemea area reflects the emergence of a new system of governance that is well-suited to modern demands for local food and to the requirements of less state-dominated production. Whatever the pessimistic opinions of some locals who profess to believe that this interprofessional association is linked to a group of 'dreamers' and does not involve the majority of the wineries, it remains a positive step towards a common label and a common marketing strategy for quality Nemea wine. Foreshadowing as it does a new type of governance, it also demands upgraded learning capacity from its participants. It represents an invitation to joint involvement in constructing a new wine industry in Nemea, to the collective benefit of the territory and, in turn, of its social and economic actors.

In recent years the Nemea VQPRD has become an area of wine tourism. Large numbers of visitors come to the area to taste the wine and learn more about the wine culture. A number of wineries have developed facilities for servicing wine tourists. More ambitious plans are in the pipeline in response to the area's emerging industry of wine tourism.

Last but not least, in summer time there has been a revival of the Nemean Games. Local people and tourists alike come together to compete in a variety of sports. The celebration is combined with other artistic events such as promotion of the local cuisine, includes Ayiorghitiko wine tasting and wine contests. Ayiorghitiko is becoming a something like a pivot of local rural development.

The local-expert knowledge nexus in the two areas

An examination of the evolution and characteristics of each of the two VQPRD wines leads to the conclusion that the local-expert knowledge nexus in the two wine producing areas is asymmetrical.

In the case of the Mavro Messenikola VQPRD, knowledge about the winemaking is shared by a limited number of people. The farmers consider themselves unable to participate in this 'scientific' and 'managerial' knowledge, identified with the operations of a 'certified' winery. They retreat to their everyday, less risky, practice of making wine that is good only for locals and only for short-term consumption. This is neither demanding for winemaking farmers nor particularly rewarding. The 'leap to a higher scale' that would be entailed in opening a winery would mean cooperating with other farmers and learning to handle problems of economic organisation and programming. It would also presuppose technical knowledge of professional winemaking. In the Lake Plastiras area it is clearly difficult for a farmer to acquire either the social capital or the motivation to pursue a territorial identity through wine certification. Farmers' dependence upon the policies of the local Union of Wine Cooperatives to sustain their livelihoods leads to a gradual erosion of their income. Their 'free-rider' attitude of seeking to gain the highest possible price for a minimum of quality is in keeping with stance of the Cooperative. It is economically indefensible but consistent with the productivist orientation of most domestic agricultural policy and with the widely held view that the higher the yield the higher the income. Farmers thus remain firmly locked into the agro-industrial logic.

In the past vineyard yields were smaller but the wine was of much higher quality. And other products, for example *tsipouro*, were also made from the remains of grapes. The new tacit knowledge based on agro-industrial logic has displaced the former repertoire of practices and of experiential knowledge. There was a break with former knowledge repertoires, justified on the basis of the higher incomes and guaranteed prices that came to be associated with the seemingly endless need for conventional, mass consumption wine. The trend is common to both areas but it is more dramatic in the case of the Lake Plastiras area, where human and natural resources are limited and all the action is on a small-scale. Smallness of scale, proximity, interrelatedness, geography: these are all important factors when examining knowledge formulation, transfer and change (Howells 2002).

The cost that farmers would be required to pay if there were a shift 'back' to their being wine connoisseurs, reverting to an *art de la localité*, is in reality very high. This shift would be highly uncomfortable for those who have been accommodated as specialists or professionals and who would need to be re-trained and virtually re-educated under the new conditions. The pride of being a farmer in the area would have to be re-negotiated with a 'specialist', that is a non-farmer with expert knowledge and in a position of dominance. This is simply not acceptable to the majority of vine-growers in the Lake Plastiras area, who appear to be attached to their present self-image and would be embarrassed to see it eroded. The strategy of the locals is to confine themselves to their familiar livelihood practices and go about their business.

An additional obstacle to the construction of a local-expert knowledge nexus in the area is the fact that the Mavro Messenikola VQPRD was found by experts to be something that needed adjustment. By 1994 it had been concluded that two French red wine varieties (Syrah and Carignian) had to be added to the local Mavro Messenikola VQPRD wine variety (in a proportion amounting to 30 per cent of the total) to improve its quality.

The Karamitros winery is an exception to the conditions prevailing in the Lake Plastiras area. There is a widespread impression that quality wine production is something vastly difficult, but the Karamitros winery is a living testimony to the fact that territorial development on the basis of the wine zone is possible. It is in no way surprising that George Karamitros should be in the habit of citing both the traditional knowledge that his family has maintained from older times and the new scientific and technical knowledge that is embodied in the modern machinery and equipment used in the vinification process. He has received vocational training in winemaking, he is proud of the his own family's strong winemaking traditions, the maintenance of which in no way prevents him from also working as a professional musician. These attributes are part and parcel of what amounts, as indicated, to an exceptional case.

By contrast, in the Nemea VQPRD knowledge about winemaking is no well-kept secret but rather something widely known. It has been transported from place to place within the area and also imported into the area from elsewhere. There has been a knowledge transfer into the area through the wineries that have relocated there as well as through the operation of certain institutions such as the local interprofessional association and/or the cooperation with large wine companies (Howells 2002: 877–878).

The re-discovery of local knowledge (Kloppenburg 1991, Laureano 1999), which serves both as a marketing tool and quality assurance guarantee, seems to have become a major issue among the wineries in the Nemea wine appellation. In terms of rhetoric everyone claims to be using traditional knowledge, and there is nothing surprising about that. Nobody of course denies that modern scientific knowledge and expertise is vital for the production of quality wine and for diversification of wine production. But to redress the balance many of the wine markers claim, albeit unconvincingly, that: '*good wine is produced in the vineyards*'.[14]

Because of its specific attributes but also because of its capacity for continuous readjustment, Ayiorghitiko is regarded as a quality wine variety. The Nemea wine zone is currently a dynamic area which has attracted a large number of wineries with sister wineries elsewhere.[15] These wineries came to this relatively out-of-the-way place in search of the relatively remote Ayiorghitiko vineyards that would provide them with a valuable natural resource for making wine. But this seeking out of vineyards is not unrelated to the existing tacit knowledge in the Nemea area, which geographically is highly differentiated. The tacit knowledge (Gertler 2003), a large part of which has been lost over the years, has been linked to specific 'vine-topes' and their cultivation. It has had to be regained and reconstructed on the basis of the new standardised and technical solutions that have become available to winemakers. The wineries that have relocated aim at formulating new tacit

---

14 This claim was made by many interviewees.
15 A number of such wineries have recently been built in the Nemea area: for example, Domain Lafazani, Domain Semeli, Spyropoulos, Gaia, Katogi-Strofilia, Harfaftis.

knowledge alongside the expert and scientific knowledge they bring to the area. The local vine growers provide significant dynamism simply due to the fact that they have preserved a certain tradition of winemaking by remaining in the Nemea area and by maintaining their production.

The result is that local farmers have had to utilise their lay knowledge while at the same time acquiring another form of tacit knowledge with better connections to scientific and expert knowledge, making possible a vast improvement in the quality of the original product. A new type of tacit knowledge is evidently being sought with a view to re-embedding it in the existing corpus of reliable formal-scientific-expert knowledge. The main reason for this is that it is not possible to dissociate tacit knowledge (or local knowledge) from codified knowledge (Agrawal 1995, Nygren 1999). In the case of the Nemea VQPRD the requirement is thus a new synthesis of tacit and formal knowledge with the capacity to generate an economy of means and ends.

Local and/or traditional knowledge is of central importance in winemaking from the viewpoint of justifying a linkage between wine quality and the specific locales (*terroirs*) in which it is produced. And such *terroirs* are actually being reconstructed in the case of Nemea out of a struggle between winemakers to acquire specific 'vine-topes'. Both the rhetoric over the existence of local/traditional knowledge and the local-expert knowledge nexus itself play a vital role in reclaiming local wine and challenging its conventional industrialised counterpart.

But consumers themselves are not yet ready fully to accept that scientific/expert/technical knowledge is of primary significance in winemaking. With local food what is important is its identity and the geographical differentiation from which that identity is derived. Local/traditional knowledge is of relevance to this. Even if it has lost much of its specificity, it retains its rigour from the reappraisal of role to which it has been subjected and rediscovery of features that had fallen into neglect.

The value of technical knowledge is increasingly recognised, though not at the expense of proper management of local natural resources. Certain modern wineries have supported precisely the view of winemaking that, while recognising the enormous advantages offered by technical knowledge also stresses the importance of vineyard management, location and stewardship for manufacturing a range of quality wines (such as those of Paraskevopoulos from the Gaia Estate). Certain local wineries in the Nemea wine zone (for example Lafkiotis Winery) aim at striking a balance between the identity and geographically-based variety of their wines with their level of scientific, expert knowledge that goes into making it. One major problem in Nemea wine zone is that the lay knowledge of farmers is not always useful to winemakers and indeed may even pose problems of that can be reflected in the quality of a wine and the mode of operation of a winery (Callon 1999).

It is ultimately from its twofold and contested character that knowledge derives its vitality in local wine production networks. The interrelationship between tacit and formal knowledge is at the basis of the discussion on quality and its social construction. In the Nemea wine zone a new tacit knowledge has been under construction with a view to being re-connected to the available formal knowledge. Such a re-embedding of tacit knowledge provides the basis for local construction of wine quality, and it does seem to have earned some recognition on the market. But there is still a long way to go for a territorial identity to be elaborated for quality wine in the Nemea wine zone. The role, attitudes and ideology of farmers are of crucial importance, as are those as the winemakers that are co-creators of local wine quality. Empowerment of the existing institutions can provide further motivation for re-embedding of local wine in its social, economic and territorial context.

Conclusion

Comparison of the two wine-producing areas is not easy due to the fact that a variety of different issues have to be taken into account. But the similarities and the differences can in general be enumerated as follows.

Starting with similarities, both study areas are in the vicinity of a large city that is the administrative and socio-economic centre of the wider region. They are however also both peripheral, geographically and socio-economically. Both areas can moreover be considered somewhat marginal in location due to their socioeconomic and developmental problems. The ageing of the local population, the heavy reliance on farming, the lack of off-farm economic opportunities: all these factors imply similar types of problems for both areas. There is also in both cases a tradition in vineyard cultivation and winemaking that goes far back in time, embodying local/traditional knowledge both in farming and in winemaking. There are many similarities in the attitudes of the local population and of farmers to the wider socioeconomic situation. The perception is that it is having a negative impact on their income, their farming and their perception that they are farmers and not winemakers. In both areas there are many similarities in the way that farmers perceive the role of local wine cooperatives: their 'free rider' attitude when it comes to selling their produce to wineries, their insatiable need for increased production and for maximum prices with minimal attention to quality. All in all, the cooperatives and the large wine companies have played a negative role for establishment of a local wine production. Each for different reasons has reinforced the loss of local/traditional knowledge but has not introduced expert and/or managerial knowledge to both areas.

But there are differences between the two study areas. To begin with, the Nemea area is relatively close to Athens (one hour away) while the Lake Plastiras area is more than a four-hour drive away. Nemea is well known as a wine zone, while only the well-informed know that there is a small wine-producing zone in the Lake Plastiras area. Nemea is renowned for its certified wine and to a lesser extent for its identity as an archaeological site. Lake Plastiras is known mostly as an agro-tourist area and somewhat less so as an ecological site. The territorial identity of the two areas has been constructed on different bases (Tregear et al. 2007). As wine zones they are also very different in terms of scale. The Nemea wine zone covers a farming area of nearly 10,000ha; the Lake Plastiras zone covers no more than 1,000ha). The size of the vineyard area in the case of Nemea is 4,500ha; at Lake Plastiras is just 100ha. Given its large-scale, the Nemea zone has retained a large proportion of its human resources in farming and above all in vineyard cultivation. The existence of this relatively skilled labour force, working in their own vineyards, has been a major factor behind the establishment of new wineries in Nemea. Due to land scarcity, the development of agro-tourism in the surrounding area and the dominance of intensive industrial crops (cotton) in the nearby plain zone, there has been no dynamic expansion of vineyard cultivation in the Lake Plastiras area.

There are moreover differences in terms of the institutional barriers to economic development in the two study areas. The Nemea wine zone is not faced with such barriers, with the exception of the Stymfalia archaeological site (NATURA 2000 site) and its environs. By contrast, the wine zone in the Lake Plastiras area faces the problem of a land use control mechanism[16] that imposes very notable restrictions on wine processing activity. There are no institutional limitations within settlements but there are significant limitations for building wineries on farmland. It is therefore not surprising that the 'motivational and social capital role of cultures' (Jenkins 2000: 307) in the two areas differ significantly. There are, in other words, different levels of motivation and activation and different perceptions of, and attitudes towards, individuals' capacity to overcome institutional and/or economic barriers and the resultant feelings of powerlessness. Nemea farmers have a large choice of wineries to which they can sell their produce. Quality becomes critical in determining the price at which they may sell. Lake Plastiras farmers by contrast have no other option than to sell their produce to the local cooperative winery.

One important point to make is that in both areas the prevailing agro-industrial logic led to the loss of a great part of the local/traditional knowledge from the collective memory of the local communities. The key question for this discussion is whether the local communities can re-invent local/traditional knowledge and find ways of merging it with expert and managerial knowledge. The Lake Plastiras area is evidently lagging behind in this process whereas the Nemea area is going ahead.

The most tangible comparative advantage of Nemea is that it has attracted considerable outside investment which may be attributed to its scale. Geographical and socioeconomic factors certainly play a role. The advantages are summarised as follows: a) the location of the area (its proximity to the metropolitan area of Athens); b) the quality of the wine variety (its flexibility as a wine variety, the

---

16 Due to the lake there is an institutionalised 'Zone for Housing Control' (ZOE) which poses significant limits to any productive activity in the area for conservation purposes.

availability of adequate wine quality, the tradition it reflects); and c) the specific socioeconomic characteristics of the area (professionalisation of local farmers, adaptation to new cultivation techniques, availability of semiskilled and skilled labour power, availability of technical staff and skills).

The Nemea wine production cluster is a reflection of the existence of a local-expert knowledge nexus. The breadth and the scale of the cluster are critical parameters of the local capacity to overcome the dominance of the agro-industrial logic. The local capacity to re-construct local/traditional knowledge and negotiate knowledge forms remains a critical element for the territorial development of Nemea.

Such territorial development is contingent on successful re-construction of Nemea's '*terroir*'. The role of wineries, local government and vine-growers is important in framing the local-expert knowledge nexus that is of pivotal importance for the governance of this '*terroir*'. The actions of 'dreamers' (such as Papaioannou, Paraskevopoulos, Palyvos, Skouras), the bearers of both expert and managerial knowledge, have been a motivating factor behind the success of the Nemea wine zone. Reclaiming of local food production on the basis of local-expert knowledge nexus involves a re-invention of local identity, and it is being shown to be feasible.

References

Agrawal, A. 1995. Dismantling the divide between indigenous and scientific knowledge. *Development and Change*, 26(3), 413–439.
ANEM (2003), *Diagnostic Report for Vines and Wines*, Unpublished report, Region of Thessaly (in Greek).
Barham, E. 2003. Translating terroir: The global challenge of French AOC labeling. *Journal of Rural Studies*, 19(1), 127–138.
Broude, T. 2005. Culture, trade and additional protection for geographical indications. *Bridges*, 9, 20–22.
Callon, M. 1999. The role of lay people in the production and dissemination of scientific knowledge. *Science, Technology and Society*, 4(1), 81–94.
Centre for Strategic Planning in the Peloponnese (CSPP) (2005a), *Local Development Programme for the Nemea Municipality area*, Unpublished Study, COHESION, INTERREG IIIB CADSES (in Greek).
Centre for Strategic Planning in the Peloponnese (CSPP) (2005b), *Study for the Characteristics of the Wine Sector in the Nemea area*, Unpublished Study, COHESION, INTERREG IIIB CADSES (in Greek).
DuPuis, E.M. and Goodman, D. 2005. Should we go 'home' to eat?: Toward a reflexive politics of localism. *Journal of Rural Studies*, 21(3), 359–371.
Feagan, R. 2007. The place of food: Mapping out the 'local' in local food systems. *Progress in Human Geography*, 31(1), 23–42.

Fonte, M. 2008. Knowledge, food and place: A way of producing, a way of knowing. *Sociologia Ruralis*, 48(3), 200–222.

Fonte, M. and Grando S. 2006. *Local food production and knowledge dynamics in the rural sustainable development*. Input paper for the Working Package 6 (Local food production) of the CORASON Project. Unpublished manuscript.

Gade, D.W. 2004. Tradition, territory, and terroir in French viniculture: Cassis, France, and Appellation Contrôlée. *Annals of the American Association of Geographers*, 94(4), 848–867.

Gertler, M.S. 2003. Tacit knowledge and the economic geography of context, or the undefinable tacitness of being (there). *Journal of Economic Geography*, 3, 75–99.

Goodman, D. 2004. Rural Europe redux? Reflections on alternative agro-food networks and paradigm change. *Sociologia Ruralis*, 44(1), 3–16.

Goodman, D. and DuPuis, E.M. 2002. Knowing food and growing food: beyond the production-consumption debate in the sociology of agriculture. *Sociologia Ruralis*, 42(1), 5–22.

Guthman, J. 2004. *Agrarian Dreams? The Paradox of Organic Farming in California*. Berkeley: University of California Press.

Guthman, J. 2007. The Polanyan way? Voluntary food labels as neoliberal governance. *Antipode*, 39(3), 456–478.

Gwynne, R.N. 2006. Governance and the wine commodity chain: Upstream and downstream strategies in New Zealand and Chilean wine firms. *Asia Pacific Viewpoint*, 47(3), 381–395.

Harvey, D. 2002. The art of rent: Globalization, monopoly and the commodification of culture, in *The Socialist Register 2002: A World of Contradictions*, edited by Panitch, L. and Leys, C. London: Monthly Review Press.

Hendrickson, M.K. and Heffernan, W.D. 2002. Opening spaces through relocalization: Locating potential resistance in the weaknesses of the global food system. *Sociologia Ruralis*, 42(4), 347–369.

Hinrichs, C.C. 2000. Embeddedness and local food systems: Notes on two types of direct agricultural market. *Journal of Rural Studies*, 16(3), 295–303.

Howells, J.R.L. 2002. Tacit knowledge, innovation and economic geography. *Urban Studies*, 39(5–6), 871–884.

Ilbery, B., Morris, C., Buller, H., Maye, D. and Kneafsey, M. 2005. Product, process and place: An examination of food marketing and labelling schemes in Europe and North America. *European Urban and Regional Studies*, 12(2), 116–132.

Jenkins, T.N. 2000. Putting postmodernity into practice: Endogenous development and the role of traditional cultures in the rural development of marginal regions. *Ecological Economics*, 34(3), 301–314.

Jones, A. and Clark, J. 2000. Of vines and policy vignettes: Sectoral evolution and institutional thickness in the Languedoc. *Transactions of the Institute of British Geographers*, 25(3), 333–357.

Josling, T. 2006. The war on terroir: Geographical indications as a transatlantic trade conflict. *Journal of Agricultural Economics*, 57(3), 337–363.

Kloppenburg, J. 1991. Social theory and the de/reconstruction of agricultural science: Local knowledge for an alternative agriculture. *Rural Sociology*, 56(4), 519–548.

Lagendijk, A. 2004. Global 'Lifeworlds' versus Local 'Systemworlds': How flying winemakers produce global wines in interconnected locales. *Tijdschrift voor Economische en Sociale Geographie*, 95(5), 511–526.

Laureano, P. 1999. *The System of Traditional Knowledge in the Mediterranean and its Classification with Reference to Different Social Groupings*, Report prepared for the Secretariat of the Convention to Combat Desertification, UNCC, ICCD/COP(3)/CST/Misc.1.

Lazarakis, K. 2005. *The Wines of Greece*. London: Octopus Publishing Group.

Marsden, T. 2000. Food matters and the matter of food: towards a new food governance? *Sociologia Ruralis*, 40(1), 20–29.

Marsden, T., Munton, R., Ward, N. and Whatmore, S. 1996. Agricultural geography and the political economy approach: A review. *Economic Geography*, 72, 361–375.

Marston, S.A. 2000. The social construction of scale. *Progress in Human Geography*, 24(2), 219–242.

McMichael, P. 2000. The power of food. *Agriculture and Human Values*, 17(1), 21–33.

Moran, W. 1993a. Rural space as intellectual property. *Political Geography*, 12(3), 263–277.

Moran, W. 1993b. The wine appellation as territory in France and California. *Annals of the Association of American Geographers*, 83(4), 694–717.

Morgan, K., Marsden, T. and Murdoch, J. 2006. *Worlds of Food: Place, Power and Provenance in the Food Chain*. Oxford: Oxford University Press.

Murdoch, J., Marsden, T. and Banks, J. 2000. Quality, nature and embeddedness: some theoretical considerations in the context of the food sector. *Economic Geography*, 76(2), 107–125.

National Statistical Service of Greece (NSSG) 2005. *The Wine Sector in Greece: National Trends and Regional Distribution*, Working paper No 15, UNECE/EUROSTAT/FAO/OECD.

Niles, D. and Roff, R.J. 2008. Shifting agrifood systems: the contemporary geography of food and agriculture; an introduction. *GeoJournal*, 73(1), 1–10.

Nollas, I. 2005. *The Greek Vineyard: Evolution and Prospects*, unpublished manuscript (in Greek).

Nygren, A. 1999. Local knowledge in the environment-development discourse: From dichotomies to situated knowledges. *Critique of Anthropology*, 19(3), 267–288.

Parrott, N., Wilson, N. and Murdoch, J. 2002. Spatializing quality: Regional protection and the alternative geography of food. *European Urban and Regional Studies*, 9(3), 241–261.

Pratt, J. 2007. Food values: the local and the authentic. *Critique of Anthropology*, 27(3), 285–300.

Scott, J.C. 1998. *Seeing like a state: How certain schemes to improve the human condition fail.* New Haven: Yale University Press.

Smith, E., Marsden, T., Flynn, A. and Percival, A. 2004. Regulating food risks: rebuilding confidence in Europe's food?. *Environment and Planning C: Government and Policy*, 22(4), 543–567.

Tregear, A., Arfini, F., Belletti, G. and Marescotti, A. 2007. Regional foods and rural development: the role of product qualification. *Journal of Rural Studies*, 23(1), 12–22.

van Leeuwen, C. and Seguin, G. 2006. The concept of terroir in viticulture. *Journal of Wine Research*, 17(1), 1–10.

Watts, D.C.H., Ilbery, B. and Maye, D. 2005. Making reconnections in agro-food geography: alternative systems of food provision. *Progress in Human Geography*, 29(1), 22–40.

Whatmore, S. and Thorne, L. 1997. Nourishing networks: alternative geographies of food, in *Globalizing Food: Agrarian Questions and Global Restructuring,* edited by D. Goodman and M. Watts. London: Routledge, 211–224.

Winter, M. 2003a. Geographies of food: agro-food geographies – making reconnections. *Progress in Human Geography*, 27(4), 505–513.

Winter, M. 2003b. Embeddedness, the new food economy and defensive localism. *Journal of Rural Studies*, 19(1), 23–32.

Conclusion

Europe's Integration in the Diversities of Local Food and Local Knowledge

Maria Fonte

The analysis of local food initiatives presented in this volume is part of the larger EU-funded CORASON project. This was carried out in 12 European countries selected according to the 'Green Ring' hypothesis, (Granberg, Kovach, Tovey 2001: xiii) and contains those countries located in the geographical rim of Europe, characterised by a strong agrarian presence, that have followed pathways into modernity different from those of the earlier industrialising core. The objective of the project was to analyse how rural areas in these Green Ring countries are related to the global economy and society from the perspective of knowledge processes. In particular, are they experiencing a process of homologation or further marginalisation? Do they follow a common trajectory or are they diversifying their adaptation strategies to a changing world? Are they able to play a role in the knowledge dynamics, especially in relation to global challenges, like climate change, biodiversity conservation, sustainable management of natural resources?

Bruckmeier and Tovey (2009) have pursued these questions analysing initiatives related to non-agricultural and natural resource-use practices. They highlight the processes through which rural inhabitants revitalise and strengthen their knowledge and experience in cooperation with experts. The main objective of this volume has been to show how initiatives aimed at place-embedding of food offer diverse, innovative solutions to the knowledge problems of an integrating European society.

Despite the variety of theoretical references of each author, the book responds to a unitary design that is manifested in

- the selection of case studies, responding to an agreed reference framework
- the methodological approach to the analysis of case studies which focuses on the importance given to the context in which knowledge dynamics emerge, notably the dynamic of social networks and the knowledge processes these networks mobilise.

The following sections summarise the most important findings from the case studies analysis, with respect to agro-food contexts, social networks and knowledge processes.

The relevance of the agro-food context: local and locality food

Since the early discussions in the CORASON research team, differences arose among researchers in the way local food can be conceived and organised. Especially relevant is the distinction between 'local' and 'locality' food (Maye and Ilbery 2007; Watts, Ilbery and Jones 2007), that we called 'local food for local consumers' and 'local food for distant consumers'. The former refers to food produced and consumed within a close distance, that makes possible a face-to-face relation between producers and consumers; the latter to food with an identifiable geographical provenance, sold to distant consumers. Watts and colleagues (2005) consider the first a 'strong', the second a 'weak' alternative to the conventional food provision system.

Our case studies point to the different agro-food contexts in which the two models of local and locality food emerge, which can be characterised as 'food desert' and 'marginalisation', as shown in Figure C.1. These two categories may justify different strategies for the re-localisation of the food system, with a 'reconnection strategy' for food deserts category, and the 'valorisation of the origin of the product' in the marginalisation category (Fonte 2008).

The research areas in Ireland, Scotland, Sweden and Germany are part of national agro-food economies that are since longer time export oriented or inserted in global trade, although there may be differences between the regions in these countries. Food production is disconnected from local consumption and food distribution is based on large retailers, who have out-competed the local market for food. Small producers can not find an outlet for their products, while consumers can not buy locally produced food. The initiatives of food re-localisation analysed in these areas aim to re-establish both a local market for food and a local culture of food growing and processing.[1] Place re-embedding of food and food-markets is pursued through the social and economic re-connection of producers and consumers, as a strategy able to reverse direction, from 'foodless landscape' toward sustainable food communities. In this perspective the product qualities relate mainly to its being fresher, healthier and environmental sustainable.

The research areas of the Mediterranean countries (Portugal, Italy, Spain, Greece), but also of Norway and Poland, are marginalised regions, characterised by an agro-food system based on the persistence of traditional small-scale family farms. Food has often a place-based identity and is still place-embedded, but local food markets are under the threat of depopulation and migration. A special case of interest is the case-study of Poland, a country that is recently experiencing an explosion of mass tourism.

In these areas the valorisation of local food is pursued as a territorial strategy of integrated rural development, aiming to re-vitalise marginalised food economies.

---

[1] The initiatives considered were: C-Farmers' Market in Ireland, the Netzwek Verpommern in Mecklenburg, the Food Link Van and the Skye and Lochalsh Horticultural Development Association in Skye, Scotland and, finally, the Eldrimner initiative in Sweden.

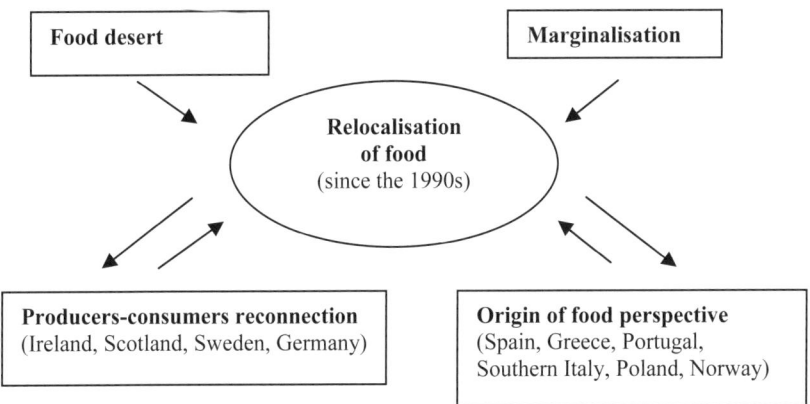

Figure C.1   Agro-food contexts and strategies of re-localisation

Here the most urgent problem of the rural community is to revitalise and strengthen, rather than to construct, place-embeddedness of food. Two are the pillars of an integrated rural development strategy in marginal areas: the valorisation of local typical products, which generally implies the construction of a certification system, and the promotion of collective initiatives of diversification of the rural economy, especially through rural tourism (organisation of fears and festivals, for example). The main characteristics of the two models are summarised in Table C.1 and will be further discussed in the following sections.

Actors, markets and social networks

The social dimension of sustainability is pre-eminent in the 're-connection' strategy, where social movements and civil society associations are important actors in the re-vitalisation of a rural community. In Jämtland, Sweden, new rural movements, developing from traditional forms of rural economy and livelihood, become very active and influential in the 1980s and 1990s, 'experimenting with pioneering forms of a new sustainable local economy and society as in projects of organic farming, eco-village movements or LETS-projects' (Bruckmeier et al. 2006). Similarly Tipperary was one of the heartlands of the early co-operative movement (1890s to 1920s) in Ireland. The Food Link Group in Skye, Scotland, is a voluntary association of local businesses, chefs, citizens with an interest in local produce (meat, fish, dairy, vegetables). Netzwerk Vorpommern, in Germany, is a spin-off of the organic movement, promoted by a group of highly committed organic consumers, willing to provide local market channels for the organic producers of the area.

Table C.1 Main characteristics of the two strategies of food re-localisation

	Re-connection (Local food)	Origin-of-food (Locality food)
Context	Food desert	Marginalisation
Limiting dimensions in the food system sustainability	Social, environmental	Economic
Resources (cultivated varieties, breeds, etc.)	Place specificity needs to be re-built	Specific to the place
Product quality	Fresher, healthier, more environmentally sustainable	Place-identity
Producers	Post-modern farmers	Post-traditional farmers
Place of exchange	Local market	Local and extra-local markets
Consumers	Local (also the tourist, but it is not the initiatives' target)	Local and distant (migrants, rural tourists, responsible / aesthetic trans-local consumers)
Actors	Civil society and social movements	Local institutions and producers associations
Certification	Not so important (also contested)	Important
Limiting factors in the knowledge system	Local lay knowledge need to be re-built	Local lay knowledge need to be re-vitalised through connections with managerial and scientific knowledge

Source: Own elaboration from case studies.

In the 'origin-of-food' strategy, economic dimension is the limiting factor in the food system sustainability; by consequence the marginalised rural communities look for new sources of income to survive. Especially when the initiatives set in motion a process of origin certification, the key actors are often linked to institutions and firms association, rather than social movements.[2]

With a great difference of actors involved and objectives to pursue, conflicts may arise around the appropriation of the 'local food project' (see in particular Tovey, this volume): local institutions, LEADER GALs and other development agencies are often more interested in the valorisation of local food system as contributing to

---

2   This is the case of the certification of Barranco cured ham in Portugal, the Oscypek neo-traditional network in Poland, the Raquena sausage PGI in Spain, the valorisation of Nemea wine production VQPRD in Greece, the valorisation of fermented fish in Norway and the construction of Aspromonte National Park certification on southern Italy.

rural development; experts and business capital in the appropriation of commercial profit emerging from new niche markets; social movement in the defence of family farms or the construction of an alternative project to the conventional food system. Nonetheless, actors are not strictly identifiable with their specific function. They are often hybrid actors: they may represent an institution, but adhere to a social movement, for example. While Tovey (this volume) stresses the risk of co-optation in the overlapping of social networks, we point to the possibility that well designed, initiatives of local food valorisation may become a collective endeavour, in which a process of social learning and participative experience building may be set in motion. In the unfolding of the learning process, initially different interests or position may converge towards new commonly agreed upon objectives and a process of cooperation may be stimulated for the promotion of the community welfare. Specifically with respect to certification processes, we find examples that lead to the expropriation of the traditional knowledge and the economic benefits deriving from the valorisation of the product (Barrancos cured ham in Portugal), but also participative certification processes aiming to re-vitalise the sense of place of the local community as a way for promoting civic values of citizenships and sustainability (south of Italy).

The Polish study is an exemplary case of how the reputation of a local product can be jeopardised by the pressure of mass tourism and an expanded market. It is interesting to note that here expropriation of knowledge and benefits occurs by local 'fake' producers rather than extra-local actors. Better than other initiatives, this case is useful to illustrate the different functioning and different effects on local community of 'mass' and 'local' markets. Mass (capitalist, self-regulating) markets lead to loss of the local breed (biodiversity), standardisation of the product and production process (commercialisation of tradition), on one side, and de-linking of producers and consumers (one-shot market exchange) to the other.

The challenge that local food wants to launch is quite arduous: use the (local) market exchange to re-embed food into a local place and into a local community. The concept of de-commodification analysed by Appadurai (1986) may turn very useful in order to escape a linear vision of history and economy as a progress toward a global market economy, purged of any social link. In asserting the importance of the social interaction between producers and consumers, local food initiatives refuse the dichotomy between market and sociality and assert that relation of exchange need not to be purely instrumental, deprived of any social implication, but may be re-embedded or nested into a locality and into social relations.

Knowledge processes

The privileged perspective for considering the place-embeddedness of food in the CORASON project is the analysis of the knowledge dynamics. We started from the assumption that the 'knowledge of the place', or local knowledge, is a constitutive element of local community, based on a specific method of knowing things and

Table C.2   Local food and knowledge dynamics

Knowledges: Forms	Contexts	
	Food desert	**Marginalisation**
Local lay knowledge	Local knowledge is lost: farmers need to learn how to grow with not conventional (chemical) methods.	Local knowledge is the traditional knowledge of a place and concerns how to grow and consume local food. It is elaborated, conserved and transmitted in the rural community.
Managerial/ commercial	Is diffused among farmers: their trading and selling skills are regarded as part of their larger repertoire of skills in social interaction, not needing to be formally taught.	Not diffused among farmers: it is a limit to the valorisation of traditional products.
Scientific and technical	It belongs to the extra-local technocratic structure.	It belongs to the extra-local technocratic structure.
Knowledges: Dynamics		
Local-lay knowledge and tacit dimension	Local-lay knowledge is rebuilt through experience and exchanges with local knowledges in other (national and international) regions. Tacit knowledge is important in facilitating communication.	Traditional knowledge interacts with managerial and expert-scientific knowledge. Tacit knowledge is important in facilitating communication.
Experts	They are not the 'scientific' experts, but 'somebody who has already done it' (practical hands-on expertise).	'Experts' are the scientists and technicians (especially in wine and oil sectors) or the experts on regulations and certifications schemes.
Scientific knowledge	Not important. It may a starting point, but it need to be adapted, integrated.	Interaction between scientific and local knowledge is important (wine, oil, ham, fermented fish). Scientific knowledge may appropriate local/oral knowledge (ham, wine). Local knowledge may stimulate new research.
Managerial knowledge	Farmers update their managerial skills interacting with institutions, managerial and scientific expert.	Interaction with managerial knowledge is crucial in the process of valorisation of local food, when applying to certification or in strategies of rural development through tourism.
Transfer of knowledge	Mainly through shared experiences; also from conversation and observation.	Through 'observation' and compilation of traditional knowledge.

Source: Own elaboration from case studies.

relations among things. If the local food project aims to construct an alternative food systems, then, as Kloppenburg (1991) noticed about 20 years ago, this needs a re-building of knowledge processes, starting from local knowledge.

Dealing with agriculture practises, knowledge of place involves knowledge of natural processes. We found it useful to distinguish between 'lay' and 'tacit' local knowledge: while the last refers to unspoken, pre-discursive knowledge used in the communication among people, lay knowledge refers to the views local people elaborate about how things work; for example, in the case of agriculture, when it is better to sow, which plant varieties to grow in a particular soil, etc.

The case studies in this volume highlight the different effects sorted by the diffusion of the agro-industrial model of food production in the two contexts: in the areas characterised as 'food desert', technical knowledge of how to grow food with not-conventional practices and consumers' culture about local food is lost or lacking; in the context of 'marginalisation' local knowledge about food is maintained by farmers and consumers in the form of 'traditional knowledge' (Table C.2).

By consequence, the re-construction of a local economy of food entails a process of re-construction of local knowledge in one situation (the 're-connection perspective'), and a re-vitalisation and valorisation of 'traditional knowledge', in the other.

In the 're-connection perspective', scientific knowledge is not reputed useful or sufficient for the reconstruction of local food. There is actually an attitude of precaution towards scientific knowledge, while experience and practical hands-on expertise is looked for. Local-lay knowledge is considered of the utmost importance and is rebuilt through experience and exchanges with local knowledges in other (national and international) regions. The 'expert' is, in this perspective, re-defined as 'one who has already done it'. The interaction with scientific knowledge is sporadic and occasional, while the priority is seen as the construction or re-construction of a local knowledge of the place. The two ways of knowing (local and scientific) remain separate (a post-modern perspective? Fonte, this volume: 164).

In the marginalised areas, the local food project sets in motion a process of recuperation and valorisation of traditional knowledge. Here, though, knowledge processes lead frequently to an interaction with experts, who are considered to be managerial or scientific experts, the scientists who know how processes work or the ones who knows norms and regulations. Managerial knowledge is lacking to the (post-)traditional farmer, while it is needed to manage the complex procedures necessary for the certification of a product or even as organisational skills when promoting the local territory through fears and festivals. The interaction with scientific knowledge is frequent, especially when dealing with complicated processes of food transformation: fermented fish, wine or oil production. But while in many food processes (production of cheese, ham, salami, fermented fish) traditional artisanal knowledge still remains the hallmark of the product excellence, in wine production the contribution of scientific knowledge is reputed essential for the attainment of high quality standards. The cases studies from Spain and Greece highlight how in this sector there has been a professionalisation of technical knowledge (led by the State in Spain – the *Requena School of Viticulture and Oenology* – by big private enterprises in Greece) which has in the best cases relegated local knowledge to the vineyard and the vine cultivation, missing important opportunities for the enhancement of local knowledge. Actually the case of Spain highlights how professional oenologists were even disregarding the local variety of grape, interpreting designation of origin as a devise for selling a product, rather than valorising a territory. Nonetheless, the development of labels of origin and the rhetoric of the *terroir*, even when used only as a marketing devise, have the potential of re-embedding wine in the territory, pushing in this case towards attempts of integration between local and professional forms of knowledge, that may lead to a *Mode 2* science (Fonte, Introduction, this volume: 16).

Concluding remarks

The case studies gathered in this volume show how in the most dispersed and different regions of Europe people act together to promote initiatives for both, the re-localisation of food and the valorisation of local food. Agriculture, local food and rural areas have then become not only an arena of passive resistance, but also a field of action for small farmers, producers and citizens-consumers interested to build the sustainability of rural areas, sometimes extending the food networks to include new, often urban, actors, sometimes constructing new food communities.

The specific approach of this volume – based on the study of knowledge dynamics – shows the importance of re-localising knowledge while re-localising food. There is still a wealth of local knowledge in the rural areas of Europe that needs to be recognised, defended, used and valorised. But also where local knowledge has been lost, it needs to be re-built anew if the local food project wants to sustain an alternative agriculture and food economy. Re-asserting the value of local knowledge can stop the deterioration of natural resources and biodiversity, that it is nowadays an ineluctable task, can help to build a more robust way of knowing, where experience enters full title in the generation and transmission of knowledge, and, last but not least, can re-locate rural people at the centre stage in the construction of the European knowledge society.

References

Appadurai, A. 1986. Introduction: commodities and the politics of value, in *The Social Life of Things: Commodities in Cultural Perspective*, edited by Appadurai, A. Cambridge: Cambridge University Press, 3–63.

Bruckmeier, K. and Tovey, H. (eds.) 2009. *Rural Sustainable Development in the Knowledge Society*. Aldershot: Ashgate.

Bruckmeier, K., Engwall, Y. and Höj-Larsen, C. 2006. *Local Food Production and Knowledge Dynamics in Rural Sustainable Development*. Manuscript: CORASON-project, report work package 6; Gothenburg University, Human Ecology Section. Available at: http://www.corason.hu [accessed: 25 November 2009].

Fonte, M. 2008. Knowledge, food and place: a way of producing, a way of knowing. *Sociologia Ruralis*, 48(3), 200–222.

Granberg, L., Kovách, I. and Tovey, H. (eds.) 2001. *Europe's Green Ring*. Aldershot: Ashgate.

Kloppenburg, J. Jr. 1991. Social theory and the de/reconstruction of agricultural science: local knowledge for an alternative agriculture. *Rural Sociology*, 56(4), 519–548.

Maye, D. and Ilbery, B. 2007. Regionalism, local food and supply chain governance: a case study from Northumberland, England, in *Alternative Food Geographies: Representation and Practices* edited by Maye, D., Holloway, L., Kneafsey, M. (eds.). Oxford: Elsevier, 149–167.

Watts, D., Ilbery, B. and Jones, G. 2007. Networking practices among 'alternative' food producers in England's west midland region, in *Alternative Food Geographies. Representation and Practices* edited by Maye, D., Holloway, L., Kneafsey, M. (eds.). Oxford: Elsevier, 289–307.

Watts, D.C.H., Ilbery, B. and Maye, D. 2005. Making re-connection in agro-food geography: alternative systems of food provision. *Progress in Human Geography*, 29(1), 22–40.

Index

'a form of social contingency' (local food) 129
Actors
   'local food projects' 268
   'origin-of-food' strategy 268–269
African Swine Plague 136
'Agro-smak' project (oscypek cheese, Poland) 179
agro-foods, economy 1
   re-localisation 266–267
Alentejano-breed pig (*Sus ibericus*) 135, 141, 144
Alternative Agriculture Food Networks (AAFN) 175, 191
'alternative' food movements 55
Alternative Trade Organisations 40
Anglo-Irish meats 42
ANKA S.A. (local development agency, Greece) 246–247
Appellation d'Origine Contrôllée (AOC) 240
'apple man', Ireland 53
Art de la localité (Greek wine) 256
Aspromonte National Park (ANP), Italy
   certification 25–26, 161–162, 162–167
   Cittanova stockfish 159
   conclusions 167–168
   hygiene 159
   IGEA 150, 158–160, 165
   label for local products 156–162, 164
   local food culture 153–162
   local knowledge 151–153
   logo 160–161
   model farms 159
   Palizzi wine 27, 165–167
   traditional agriculture 153–162
   typicality 160
Assoçiacão de Criadores de Porco Alentejano (ACPA) 138, 141

Assoçiacão Nacional dos Criadores do Porco Alentejano (ANCPA) 138, 141
Association of Sheep Breeders (ASB) and oscypek cheese, Poland 187–189, 192–193
Ayiorghitiko grape/wine (Greece) 27, 29, 251–252, 254

Baixo Alentejo, Portugal
   certification 130, 133–134
   traditional farming 135–136
Barrancarnes-Transformação Artesanal Lda 139–142
Barrancos Cured Ham, Portugal
   certification 25, 27
   conclusions 144–145
   local actors/networks 142–144
   local food for distant consumers 137–142
   local to global production 130
   study area 134–135
   traditional farming 135–137
Barrenquenho (local dialect in Portugal) 135
BIOLAND (organic association, Germany) 68
bobal grape and wine production 27, 219, 225, 227
Bord Bia (National Food Promotion Board), Ireland 44, 51
Bovine Spongiform Encephalopathy (BSE) 90, 109
Bruckmeier, K. and Tovey, H. 2–3, 22, 265

C— Development Association, Tipperary, Ireland 45
C— farmers market, Tipperary, Ireland 44, 45–47, 50, 53, 57
Calabria region, Italy 153–154

Caledad Valenciana (CV) 218
California, USA counter-cultural organic movement 56
Camino (lifestyle journal, Sweden) 109
Cassa per il Mezzogiorno (development programme) 153
Central and Eastern Europe (CEE), post-socialist ruralities 62
'Central Union of Vines and Wine Co-operatives, Greece (KEOSOE) 254
Centre for Product Development in the Food Processing Industry (Norway) 204
certification
  agro-food system 131–134
  Aspromonte National Park, Italy 25–26, 161–162, 162–167
  Baixo Alentejo, Portugal 130, 133–134
  Barrancos cured ham (Portugal) 25, 27
  civic values 269
  distant consumers 176
  hygiene and food production 25
  KRAV (Sweden) 111, 115
  local foods 133
  origin 149–172
  oscypek cheese, Poland 26–27, 188, 193
  process 25–26
  rakfisk 204
  salami, Valdres, Norway 208–209
  Scotland 83
  semi-fermented fish, Valdres, Norway 203–207
  Spesialitet (Norway) 200–201, 204
Certificatión y Validación (CV) 218
Cittanova stockfish in Aspromonte National Park, Italy 159
Civiltà Contadina (seed savers' association) 155
Co-operative movement, Tipperary, Ireland 42
COHESION programme (Nemea, Greece) 253
Cold Meat Fair, Requena, Spain 228, 230–231, 234
'colouring wine' 219–220, 221
commodification of food 1

Common Agricultural policy (CAP) 79, 86, 127, 142, 144, 219
conclusions
  actors 267–269
  agro-foods 266–267
  knowledge 269–272
  markets 267–269
  social networks 267–269
Conselleria de Agricultura, Pesca y Alimentacion, Valencia, Spain 218
'consumer sovereignty' concept 103
CORASON ('A cognitive approach to rural sustainable development' ...)
  aims 30
  Aspromonte National Park, Italy 150, 153
  case studies 18–19
  countries 3–5
  description 2
  Eldrimner, Sweden 99, 101
  European Union 265
  Greece and wines 238
  knowledge processes 269–270
  local food concept 266
  local knowledge role 17
  Norway 199
  organisation 2–3
  originality 17
  oscypek cheese 173
  'polycentric management' 3
  Skye, Scotland 77, 79
Corinthian currants vinification 251
COVINAS co-operative (Valencia region) 220, 222–223
Craft Butchers' Association 53–54
CV *see* Certificatión y Validación

'de-commodification' concept 11
DEMETER (organic association, Germany) 68
Designation of Origin (DO)
  Barrancarnes 139
  certified agro-food 131
  Greek wines 245
  PGI comparison 228
  Regulatory Board (Spain) 233, 234
  Requena, Spain cold meat 232

Utiel-Requena, Spain wine 218–227, 235
doble pasta (double pomance) wine, Spain 28, 219, 221, 226
dried cured salami Valdres, Norway 27, 208–210

East Germany 're-localisation' concept 22
economic restructuring of local food 127–128
Eldrimner, Sweden
   conclusions 121–124
   'consumer sovereignty' 103
   development
      consumers 118
      knowledge 116, 118–121
      national resource centre 117
      networking 114
      project 111–114
      quality aspects of local food 113, 114
      success 116–117, 121
      support structures/new networks 117–118
   European Union funding 25, 115
   fishing 104
   'household fishermen' 104
   introduction 99–100
   local food production 100–102
   'maintaining one's rural ties' 104
   meaning 99
   social context 102–110
Embutido de Requena (Requena cold meat) 228
Encinasola, Spain 135
Environmental Health 91
environmental sustainability 7–9
Epitrapezios Oenos wine (Greece) 243
Erosion, Technology and Concentration (ETC) Group 24
Escuela de Vitacultura y Enolgia de Requena (School of Viticulture and Oenology) 222
Estación de Enológica de Requena (Requena Oenology Station) 222
European Union (EU)
   agricultural policies 154
   agro-environmental programme 132
   approved control 188
   certification of origin 151
   Common Agricultural Policy 79, 86, 127, 142, 144, 219
   CORASON project 265
   Designation of Origin and Greek wines 245
   Eldrimner, Sweden project 25
   food
      governance 128
      quality standards 109–110
      re-localisation 240
   Greece 242
   hygiene regulations 155, 228
   industrial fragmentation 215
   Nemea VQRD wine 254
   northern/southern distinction 128–129
   organic production 111
   oscypek cheese, Poland 189, 192
   pig-breeding 142
   Poland 174, 188
   Protected Designation of Origin 129, 134, 135
   Protected Geographical Indication 134, 135
   quality production system 128
   rural areas 127, 144–145
   sustainability 3
   Sweden 25, 105, 106, 115
   VI Framework Research Programme 2
   vine growing 220, 242
   see also LEADER
European 'Welfare Quality' project 108, 110
'expert' concept 271

Fair Trade movement 40–41, 55
Farmers Market, C— , Tipperary, Ireland 44, 45–47, 50
farmhouse sheep's milk cheese 52
'fashion' concept, Ireland 43
fermented fish production in Valdres, Norway 27
FLG *see* Food Link Group
flora (honey with protected names) 132
'food desert'
   food production model 270–271
   Tipperary, Ireland 41–43
'food from nowhere' concept 1

'food from somewhere' concept 1
food governance model 186
Food Link Group (FLG) Skye, Scotland 87, 93, 94, 267
Food Link Van Skye, Scotland 87, 94
Food Project Development Officer, LEADER, Tipperary 49, 52
food re-localisation 2, 267, 268
food-communities and neo-liberalism 12–13
'foodless landscape' concept 266
forestry in Jämtland region, Sweden 100
'Forward Strategy for Scottish Agriculture' 83
France and *produits du terroir* 131
fruit (products with protected names) 132
'fuel' concept, Ireland 43

garnacha grape (Utiel-Requena, Spain) 219
GDP (gross domestic product)
    Calabria 153
    Scotland 81
Generic Designation (DG) 216
Geographical Indication (GI) 131, 151
German Agricultural Co-operative (DLG) 65
German Democratic Republic (GDR)
    history 62–63
    'organic farming' 65
Germany
    border with Poland 63
    'organic farming' concept 65, 67, 72
    see also northeast Germany
GHG *see* greenhouses gases
'gifts' description 10
globalisation 1
GRAIN (biodiversity and sustainable agriculture NGO) 24
Greece
    European Union 242
    phylloxera epidemic 242, 251
    quality wines 243, 246–249, 250–255, 256, 258
    valorisation of wine/olive oil 27
    vins de pays wine 243
    winemaking 27–29, 237–264
    see also Lake Plastiras; Nemea
'Greek of Calabria' (language) 156
'Greek Wine Association' (SOA) 254

'Green Ring' hypothesis 3, 265
greenhouses gases (GHG) 8
Griefswald, Germany 65, 69
Guaranteed Traditional Speciality (GTS) 128, 131
Guild of Growers of Utiel (*Gremio de Cosecheros de Utiel*) 219

Hazard Analysis and Critical Control Points (HACCP) 128, 150
Helle Slakteri abbatoir, Valdres, Norway 209
Hellenic Organization for Standardization (ELOT) 247
Highland culture in Podhale, Poland 179, 182, 192
'household fishermen' (Sweden) 104
hygiene
    Aspromonte National Park (ANP), Italy 159
    food knowledge 25, 53

'industrial wines' 29
Inspection and Control for the Ecological Guarantee of Agrofood Processing (IGEA) 150, 158–160, 165
Institute of Agro-Food Quality (Spain) 230
International Organization for Standardization (ISO) 128, 150
Ireland
    IOFGA (organic certification body) 53
    McDonalds 43
    National Small Food Co-ordinator 51
    'Privately Run Farmers Markets' 45
    Teagasc (development institution) 44, 51
    see also Tipperary
Irish Farmers' Association (IFA) 43
Italy
    National Register of Traditional Products 155
    valorisation of wine/olive oil 27
    winemaking 27–28
    see also Aspromonte National Park

Jämtland region, Sweden 100, 106, 118, 267

Kalinowski Decree, Poland 183

Kalinowski, Jaroslaw 183
Karamitros, George 257
Karamitros Winery (Greece) 249, 257
Karditsa, Greece
    Union of Co-operatives 249
    wine growing are 244–245
Keimblatt (organic food shop) 68
knowledge
    Aspromonte National Park, Italy 151–153
    conclusions 269–272
    economic restructuring of local food 130
    Eldrimner, Sweden
        expert 120
        lay 119
        management 116
        scientific 120
        tacit 119
        types 118–119
    Greek wine 255–259
    'Green Ring' countries 265
    lay 17, 20, 25, 28, 119, 153, 229
    local food
        economic restructuring 127–148
        politics 18–21, 25–27
    Mavro Messenikola wine 255–256
    northeast Germany 62, 70, 72
    oscypek cheese, Poland 185, 194
    Requena, Spain, cold meat PGI 229, 229–230
    rural areas, 3
    scientific 62, 120
    Skye and Lochalsh, Scotland 80–81, 88–93
    tacit 29, 119, 153, 199–200, 229, 235, 256, 258–259
    technical 53
    Tipperary, Ireland 53–55
    traditional artisan 23–25, 27
    Utiel-Requena, Spain 227, 233–234
    Valdres, Norway 205–206, 210–211
KRAV certification (Sweden) 111, 115
kurv *see* salami

La Matanza (killing of pigs in Spain) 228
La Plana de Utiel-Requena (comarca) 218
Lafkiotis Winery (Greece) 258

LAG *see* local action group
Lake Plastiras, Greek wine
    certification 256
    conclusions 259–261
    introduction 244–246
    Mavro Messenikola 27, 245–249, 246–249, 255–256
    quality 254
    Union of Wine Co-operatives 29
lay knowledge
    certification 25
    Eldrimner, Sweden 119, 229
    growing and preparing 20, 153
    'how things work' 17
    product manufacture 229
    Utiel-Requena, Spain winemaking 28
LEADER (EU initiative)
    commercial knowledge 53
    grants 46, 49–50
    'local food project' 268
    Mavro Messenikola VQPRD wine 246–247, 249
    'micro-food' strategy 53
    programme 132
    technical (hygiene/safety) food knowledge 53
    Tipperary
        courses 57
        food entrepreneurs 57
        Food Project Development Officer 49
        local action group 45, 47–50, 51–52, 56–57
        'local food' 44–45
        networks 51–22, 55, 57
Less Favoured Areas (LFAs)
    Plastiras reservoir, Greece 245
    Scotland 81
local action group (LAG), Tipperary, Ireland 45, 47–50, 51–52, 56–57
'local embeddedness' in Tipperary, Ireland 50, 57
local exchange and trade systems (LETS) 107, 267
local food for distant markets 129–130, 134–142
local food for local markets 129–130
local food and politics

conclusions 29–30
environmental sustainability 7–9
farmers/consumers in new local food
    economies 21–23
introduction 6–7
knowledge 18–21, 25–27
provenance certification 25–27
recovering traditional knowledge
    23–25
socio-economic dimension of localness
    9–18
spatial dimension of the ' local' 7–9
valorising traditional knowledge 23–25
wine making 27–29
'local knowledge'
    concept 17
    northeast Germany 62
*Local Knowledge for an Alternative
    Agriculture* 162
'local production for global markets' 175
'local production for local markets' 175
Local Quality Convention (LQC) 246
'*local*' term and *socio-spatial proximity* 3
Lochhead, Richard (Scottish Agriculture
    Minister) 82

McDonalds in Ireland 43
'maintaining ones' rural ties' 104
Malopolska region, Poland 173, 179
marginalised food production
    model 270
    regions 266–267
markets, conclusions 267–269
Marx, Karl and commodities 9–10
Matforsk (Norwegian Food Research
    Institute) 204
MATKULT ('food culture') 117
Matmerk (Norwegian Agricultural
    Quality System/Food Branding
    Foundation) 204
Mavro Messenikola wine
    agro-natural habitat 27–28
    knowledge 255–256
    VQPRD (quality wine) 245–249, 256
'meaningful commodities' exchange 11
Mecklenberg-Ostvorpommern (MV),
    Germany 65–66
Messenikola, Greece 245

model farms in Aspromonte National Park,
    Italy 159
montado (agrosilvo-pastoral system in
    Portugal) 135, 136, 141
MV *see* Mecklenberg-Ostvorpommern

Naeringsutvikling, Valdres, Norway 201,
    203
National Register of Traditional Products
    155
National Small Food Co-ordinator
    (Ireland) 51
Naturtkost Vorpommern GmbH (Biofood
    West Pomerania Ltd) 69
Nemea, Greece wine
    Ayiorghitiko 27, 251–252, 254
    COHESION programme 253
    conclusions 259–261
    European Union 254
    introduction 249–250
    network 253, 255
    quality 254
    Savatiano (vine variety) 251
    tourism 255
    VQPRD 250–255, 258
'Nemea Union of Winemakers and
    Viticulture' (ENOAN) 253–254
neo-liberalism and new 'food communities'
    12–13
Netzwerk Vorpommern, northeast
    Germany 68–71, 72, 267
non-governmental organisations (NGOs)
    GRAIN 24
    Slow Food Poland 179
Norsk Designråd (Norwegian Design
    Council) 204
'Norsk *Rakfisk* Festival' 202
northeast Germany
    agriculture 64
    conclusions 71–73
    food and knowledge 61–75
    international companies 67
    introduction 61–65
    local knowledge 62, 70, 72
    Netzwerk Vorpommern 68–71, 72, 267
    quality management 71
    rural settlements 63
    scientific knowledge 62

socialism 63, 65
study region 65–68
Norway
　agriculture 198–199
　agro-food systems 197
　certification 200–201
　local food 197, 198
　rural development 198–199
　Spesialitet (certification) 200–201
　see also Valdres, Norway
Norwegian University of Life Sciences 205

Odermündung region, Germany 65–66
olives (oils with protected names) 132
one church, one co-operative doctrine 221
Oppland, Norway 201–202
'organic farming' concept 65, 67, 72, 149
origin certification 149–172
origin-of-food perspective 5, 268–269
Origin Designation (DO) 216, 218
oscypek cheese, Poland
　'Agro-smak' project 179
　'area of origin' 191
　Association of Sheep Breeders 187–189
　certification 26–27, 188, 193
　'commercialisation of tradition' 180
　'common norm' 189
　conclusions 191–195, 266, 269
　description 174
　'elite of herdsman' 187
　European Union 189, 192
　'fake' version 181, 185
　food
　　chains 173
　　production–consumption 175–178
　food governance 186
　Highlanders 187
　introduction 173–175
　knowledge 185–191, 194
　local food concept 194
　'market conqueror' concept 183
　mass production 187
　networks 180–185, 191, 194
　'oscypek boom' 182
　'the oscypek festival' 184
　Oscypkowy szlak 184
　Podhale development 192

Polish Patent Office 188
'popular oscypek' sector 193
'quality turn' concept 194
recipe 180–181
registered pattern' 189
'skypek' label 182
Slovakia 190
souvenir for mass tourism 181, 186
temporal/spatial aspects 178–180
'The Taste of Malopolska' 179
tourism 184
variety 185
Ostvorpommern (OVP), Germany 65–66, 67

Palizzi wine, Italy 27, 165–167
phylloxera epidemic, Greece 242, 251
place-embeddedness of food 9
Podhale region, Malopolska, Poland 174, 176–184, 189–190, 192
Poland
　border with Germany 63
　European Union 174, 188
　Kalinowski Decree 183
　Ministry of Agriculture 183
　oscypek cheese 26–27
　skiing areas 180
　'social equality' 178
　tourism 266, 269
　traditional cuisine 183
　'Western' foodstuffs 178
　see also oscypek cheese
Polish Patent Office and oscypek cheese 188
Polish Red Cow breed 192
'polycentric management' 3
'Popular Movement Council' (Sweden) 106
Portugal
　certification 130
　wine 131
　see also Baixo Alentejo; Barrancos Cured Ham
Potsdam Treaty 63
Presunto de Barrancos see Barrancos Cured Ham
'Privately Run Farmers Markets', Ireland 45

production and exchange in food
    economy 9–12
produits du terroir (France) 131
Protected Designation of Origin (PDO)
    cured ham (Portugal) 134, 138,
        140–145
    Feta Agrafon (cheese) 245
    food accreditation 128
    Graviera Agrafon (cheese) 245
    oscypek cheese (Poland) 26–27
    policy 12
    rakfisk (Valdres, Norway) 205
    Requena, Spain cold meat 232
    Spain 216
Protected Geographical Indication (PGI)
    12–13, 134
    Cold meat from Requena, Spain
        228–232
    DO comparison 228
    rakfisk from Valdres, Norway 204
    salami from Valdres, Norway 210
    Valencia region 216, 218, 229
provenance certification 25–27

quality wines in Greece (VQPRD) 243,
    246–249, 250–255, 256, 258

rakfisk (fermented fish)
    Protected Designation of Origin 205
    Valdres, Norway production 203
RDAs *see* recommended daily allowances
recommended daily allowances (RDAs) 90
're-connection perspective' case studies 3,
    18, 266, 271
recovering traditional knowledge 23–25
'reflexive localism' 237
Regulatory Board (DORC) and label of
    origin 224
're-localisation' concept in East Germany 22
Requena Cold Meat Association 230
Requena School of Viticulture and Oenology
    272
Requena, Spain
    cold meat PGI
        Cold Meat Fair 228, 230–231, 234
        introduction 228
        knowledge 229–230
        marketing 234

social/ecological sustainability
    230–231
strategy 232
cold meat/wine comparison 232–235
res nullius (a thing belonging to nobody) 13
res universitatis (a thing owned collectively)
    13
'resistance' concept 11
Ribera del Duero, Spain wine region 223,
    226
Rioja, Spain wine region 223, 226
River Oder (Germany–Poland border) 63
Roman Catholic Church in Podhale, Poland
    180
Rostock, Germany 65
Royal Society for the Protection of Birds
    (RSPB) 82

salami (kurv) from Valdres, Norway
    artisan knowledge 27
    certification 208–209
    production 208–210
SALE (Skye and Lochalsh Enterprise) 86,
    92–93
Sami people, Sweden 100
Särimner fair, Eldrimner, Sweden 105, 115
Savour Tipperary Food Guide 48
School of Viticulture and Oenology,
    Requena 220, 227
science and knowledge in post-positivist era
    14–18
scientific knowledge in northeast Germany
    62
Scotland
    agricultural production policy 81–85
    Department for Environment and Rural
        Affairs 83
    Food Certification Agency 83
    Food Standards Agency 83
    GDP 81
    Less Favoured Areas 81
    National Food Policy 83, 94
Scottish Agricultural College (SAC) 89–90
Scottish Food and Drink 83
semi-fermented fish in Valdres, Norway
    203–208

Senter for Produktutvikling in aeringsmiddelindustrien (SPIN) 204
'singularisation' concept 11
Skye and Lochalsh Horticultural Development Association (SLHDA)
   case study 77
   description 85–87
   farmers' market 92
   food quality 91
   knowledge 89
   locally led initiative 85
   theoretical frameworks 87–88
Skye and Lochalsh, Scotland
   Food Link Group 87, 93, 94
   Food Link Van 87, 94
   knowledge 88–93
   region 84–85
   SALE 86, 92–93
Skye, Scotland
   agricultural production policy 81–85
   alternative food politics and knowledge 78–81
   branding 94
   conclusions 93–94
   disconnection and disembedding 78–79
   'dual perspective' on local food networks 79
   introduction 77–78
   knowledge 80–81, 88–93
   local foods 85–88
   tourism 92–93
Slovakia and oscypek cheese 190
Slow Food Congress, United Kingdom 94
Slow Food Convivium, Tipperary, Ireland 43, 48
Slow Food Poland (NGO) 179
social networks 267–269
socio-economic localness
   neo-liberalism/new food communities 12–14
   production and exchange in food economy 9–12
   science/knowledge in post-positivist era 14–18
socio-spatial proximity and 'local' term 3
Soil Association 91

'space and place' concept 151
Spain
   La Matanza (killing of pigs) 228
   Union of Wine Co-operatives 29
   valorisation of food products 27
   winemaking 27–29
   see also Requena; Utiel-Requena; Valencia
spatial dimension of the ' local' 7–9
Specific Designation (SE) 216
Spesialitet (certification in Norway) 200–201, 204
Stymfalia, Greece archaeological site 260
supply chain strategy 216
Sweden
   Camino (lifestyle journal) 109
   European Union
      entry 107
      funding 25, 105, 115
   'Household fishermen' 104
   Jämtland region 100, 106, 118
   KRAV certification 111, 115
   local exchange and trade systems 107
   MATKULT ('food culture') 117
   Ministry of Agriculture 117
   national food production 123–124
   organic production 111–112
   'Popular Movement Council' 106
   Rural Development Program 105, 111
   Sami people 100
   University for Agriculture 108
   see also Eldrimner, Sweden
Szczecin, Poland (Stettin) 63, 65

tacit knowledge
   Ayiorghitiko grape 29
   cold meat 235
   CORASON project 153
   Eldrimner, Sweden 119
   'interactions with people' 17
   Nemea, Greece wine 256, 258–259
   Norwegian study 199–200
'traditional/hand-made/natural' concept 229
Tasting or Qualification Committee (Spain) 224
Tatra Mountains region, Poland 26, 174, 179–181, 184, 190

Teagasc, Ireland (development institution) 44, 51
tempranillo grape (Utiel-Requena, Spain) 219
territorial quality 216
territorialisation and food products 176
territory (*the origin of food perspective*) 5, 18–19
terroir wines 29, 240–242, 258, 261, 272
'The National Interprofessional Association of Vines and Wines' (EDOAO) 254
'The Taste of Malopolska' (oscypek cheese, Poland) 179
Tipperary, Ireland
 co-operative movement 42, 267
 farmers market, C— 44, 45–47, 50
 'food desert' 41–43
 history 44–45
 introduction 39–41
 knowledge 53–55
 LEADER 44–45, 46
 local action group 45, 47–50, 51–52, 56–57
 meat production 42
 nature 54–55
 networks 51–52
 power in rural development 55–58
 Slow Food Convivium 43, 48
Topikos Oenos wine 243
tourism, oscypek cheese, Poland 266, 269
tourism and oscypek cheese, Poland 180–181, 184, 186
tsipouro (Greece) 256

Uecker-Randow (UER) district, Germany 65–66
UNCTAD *see* United Nations Conference on Trade and Development
União das Assoçiacões de Criadores do Porco Alentejano (UNIAPRA) 138–139, 141
Union of Co-operatives, Karditsa, Greece 249
Union of Wine Co-operatives, Spain 29
United Kingdom Congress on Slow Food 93
United Nations Conference on Trade and Development (UNCTAD) 24

University of Applied Science, Neubrandenburg, Germany 71
University of Evora, Portugal 25, 137–138, 141
'Utiel-Gastronomic' Quality Label 225
Utiel-Requena DO 223
Utiel-Requena DO Wine Route 225
Utiel-Requena, Spain wine
 cold meat comparison 232–235
 lay knowledge 28
 production 27–29, 234–235
 technical knowledge 233–234
Utiel-Requena, Spain wine DO
 actors/networks 220–221
 bobal grape 219, 225, 227
 conclusions 227
 context 228–227
 garnacha grape 219
 knowledge 227, 233–234
 new/traditional structures 221–223
 policy 234–235
 role 223–225
 strategies 225–226
 supply chain 216
 tempranillo grape 219

'Valdres Kurv BA' 209, 211
Valdres, Norway
 conclusions 211–212
 dried cured salami 208–210
 fermented fish/salami production 27
 Food Security Authority 209
 knowledge 210–111
 local food development 201–210
 Matforum (Food Forum) 202
 Næringsutvikling 202, 203
 Protected Geographical Indication 204
 rakfisk 203, 205
 salami (kurv) 27, 208–210
 semi-fermented fish
  barriers to further development 207–208
  certification 203–207
  competitors 207
  knowledge 205–206
  product differentiation 206
 traditional food 199
'Valdres Rakfisk BA' 204, 207, 209, 211

Valencia region, Spain
    cold meat/wine comparison 232–235
    context 217–218
    COVINAS co-operative 220, 222–223
    introduction 215–216
    Requena cold meat PGI 228–232
    Utiel-Requena label of origin 218–227
valorisation
    Barrancos cured ham, Portugal 269
    cold meat in Requena, Spain 27
    food in Malopolska, Poland 179
    local food and politics 23–25, 266, 269
    olive oil (Greece, Italy and Spain) 27
    oscypek cheese, Poland 26
    traditional knowledge 23–25
    Valdres, Norway products 210
    wine (Greece, Italy and Spain) 27–29
vins de pays wine 243
Vorpommern region, Germany 68–71
see also Mecklenburg–Ostvorpommern; Ostvorpommern

'Welfare Quality' project (Europe) 108, 110
wine
    Greece
        context 242–244
        Epitrapezios Oenos 243
        OPAP category 243
        OPE category 243
        theory 238–240
        Topikos Oenos 243
        traditional/expert knowledge 27–29
        vins de pays 243
        VQPRD category 243, 246–249, 250–255, 256, 258
        see also Lake Plastiras; Nemea
    'industrial wines' 29
    Italy
        traditional/expert knowledge 27–29
        see also Aspromonte National Park
    local knowledge 27–29
    Portugal demarcation 131
    Spain
        traditional/expert knowledge 27–29
        see also Utiel-Requena; Valencia
    terroir concept 29, 240–242, 258, 261, 272
    traditional/expert knowledge 27–29
Wismar, Germany 65
World Intellectual Property Organization (WIPO) 24
World Trade Organization (WTO) 24, 150

Zakopane, Tatra Mountains, Poland 181, 184